Figures

Conservation in the Anthropocene

This book provides a critical assessment of conservation in the Anthropocene grounded in the personal, historical, and cultural development of human interaction with nature. The author argues that conservation can no longer be primarily about preserving nature but must adapt its efforts to promote changes through which humans create a landscape that is neither abandoned nor degraded but used well by humans and non-humans alike. The book first reviews the origin of ideas and conditions that have led to the concept and classification of the Anthropocene and explores how the author's own interactions with nature were shaped through his experience as a conservation biologist. Next, it considers how humans have come to be the primary drivers of ecological activity, geological events, and climate change. Chapters then focus on the need for new conservation thinking regarding novel ecosystems, urban conservation, the role of Indigenous Peoples in conservation, and the value of protected areas (PAs), parks, and wilderness. The book concludes by identifying strategies for effective conservation and argues for a new formulation of conservation values that redefine human relationships and interaction with nature. Chapters are enlivened by the personal experiences of the author and the first-person narratives of conservation activists and scientists throughout the world who are learning to practice and succeed in conservation efforts under Anthropogenic conditions.

Drawing on global examples, this book will be of great value to students and scholars of biodiversity conservation and environmental science ready to consider a new way of looking at the care and nurture of nature in the Anthropocene.

Fred Van Dyke is an independent consultant providing guidance in conservation research and education, professional development, writing for publication, and organizational leadership. Previously he was Executive Director of the Au Sable Institute and Professor of Biology at Wheaton College, USA. He has also worked as a wildlife biologist for the Montana Department of Fish, Wildlife, and Parks, a contract biologist for the US Fish and Wildlife Service and the US Forest Service, and as a consultant to the US National Park Service and private industry. He is the co-author of *Conservation Biology: Foundations, Concepts, Applications*, 3rd edition (2020).

Routledge Studies in Conservation and the Environment

This series includes a wide range of inter-disciplinary approaches to conservation and the environment, integrating perspectives from both social and natural sciences. Topics include, but are not limited to, development, environmental policy and politics, ecosystem change, natural resources (including land, water, oceans and forests), security, wildlife, protected areas, tourism, human-wildlife conflict, agriculture, economics, law and climate change.

Creating Resilient Landscapes in an Era of Climate Change
Global Case Studies and Real-World Solutions
Edited by Amin Rastandeh and Meghann Jarchow

Species, Science and Society
The Role of Systematic Biology
Quentin Wheeler

Conservation Concepts
Rethinking Human-Nature Relationships
Kurt Jax

Japan's Withdrawal from International Whaling Regulation
Edited by Nikolas Sellheim and Joji Morishita

Conservation Leadership
A Practical Guide
Simon Black

Positive Psychology and Biodiversity Conservation
Health, Wellbeing and Pro-Environmental Action
Jolanta Burke, Sean Corrigan, Jimmy O'Keeffe, and Darren Clark

Conservation in the Anthropocene
Reshaping Interaction with Nature
Fred Van Dyke

For more information about this series, please visit: www.routledge.com/Routledge-Studies-in-Conservation-and-the-Environment/book-series/RSICE

Conservation in the Anthropocene

Reshaping Interaction with Nature

Fred Van Dyke

Routledge
Taylor & Francis Group
LONDON AND NEW YORK

Designed cover image: Getty Images

First published 2025
by Routledge
4 Park Square, Milton Park, Abingdon, Oxon OX14 4RN

and by Routledge
605 Third Avenue, New York, NY 10158

Routledge is an imprint of the Taylor & Francis Group, an informa business

British Library Cataloguing-in-Publication Data
A catalogue record for this book is available from the British Library

ISBN: 978-1-032-51915-9 (hbk)
ISBN: 978-1-032-51107-8 (pbk)
ISBN: 978-1-003-40441-5 (ebk)

DOI: 10.4324/9781003404415

Typeset in Sabon
by codeMantra

To Harold Zahniser Snyder
Who inspired many to the highest intentions and noblest interactions with nature –
and did it every day

Contents

Foreword

"Imagine our descendants in the year 2200 or 2500", Nobel Prize-winning atmospheric chemist Paul Crutzen, who popularized the term *Anthropocene*, urged us in 2011.

> They might liken us to aliens who have treated the Earth as if it were a mere stopover for refueling, or even worse, characterize us as barbarians who would ransack their own home. Living up to the Anthropocene means building a culture that grows with Earth's biological wealth instead of depleting it. Remember, in this new era, nature is us.[1]

In the book in front of your eyes, Fred Van Dyke answers this call with a comprehensive account of the meaning of nature conservation in an era where humans have shaped all aspects of the world and have embedded their fingerprints on practically all threats to biodiversity.

This stimulating book builds on three decades of college teaching and comes 45 years after Fred's first publication, during which he has become a respected voice as a conservation biologist. Yet his eyes see the bigger picture, his ears hear other truths, and his sharp mind brings them all together. I met Fred in 2023 while working with faith-based nature conservationists of different religions in Southeast Asia. A group of Eco-Theologians needed help with a factual check of both scientific data and theological underpinnings for a book they were writing and searched for someone with expertise in both. Fred stepped up and volunteered his decades worth of experience to empower a new variety of ground-breaking conservationists on the other side of the planet.

This illustrates Fred's belief both academically and out in the field that it takes a village to conserve nature. Not just scientists, but businesspeople, educators, health professionals, politicians of all stripes, and faith leaders of all persuasions. Only a profoundly inclusive approach can keep this Earth and all her different beings healthy in the Anthropocene. Largely free of jargon, *Conservation in the Anthropocene: Reshaping Interaction with Nature* discusses how we got to this age of human-altered ecosystems, how that impacts many aspects of nature, and what's needed to meet the moment for us as nature conservationists. We can no longer "just" protect nature but must help manage the human presence on this planet and its effect on nature.

In our interconnected world, rejuvenation of conservation starts with a joint effort to break through our collective amnesia on our broken relationship with nature. Not only do we need to build bridges to other disciplines, but also to other worldviews and different continents. That starts with an open mind and introspection on how we relate to animals and plants, and connect with the mountains, forests, rivers, and seas. While this book is mostly centered in an American context and referencing Fred's own extensive conservation field experience on this continent, he reached out to people in Asia and Africa for input during the writing process, and many chapters include examples from other countries. This has made the book the more relevant for a diverse and global audience. The humility of his effort is also reflected in the down-to-earth tone on each page, and sense of humor he applies when reflecting on his own scientific discipline, wondering if "calling something a 'novel ecosystem' [is] just a label that sounds better than 'ruined and invasive species-infested landscape?'"

While striking a serious note on the urgent need to adapt biodiversity conservation to a new geologic era, this book is a joy to read and offers up food for thought on how humans relate to the natural world. It is our collective challenge as humans to stop acting toward the Earth as "aliens" and "barbarians." An increase in experiences and knowledge can lead to a change in our hearts and mindsets – holding the key to fixing the way we do nature conservation in the Anthropocene.

For our descendants in 175 or 475 years from now, as per Crutzen's imagination, to judge that we managed to accomplish this, every piece of the conservation puzzle is important. *Conservation in the Anthropocene: Reshaping Interaction with Nature* presents one fine piece.

Bram Steenhuisen[2]
2024

Notes

1 <https://e360.yale.edu/features/living_in_the_anthropocene_toward_a_new_global_ethos>.
2 Bram Steenhuisen is an independent professional working on inclusive natural resource governance and conservation dialogue. He lives and works in Southeast Asia for a number of conservation organizations.

Preface

When I began this writing, I entered the subject of conservation in the Anthropocene with naïve and uninformed ideas. I was going to explore how the work of nature conservation was being changed, and would have to keep changing, in a world where humans had become the dominant forces of nature – the Earth's most important drivers of ecological, geological, and climatological processes. The five reviewers of my initial proposal made short work of my ignorance. Three of the five reviews were negative and, by the rules of democracy, my effort should have ended then and there. Of these, however, two offered hope that the book might be saved if I would engage the disciplines of history, political science, sociology, economics, psychology, and others that are all vital elements of the Anthropocene debate. My editor invited me to write a revision and I prepared a new proposal responsive to reviewer input. The reviewers were sufficiently encouraged to cautiously support further effort.

Thus commissioned, I began a review of the writings in a multitude of disciplines that substantively addressed the Anthropocene – this "age of humanity" in which many claimed that the human race would transform nature into whatever it wanted nature to be. What was the Anthropocene? Where did it come from? What were its causes? Where was it headed? And what did that mean for conservation? Could species be saved on a human-dominated planet and if so, how?

The debate was diverse. But in this diversity, two things emerged repeatedly. One was the deeply personal nature of the Anthropocene discussion, with viewpoints strongly influenced by individual perspective, and perspective influenced by experience. The second was that the debate was deeply moral. Because the conditions of the Anthropocene find humans in a position of unprecedented influence over nature, they are in a position of unprecedented choice regarding nature. The answer to the question, "What shall we now do with our powers over nature?" had become a question of importance because our influence had become an influence of effect. The best of these writings invariably perceived that such choice required discernment between what would be taken and what would be left, what would be valued and what would be cast off, and what was right and what was wrong. After decades of my own compositions of scientific articles and books, the ingrained habit of simply summarizing the relevant literature on these questions was strong. I did not ignore it, but I chose to subject it to two other roads to knowledge. One was experience. The other was insight.

In my interactions with others in conservation – in organizations, meetings, reviews, and administration – I found that in most meetings my colleagues did not think that I was the foremost expert in any particular branch of conservation science (and with good reason), but invariably valued my insights about what our knowledge meant and what we should do with it. In experience, I have pursued most of my scientific effort in the field – in mountains, marshes, and meadows, streams, sagebrush, and (even) suburbs, wherever the wild things were. And most recently I spent nine years as the Executive Director of the Au Sable Institute, a private, faith-based conservation organization engaged in teaching, research, service, and advocacy in the work of saving nature. Being a private concern, our donors were critical components of financial support, but they were far flung and often far removed from us. I began to write to them each month. My letters were not an appeal to send more money (which would have been unforgivably boorish and ill-mannered in any relationship), but to relate to them the events of the natural world around Au Sable in the forests of northern Michigan, USA. Those events were the ones I witnessed, events of firsthand experience. In writing these letters, many constituents began writing back. Many did send money (and I was grateful), but more often they wrote words of thanks and appreciation for taking the time to "place" them at Au Sable, to make them part of the life that was being lived in the forests and swamps and lakeshores around the Institute. And more than once I remember a correspondent telling me, "Thank you, Fred. You put me in nature."

Such correspondence made me more aware of the power of experience, experience that is shared and related with passion and insight to others. Thus, I place before the readers of this book many of my own experiences and the things I learned from them. These experiences are not so much arguments as confessions – confessions of my own interactions with nature and how they have shaped my intentions toward it, as well as my view of myself in it. So, the book begins with confessions from a period of powerful experience in my life – the life of a wildlife biologist in the Absaroka Beartooth Wilderness of south-central Montana. These confessions were not part of my original book proposal. But as I repeatedly saw the role of personal experience and perspective in the best writings on the Anthropocene, I realized that I needed to make personal admissions of my own, and I submit them here.

As our present state in a human-dominated world is ultimately a moral dilemma, I have devoted the final chapter to the question of the value of nature in the Anthropocene – how do we know it is *there* and how should we respond to it? And how is our interaction with nature to shape our own moral development? If all of our decisions in a human-dominated world are shaped and coerced by our desire for yet greater domination, how can we ever see the value of nature at all? It is not from ourselves that we will learn to be better than we are. If the light that is in you is darkness, how great is the darkness?

I extend this effort as a first step in reevaluating everything. We have often considered that conservation was the solution to our abuse of the Earth. But we find instead, if we squarely face its history, that conservation, as we have practiced it, has often been complicit with, or even the cause of, the abuses we have perpetrated on the natural world and the people in it. If there is to be conservation in the Anthropocene – conservation that "works" – it will have to be a new "species" of conservation, a repentant conservation that acknowledges the mistakes and abuses of its past missions, the shortcomings of its present efforts, and the path to its renovation of future endeavors that embarks toward a different end – not a nature sustainable *for us*, but a nature nurtured and cared for *by us*.

Fred Van Dyke
Mancelona, Michigan, USA

Acknowledgments

To say that many individuals contributed to the writing and production of this book is cliché, but true nevertheless and demanding confession. Hannah Ferguson, Commissioning Editor on the Environment and Sustainability at the Taylor and Francis publishers, extended the invitation to me to produce this book, and has suffered the consequences ever since. I am deeply grateful for her trust in me to complete this effort. Serena Watkins, Hannah's Editorial Assistant, did the work of preparing the infrastructure to make my effort publishable. Five anonymous reviewers of my initial proposal, although diverse in their assessments of my ideas, all contributed insights and guidance that proved invaluable to the book's production. Dave Warners and Noah Toly of Calvin University (Grand Rapids, Michigan, USA) extended me the honor of doing my research and writing as a Visiting Fellow of Calvin. My greatest appreciation is directed to Sara Haig Ali, a most talented personal assistant who proved essential, resourceful, and creative in every step of the location, acquisition, permission, and (in some cases) original creation of the book's illustrations. Not one of them would be in place now without Sara's skill and dedication.

Six authors – Sabrina Pania (Solomon Islands), Thabang Sibiya and Izac Smit (South Africa), Silva Namalwa (Kenya), and Hesti Tata and Bambang Promono (Indonesia) – contributed personal stories of their conservation efforts in local communities in the midst of the constraints and challenges imposed by the Anthropocene. Fifteen reviewers – of varying backgrounds and expertise, but all with a keen eye for detail, truth, and good writing – contributed guidance to my research and development of the book's contents. Any remaining errors of fact or interpretation are entirely my own.

Reviewers

Dean Biggins, *United States Geological Survey*
Sara Bodbyl Roels, *Van Andel Institute*
Steven Bouma-Prediger, *Hope College*
Bidhan Das, *International Centre for Sustainable Environmental Research & Application, North Bengal International University*
Alex Greene, *Centre national de la recherche scientifique* (Thailand)
David Hoekman, *Olivet Nazarene University*
Brian Keas, *Michigan State University*
Ben Lowe, *A Rocha USA*
Kimberly Lyle-Ippolito, *Anderson University*
Paul Mugge, *Organic Farmer*
Kristen Page, *Wheaton College*
Susan Sicard, *Remington Academy*
Bram Steenhuisen, *World Wildlife Fund*
David Warners, *Calvin University*
Rick Westerhof, *United States Fish and Wildlife Service*

1 Confessions from the Beartooths

Of Meadows and Mountain Goats

I was sitting with my research assistant, Steve, in a swale of green and growing black-topped sedge near the edge of Montana's Silver Run Plateau. It was late June. The sun was declining in the west, and shadows were growing longer. I had just begun my preparations for dinner for both of us on my alpine stove. We were both looking north toward the slopes that descended into the West Fork canyon near the town of Red Lodge, USA. At our elevation of nearly 12,000 feet, you could see a long way. The West Bench splayed out beyond the canyon, and farther west slopes climbed to the heights of the Froze to Death Plateau, the home of Froze to Death Mountain and Froze to Death Creek. Allegedly the names came from experiences of Native Americans there, but their own words for their ordeals were not preserved. Whoever survived to place the name on a map apparently did not suffer the prophecy it foretold, but I have been there. It was a fitting descriptor, even in the summer months.

The rich blanket of sedges and forbs (non-woody flowering plants) growing luxuriantly on the Silver Run in the comparative warmth of, perhaps, 50°F made us glad to be where we were. As the stove's small blue flames were just beginning their work on dinner, we noticed movement. A brilliantly white head appeared, or better, emerged from the plateau's northern edge. The head began to rise, two small black horns atop a long, narrow face. It was a mountain goat (*Oreamnos americanus*). And we knew enough now, from studied use of our topo maps, that it had climbed the upper end of Timberline Canyon, the steep, deep dwelling of Timberline Creek, which, at its upper end, was home to Timberline Lake. The goat took no notice of us – or ignored us (we were instantly alert and immediately stock still) – and continued deliberately, with no sign of exertion, to rise (its movements too serene to be called "climbing") above the plateau's edge and onto the green sedge before it. It began to feed. Head down, eyes watchful, but showing no alarm.

This sight would have been pleasure enough, but as we were studying the goat, a second appeared behind it, following the same path, and soon engaged in the same effort. Then a third, then a fourth, and in less than a minute there were not only billies (males), but nannies with their kids, all spreading out before us, rising like a white sunrise over the edge of the Silver Run. Counting is the glory of science, and I found myself, as a young and often too serious wildlife biologist, counting the goats. Their numbers rose, like their profiles, until, with the last arrival, there were 16, all grazing on the banquet of sedge and flowers before them – alert but serene.

The only movement I can remember making at that moment was to twist, as deftly and inconspicuously as possible, the small red lever on the stove to turn off the flame, ending

DOI: 10.4324/9781003404415-1

our dinner preparations. After that, a considerable time passed, but I was not conscious of its length. I recall now that I did not even think of the observation as a time interval at all. The sight was simply too absorbing. Eventually, the goats, for reasons and pleasures of their own, slowed (but did not entirely stop) their feeding, and began to move back toward the edge of the plateau. They disappeared slowly over the edge in ones and twos, effortlessly vanishing as if "sinking" into the canyon below. At last, only a single billie remained. Was he the goat who appeared first? He looked back toward us for a moment, then sank out of sight below the rim.

When the goats were gone, I became aware that the sun had descended below the edge of the Silver Run. The shadows, which had been lengthening, were now disappearing altogether and the colors were fading from the plants around us. The first stars, perhaps two or three, were beginning to appear. We were hungry, so I resumed our dinner effort. We had something to eat in a little while and were glad for it. But I don't remember either of us speaking. It was not so much that we were at a loss for words as that perhaps we realized that words themselves would be a loss – a diminishing of what we had seen by trying to frame the experience within the boundaries of language that could not contain it.

I cannot presume to speak for Steve, or guess his thoughts, but I remember my own. Or rather, I remember my feelings, which were stronger. Of all the things I could have felt, I was surprised to realize that what I was experiencing was an overwhelming sense of what I can now only call "privilege." Why had I been permitted to see this sight, ordinary enough in the telling, but so remarkable and moving in the seeing? It would be years before I would find any words that could imperfectly express my state of mind. They came from the voice of Henry David Thoreau.

> There are strange affinities in this universe – strange ties, stranger harmonies and relationships, what kin am I to some wildest point among the mountains – high up one's shaggy side… But why should I find anything to welcome me in such a nook as this?[1]

What kin indeed? And why should I find anything to welcome me? The goats did not offer a greeting. Yet I did not feel slighted, only honored I had been permitted to witness this "ordinary" event in their lives. What right had I to see such a moment of complete serenity in the life of these creatures, a serenity that seemed to them at this place and time as natural as breathing? And, in remembering these things, I also remember now with the perception of a privilege unworthily bestowed, there was a sense of overwhelming *joy*. That word, imperfect as it may be, will have to do, as it best describes my sense and state of mind. The thought that was the companion of this feeling was something like, "How good it is, and how glad I am, that there is a place in the world where mountain goats can 'enjoy themselves' in a life and place unique to them, that is their own, and where there is no one to make them afraid." And I found myself wishing that the world I knew might be more like the world I had just seen. Perhaps my feelings were like those of the essayist Wendell Berry on a spring morning upon discovering bluebells strewn across the ground beneath the trees of a great old wood. "For me," says Berry, "there will always be the sense of joyful surprise with which I found them – the sense that came suddenly to me that the world is blessed beyond my understanding, more abundantly than I will ever know."[2]

We are living in a time humans have begun to call the *Anthropocene*, the "age of humanity." It is an age when human influence has grown to surpass the strength of the Earth's "natural" forces of ecology, geology, and climatology combined, making humans the dominant shaping influence of planet Earth. Many voices proclaim, and some celebrate, this "death of nature," and exult in the revelation that all our past ideas of "the environment," especially of wild nature, were only constructs of human imagination. The natural world is now but an artifact of human activity and choice. There is some basis for these convictions. The presence of the mountain goat in the Absaroka Beartooth Wilderness is itself a product of human contrivance. Goats from neighboring mountains in this part of the state could not (or would not) cross intervening, lower-elevation prairies, and so did not come here. Human beings (specifically, biologists of the Montana Department of Fish, Wildlife, and Parks (FWP)) brought goats to these mountains in the 1950s. They are thriving and at home in the Beartooths now, and every new goat born into them will live out its life with a memory of no other place. The goats are now a part of our relations with the Beartooths. To paraphrase Thoreau, a person's relation to nature must be a personal one. Accordingly, I begin this book with personal confessions, confessions of those things that have shaped my relationship to nature in time and place, things that have remained with me even when times have passed and places embedded in those times are far away. These are confessions of experience, and memories of them. My understanding of nature has been shaped through many experiences in many settings, but perhaps none more vividly than in the Beartooths. My understanding of the Anthropocene must begin here.

The Substance of Place

Human constructs are a part of the identity of the Beartooth Wilderness (see Figure 1.1). Its boundaries and legal designation were congressional creations in the United States in 1978. Those limits eventually enclosed an area of 944,000 acres, 1,475 square miles (3,820 square kilometers), about the size of the US state of Rhode Island. Such demarcations made the Absaroka Beartooths the sixth largest wilderness in the conterminous US. The designation, dedication, and boundaries of the Beartooths are human creations.

Figure 1.1 The Absaroka Beartooth Wilderness Area of south-central Montana and northwestern Wyoming (USA). Nearly 1 million acres (400,000 hectares) in size, it was formed from roadless portions of the Custer, Gallatin, and Shoshone National Forests and designated as part of the US National Wilderness Preservation System by the US Congress in 1978. USDA Forest Service photo.

The world of the Beartooths is not. It is a world of nature, affected by humans but not determined by them.

Before they were a "wilderness," these same lands were roadless portions of three national forests – the Custer and Gallatin in Montana, the Shoshone in Wyoming. The area's two-part name acknowledges both substance and history. The land is dominated by two mountain ranges – the Absaroka, composed almost exclusively of extrusive (metamorphic) rock, and the Beartooths, which are granitic (igneous). The names also recognize people and places. The dominant tribal nation in this region called themselves the Apsáalooke (phonetically, ah psa ro gee), meaning "children of the large-beaked bird." The first white settlers mistranslated their name into "Crow," and transliterated the Indigenous word as "Absaroka." The Apsáalooke, by their own accounts and testimonies, traveled in these mountains, hunted in them, and revered them, but did not make permanent residence in them. George Reed, former Chairman of the Crow Cultural Committee, wrote in a letter to the Custer National Forest in 2007, "All the mountain ranges in the territory of the Apsáalooke (Crow) Nation are sacred because that is where First Maker travels as he watches his creation."[3] I found stone circles suggesting tipi rings in the stream basins and draws of the Line Creek Plateau's east face at the edge of the Custer Forest, but these were few. The nearest Apsáalooke village settlements were farther east, in the Bighorn Valley, although the full extent of the Apsáalooke Nation extended throughout portions of Montana, Wyoming, and South Dakota. Today it is the more proximate Bighorn Valley that is home to the present Crow Nation (the Crow Indian Reservation), where tribal members reside on 2.3 million acres (an area slightly larger than Yellowstone National Park) that include some of their traditional lands.

The Absaroka Beartooth Wilderness is part of a still larger landscape, the Greater Yellowstone Ecosystem (GYE). The Beartooths, more wilderness areas (the Lee Metcalf Wilderness), and national forest lands (the Custer and Gallatin) form the northern front. To the west are the Beaverhead and Caribou-Targhee Forests and adjoining national wildlife refuges. To the south lie the Gros Ventre Wilderness and the Bridger-Teton National Forest. To the east are the Shoshone Forest and the North Absaroka, Teton, Washakie, and Popo Agie Wilderness Areas. At the core are two national parks, Grand Teton and Yellowstone. At 22 million acres, enfolding 34,375 square miles, the GYE is one of Earth's largest remaining intact temperate ecosystems.

Standing on the Silver Run Plateau, it is impossible to forget just how large this landscape is. To the south, my guides for daily work were Pilot and Index Peak, northeast of Yellowstone National Park, distinct yet remote at the edge of the far horizon. To the west, only slightly closer, Whitetail Peak rises. To the east, just below the Plateau's nearest slopes, the main canyon of Rock Creek, and to the north, not much farther, the flowing ranch lands of the West Bench, and the town of Red Lodge, my home. The public use of global positioning system (GPS) technology was still in its infancy, and my topo maps and compass were constant companions and helpful guides. But in time I didn't need them. The land itself was constant testimony to where I was, and a trustworthy guide to my next destination.

Coming to Montana

I had come to Red Lodge from the more nondescript landscape of Indiana, where I had been a new assistant professor at a small college. My work was fruitful and enjoyable, and I probably learned more about life from my students than they learned about biology from me. But having come to Indiana from the Dixie National Forest and Escalante

Canyons of southern Utah, where I had been studying mountain lions as part of my PhD research, I longed to be learning and doing, not reading and talking. I had the privilege of teaching some wildlife and conservation courses, yet I sensed that even in these I might be getting "stale," speaking of things I had yet to study firsthand – much less "do" in any way that might actually benefit wildlife.

I accepted an invitation to take a position as a wildlife biologist in Montana with the FWP. The new biologist was expected to live in Red Lodge, home to the Beartooth Ranger District Headquarters. The pay was low (although the FWP assured me they provided a great wardrobe). There were no organizational or administrative responsibilities or supervisory powers. In addition to the standard elements of my on-site interview in Billings, the FWP's regional headquarters, I was also tested on my ability to drive a department vehicle, a Ford F-150 pickup truck. Standing in the headquarters parking lot on that snowy November day, the Regional Wildlife Manager, Charlie Eustace, handed me the keys. Nervously I dropped them into a fresh snowbank. "Well, that's just what we need," Charlie laughed, "some ice in the door lock." I would realize with time that Charlie had a great sense of humor. I didn't fully appreciate it at that moment. And I would come to love the truck. It never failed me, despite the torturous travel and difficult roads I subjected it to. "Old Blue," as I came to call it, may have been a machine, but over time it became more like a personal companion.

If I had been able to afford a career counselor when I was in Indiana, Montana would not have been on the shortlist of rational recommendations for my next job. It would certainly be an "adventure" for my wife, Linda, and my two small daughters, Rachel and Liel, who were all happy in Indiana. But I wanted to see Montana. I wanted to do "real" wildlife management. And I wanted to work with "big hairy animals." I had already done that in Utah with mountain lions, but now I wanted to study the herbivores. None of these were compelling, or even very intelligent, reasons from a family standpoint. Today I cannot imagine how I managed to persuade Linda to go, but we were both adventurous youths, and off we went.

My assignment was to create an integrated timber–wildlife management plan involving both the FWP and the Forest Service (FS). And nobody knew exactly how to do it, or even exactly what the assignment meant. With the retrospect of a long career, I understand now that perhaps the greatest thing about doing conservation is the endless opportunity for creativity. My first efforts focused on determining exactly what people *did* know about timber and wildlife in the Beartooths. The District was "timber poor," so much of it being wilderness and off limits to cutting. Thus, its timber sales were small compared to its wildlife populations, which were large and getting larger, especially among herbivores like elk (*Cervus canadensis*) and moose (*Alces alces*). Every hunter in Red Lodge (which included all of the men and most of the women) knew where to look for elk in November, but were less sure the rest of the time. There was no systematic information on habitat use, or even a completely clear idea of how many different populations were present. And there were no informed notions about exactly how elk would respond to making changes in their habitat, managed, unmanaged, or mismanaged. But the staff of the Beartooth District were cordial, supportive, and helpful. And at the end of our first meeting, one staff member reached into his pocket and handed me a small brown round-headed key. "Take this," he said. "It will open any lock on the Forest."

Over the next two years I determined that we were dealing with several distinct elk populations: three large ones – the Silver Run, Line Creek, and Picket Pin herds[4] (see Figure 1.2) – and three smaller "colonist" populations[5] arising from them and moving into

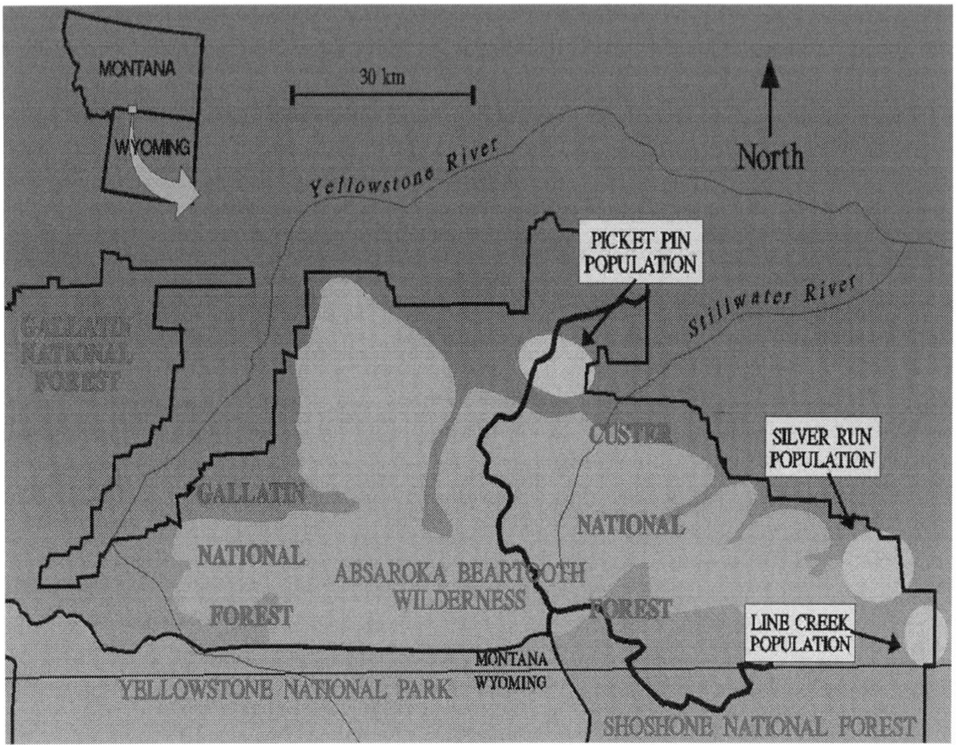

Figure 1.2 Distribution of three elk (*Cervus canadensis*) populations using portions of the Absaroka Beartooth Wilderness Area in Montana and Wyoming (USA) during the author's studies from 1988 to 1992. Figure design by Fred Van Dyke. Reprinted with permission of Creative Commons.

areas where local opinion assured me that "there's no elk there." But there were. I was able to begin comparing the habitat elk actually used to the habitat they did not ("available" habitat versus "used" habitat) and current range use to past range use (there was some now decade-old telemetry data from previous studies to permit that assessment).[4] I wondered how well elk were doing in these places, and so evaluated the relationship between plant nutrient content and elk nutritional status in different seasons on different ranges.[6] I wondered if forage availability and nutritional quality of the plants might be changed by prescribed fire, and would elk care? So I measured plant and animal responses to prescribed burning on the lower-elevation winter ranges, and found out that fire changed both.[7, 8, 9, 10] Oil companies could lease sites in the forest for exploratory drilling, so I could also evaluate how the response of elk in a population exposed to oil drilling (Line Creek) compared to that of a nearby population that was not (Silver Run).[11] In all these queries, my radio-collared animals were valuable informants, but over time I came to pay as much attention to their tracks as their radio frequencies, and often I could smell them, if the wind was right.

I got more grant money, expanded the study to include moose,[12, 13, 14] collared more elk in all the populations (large and small), and started a wildlife research internship for college students. I was having a wonderful time! Near the end of my first two years, Charlie told me, in my annual review, "Fred, you're doin' a great job. I think I can get another two years of funding for this study!" Two more years passed. At the end of the fourth year, Charlie declared, "Fred, you're still doin' a great job! I think I can get money for one more year, but after that, I think the funding's gonna run out."

In the meantime, a permanent job opened up with the FWP based in Bozeman, another regional headquarter town. I might have been able to get it. I considered applying, but, after reflection, did not. I wanted to "finish" the Beartooth work, even if I wasn't exactly sure what "finishing" meant. It was more certain the funding would run out and finish me. I liked the Department. I thought Charlie was a great boss. But I realize now that my loyalty was to the Beartooths, not to the FWP. That loyalty developed over time. It was tied to the creatures I studied and the places they traveled. The elk and moose were no respecters of the Beartooths' boundaries. They followed their food, and, in the hunting season, they followed paths to safe places. In doing so they crossed the boundaries of national forests, wilderness areas, and private land. So, I did too. The first step of knowing another life is to be present with it, to know what it knows, to understand what it needs, and to accomplish those things that might do it good.

The Breadth of a Living Landscape

Even before I could begin to understand their effects, the land was working its will on me. The FS allowed me to use one of their horses, Amigo, the better to travel the country. I remember a sky one winter day when Amigo and I traveled to the head of a drainage called Meyers Creek, one of the principal use areas of the Picket Pin herd. It was still early in what had been a mild winter. I was searching, trying to discover the paths and haunts of the elk it was my job to "manage." But it was the elk's intention to manage themselves and have nothing to do with me, their "manager." I never found them on that ride, but the day was not without reward. Amigo and I finally reached the Meyers Creek Pass, a low point ("saddle") in a long shoulder of land called Meyer Mountain. There we could see the divide, where rainfall behind us flowed south to Meyers Creek, and, ahead, north to Dry Fork. The sky was radiant. In a basin below us lay a jewel on the land called Moccasin Lake. I was tempted to ride down, not with any hope of finding elk, but because its beauty so compelled me. But Amigo was sweating, and he had been doing all the work. So, we remained a moment, then I turned him round, and we retraced our path to more familiar places.

As winter progressed to spring, I began to gain more understanding of the relationship between land and beast as I plotted the routes of elevational migrations of these elk. Winter is long in the Beartooths (often from late September through at least mid-June in the high country), but spring comes eventually, and summer follows, alternately with gentleness or violence, and always with steadfast perseverance. I recall a day in May measuring the progress of elk moving slowly up the slopes toward the alpine meadows on the plateaus, where the land was beginning to prepare a banquet of fresh sedges and grass. But the elk could not go there yet. The upslope drifts were too deep, and their forage not yet present beyond them. One herd followed the edge of a sparkling stream, Wyoming Creek, which originated in the state of its name before flowing north into Montana. Its waters patiently led the elk toward the meadows by gentler paths than they would have found in attempting direct ascent up snow-covered slopes. I followed their tracks and measured their foraging. As the snow receded, foot by foot, the elk climbed slowly upward, day by day.

Sometime in the course of this work (I cannot recall exactly when) it struck me that these creatures were measuring time by the pace at which grass grew and snow melted, things I had been taught were too slow to mark anything but idleness ("Don't let the grass grow under your feet," was a common parental reprimand if I approached a task too leisurely). They were measuring time by the pace of the needs the land was meeting. And in that moment, rather suddenly, I realized it was my clock that was broken, recklessly racing ahead of the different, more authentic pace of nature's life. "Time is told to it mutely and immediately," said Wendell Berry, referring to the pace of nature, "with perfect accuracy, as it is told by the heart in the body. Its time is the birth and the flourishing and the death of the many lives that are its life."[15] Something happened to me that day, and my own clock began to reset itself. I could still choose to measure time as other people did, and accept, for courtesy and convenience, the schedules that bound us. But I began to realize this was not the time or pace that truly mattered, or the one in which nature lives.

Because winter persists and spring tarries, summer comes to the plateaus with an almost explosive suddenness. Yet even after arrival it is ever unstable and untrustworthy in its comforts. I learned from experience that it could, and often did, snow in any month of the year in the high country, and freezing temperatures were common overnight. The growing season is usually only 45 days, or in a warm year 60 days. So, the mantra of an alpine plant is "So much to do, so little time." The growing season comes slowly, but when it does, the plant's existing nutrient reserves must be mobilized quickly, facilitated by the plant's high root-to-shoot ratios. Nutrient-use efficiencies from stored biomass and rapid rates of carbon assimilation (to make the products of photosynthesis needed for growth) are both high in alpine flora. Conservationists seem to think they can conserve mountains, animals, and flowers at the same scales of place and time. This is an illusion. Their clocks are aligned, but in different time zones, and those who would presume to manage them all must be able to move at ease among them.

As the physiology of alpine plants is complex, the results are spectacular – not only in the massive and often synchronous blooming of their flowers, but in the vividness of the colors they display. My first trip to the Silver Run left me in a state of shock. I found myself simply staring at the plants for a long time, and up close – the yellow of the alpine avens (*Geum rossii*), the purple of the sky pilot (*Polemonium eximium*), the red of the alpine paintbrush (*Castilleja nana*), the white of the American bistort (*Bistorta bistortoides*). All were dazzling. Eventually remembering my objectives, I got to the work of measurement and clipping (for later biochemical analysis). But when I was descending the plateau, I found myself thinking, "Truly, I never saw 'blue,' or 'red,' or 'green' before now. I never saw 'color' until today."

Others were affected similarly. One assistant, Chris, spent his first day with me ascending the Corral Creek Trail, one way to the heights of the Line Creek Plateau, east of the Silver Run. Chris was a varsity college soccer player in outstanding physical condition. Still, I should have started his internship with something easier. Climbing a steep trail composed mostly of rock while carrying a full backpack and ascending 3,000 feet in 4 miles is strenuous. But my timing was right. We emerged from the forest at the plateau's edge and stepped into its meadows with a sudden greeting of almost blinding brightness, a brilliance of growing and glowing color of every hue as far as the eye could see. We sat down under one of the last remaining trees at timberline. After some water and a little nourishment, we went to work, estimating the abundance and coverage of different species in this, the summer "buffet" of our study animals.

Over lunch, I asked Chris if he had seen alpine meadows before. He shook his head quietly. I probed a little further. "What do you think of it?" He was silent for a moment, and then answered, his voice trembling ever so slightly. "I almost broke down and cried," he said, almost whispering. "It was so beautiful, like nothing I..." He didn't, or couldn't, finish the sentence. Perhaps tears are the only right reaction to the depths of such beauty: "too rich for words, for Earth too dear."

The alpine plants must flourish in their short growing season and endure cold, often frosty nights. But such nights give the sky a vividness it could never know on a warm summer evening in more congenial places. And nights were as much a part of the work as days, if for different reasons. The roads Old Blue could use ended at or before the wilderness boundary. There were miles to go – on foot, and all of it uphill – before data collection could begin. If the two-to-three-hour walk ahead was added to an equally long walk back, the day would be spent with an abundance of exertion but an absence of information. If you wanted a good day or more of work on the plateau, you made plans to spend some nights there.

For all the time, over multiple decades, that I have spent outdoors sleeping in tents (or not in tents), I still find that the first night is when I have the most trouble falling asleep. So, on these Beartooth plateaus, I often climbed out of my bag, and out of my tent, to sit among the sedges and forbs, looking at the sky. There was no "light pollution," and such true darkness made the universe above a vivid, throbbing realm. There was true "progression" to the appearance of stars and planets, appearing almost one at a time, and yet, in what seemed no time at all, there were thousands.

I had seen stars and their beauty in other times and places, but I recalled past observances as being rather "inert." Not so these nights. The number of meteors I had seen flash across the sky before I came to the Beartooths could have been counted on the fingers of one hand. Here, at 10,000, 11,000, and 12,000 feet in the alpine, shooting stars seemed almost as numerous as static ones – flashing and streaking through the heavens, so constant I dared not look away from the canopy for even a moment for fear of missing even one. After the first experience, I supposed I had been fortunate enough to see a meteor shower. And then it happened again, the next night, and in many of the nights to come! I cannot tell you this happened every night, but it happened often. And I wondered at my ignorance, thinking of "outer space" as "dark." It was pulsating with light, vibrant with celestial animation.

Eventually, even on the first nights, I managed to get to sleep. Upon awakening, I learned that morning is a real "event" in the Beartooths. Not just because the sun is beginning to rise, unclouded, unhindered, above the rim of the plateau, but because of the stark and piercing coldness. On the Beartooths' plateaus, it was not merely light that was an occasion of the morning. It was also *warmth*, and the second came more slowly than the first. I often found some flakes of ice on my skin, and there might be more on the plants around me. Yet they also offered sustenance, even in such "frosty" conditions. Grouse whortleberry, or "little-leaved huckleberry" (*Vaccinium scoparium*), a low-growing diminutive shrub, was one of the few fruit-bearing plants to be found in some of the alpine meadows or beneath the white bark and lodgepole pine along their rim (see Figure 1.3). The fruits are tiny, but in a favorable year, abundant. With some persistence, and a small bowl to hold the harvest, I usually gathered a good crop. Properly mixed with oats, they could transform alpine-stove oatmeal (rather bland fare in itself) into something better, more unique (and unobtainable) than one could have ever found at the most exclusive gourmet restaurant. The picking, like the eating, was best done in the morning, permitting my fingers, vital instruments of data collection, to begin to thaw.

Figure 1.3 Grouse whortleberry (*Vaccinium scoparium*), a low-growing huckleberry often found at the edge of alpine plateaus in the Absaroka Beartooth Wilderness Area (USA). Photo by Paul Dawson.

Hikers and backpackers are forever warned to be careful and well-provisioned with liquid refreshment. The microscopic parasite *Giardia duodenalis* causes the diarrheal disease *giardiasis*, and its symptoms can persist for up to six weeks. *Giardia* can move to water from animal feces (like elk poop), and there were plenty of elk defecating in the plateaus. Yet I found the water foaming and splashing in these alpine meadow streams as compelling as the plants they nurtured. Like the colors of the plants, so vivid and alive, it seemed that the first day I saw these waters, splashing, foaming, sparkling, as if electricity was running through their currents, I would have said that I never saw water before. As a biologist I knew well that *Giardia* could be present. But the thought of turning away from it, from *them*, of *not* drinking what they offered to me, seemed an insult to nature itself, an unpardonable discourtesy of a thoughtless oaf. I did not debate with myself for long. Extracting a small cup from my pack, I filled it at the very top of a white splashing wave and took a long drink. It tasted like no liquid I had ever imbibed before. For the next five years, I drank from these alpine streams. I never suffered any ill effect.

A Wilderness Companion

As the breath of the Beartooths flowed into my professional life like the water from their streams, it respected no boundaries around anything else. I first began to notice this with my dog, Hannah. On days when I knew I would be giving attention to plants, and unlikely to encounter animals, I would take her with me. Hannah was never intended to be "my dog," but that was how it eventually worked out. Hannah was a German shepherd–lab mix. Linda and I got her to be Rachel's "birthday puppy," celebrating Rachel's first birthday in Red Lodge. We presented Hannah in a large single-handled wicker basket. Rachel and Hannah were overjoyed to meet one another. I remember the meeting vividly now, not only because of the enthusiasm of their greeting, but because this was the last time Hannah would ever fit into a basket.

Blessed with a healthy diet and vigorous (if mongrel) genetics, Hannah grew like a proverbial weed. She had the energy of an aerobics trainer on steroids, and, in the pains of her "teething" stage, used her mouth like a chainsaw stuck at high speed. So incessant was her chewing (on any object within reach) that she could well have been named "Hannah the Piranha." Carpets, furniture, and my backpack were among the victims. I knew things in the family were getting tense, and it came to a head one night at dinner.

"Fred," said Linda as she passed me the green beans, "there is something that Rachel, Liel, and I would like you to do for us."

"What's that?" I asked, my suspicions already beginning to rise.

"Well, we've made these signs, and we'd like you to put them up around town." Linda handed me one of their creations. In large letters, it began, "FREE to a Good Home…" "We can't take her anymore," said Linda. "We have to find a new place for Hannah."

I chewed my beans a little more thoroughly. "Let me get this straight," I said slowly. "You want to get rid of Hannah, and you want me to do your dirty work for you." I set my fork down on the dinner table. "Well," I continued, "I like that dog [I seemed to have either forgotten or forgiven the recent damage to my backpack] and I'm not going to do it. If you want these signs around Red Lodge, put them up yourselves."

The rest of the dinner was uncomfortably quiet. I don't know whether everyone was angry with me for the way I had put things, or ashamed of themselves for coming up with the idea in the first place. I did realize that what everyone needed was "space." Fortunately, I had a way to create it with some unused hours of annual vacation. "I'm going fishing tomorrow," I said, "and I'm taking Hannah with me!" Facial expressions around the table were hard to read, somewhere between relief, puzzlement, and resignation. But I kept my word.

I took Hannah to the Lake Fork Trailhead, the route to my favorite lake, Black Canyon. The first 5 miles were reasonably easy, a mostly level path on a good trail, and Hannah kept up admirably. The final 2 miles were harder. The trail became narrower, rougher, and steeper. Rocks and logs increasingly blocked the path. These were formidable for still- (relatively) small Hannah, but she determinedly climbed over each one, noticeably weary but with no sign of quitting. The final last half mile was a near cliff-angled slope of small- to mid-sized boulders that "dammed" the lake at its north end. There was a "trail" through the rocks that could be followed with experience and some detective skills, paying careful attention to the unpredictable appearance of primitive "trail markers" (small rock piles placed on top of some of the larger boulders by thoughtful past hikers). And there were gaps (between the boulders) that had to be jumped (actually, leaped). Courageous creature though she was, the first such gap stopped Hannah cold. She looked up at me with a pitiable expression, wanting to follow, but fearing the deep crevice between the stones. I bent down, picked her up (my fishing rod and gear were strapped over my back, so my hands were free), and carried her, by various leaps and jumps, through the final 100 yards of the ascending escarpment, setting her down on the other side. The beauty of Black Canyon Lake was now in full view.

The daily limit of cutthroat trout (*Oncorhynchus clarkii*) in the Beartooths was ten, and Black Canyon rarely disappointed. After two hours the stringer was full. We set out for home. Emboldened by the experience, and aided by the pull of gravity on a downhill slope, Hannah now attacked the gaps with the enthusiasm of a foot soldier on a bayonet charge, bounding enthusiastically (dare we say "gleefully"?) from boulder to boulder without hesitation. We reached a better trail soon and enjoyed a quiet return, delivering our trout to the family kitchen. That was the last time I ever had to carry Hannah over anything. Within a year she had grown to such proportions that I believe she could have claimed the Alpha Female position in any respectable wolf pack (a role that, with age, she increasingly asserted in my home, much to Linda's consternation). Her teething stage now passed, she was no longer Hannah the Piranha. But to our family, and to some of the locals who often saw her heading out of town with me in Old Blue, she became Hannah Montana. Hannah traveled with me across forest, wilderness, district, and administrative boundaries that defined my work, but took no notice of any. She filled her lungs gladly with the air that crossed all borders and gazed on all their vistas with no concern for their limits.

The Human Imprint – Fishing, Herding, and Mining

As the elk and moose I studied led me to many secret places of the Beartooths in my professional life, so fishing drew me to new and different sites in my personal one. The Beartooth Plateau, according to some sources, has the highest density of high-elevation lakes in the world. There are 948 high mountain lakes in the Beartooths and its adjoining lands – 318 have fish, 630 are barren. Of the former, 204 have self-sustaining populations, 114 are stocked by the FWP. Historically the number of fishless lakes was much higher. Many of the high lakes, especially those nestled in glacier cirques (steep-sided, narrow, hollowed-out areas at the head of a mountain valley, formed by glacial erosion), had no inlet or outlet streams that fish could use to reach them. And many that had streams had these waters interspersed with high waterfalls, roaring cascades, and steep gradients that no fish, no matter how determined, could overcome.

In their own homelands, First Nations people in Canada often stocked lakes that lacked fish, but there is no evidence that the Apsáalooke or other Native Americans did so in the Beartooths. FS staff and long-time residents of Red Lodge told me that some lakes were stocked in the early 1900s by locals who physically hauled trout to the lakes in milk containers borne by mule-drawn wagons (and some by hand, carrying the fish in buckets!). By the 1960s the FWP had taken over the effort using a less strenuous but more systematic method of dropping fish into the lakes from an airplane. The waters received their new residents but did not always welcome them. The high-elevation cirque lakes can plunge steeply to great depths within a few yards of the shoreline, offering none of the shallow water areas that trout need to spawn.

The FWP's own actions in lake stocking were themselves targets of criticism. Pat Marcuson, who made the earliest and most comprehensive studies of Beartooth Lakes beginning in the 1960s, found that FWP stocking placed aggressive brook trout (*Salvelinus fontinalis*) in high lakes that outcompeted other species of trout for food, but then failed to reach large sizes themselves, producing teeming populations of stunted brookies where there had once been large native fish. Yellowstone cutthroat trout reared at a state fish hatchery in Big Timber, Montana, and then stocked in other mountain lakes, suffered from a lack of genetic diversity. The inbred fish hybridized with native cutthroat, damaging the genetic integrity of the latter (see Figure 1.4). Worse, pilots sometimes dropped the wrong species into lakes targeted for a different species. The results were predictably bad[16] and a great personal embarrassment to the pilot!

Figure 1.4 Cutthroat trout (*Oncorhynchus clarkii*) are now the most common, and most commonly stocked, fish species in the lakes of the Absaroka Beartooth Wilderness Area in south-central Montana and northwestern Wyoming (USA). Public domain.

Over time, FWP stocking improved with better research and practiced technique. Different lakes were stocked in different years, but in those without spawning habitat, the planted cohort grew, lived, and died together, usually in a period of four to five years. During my time in the Beartooths, I learned that if one knew (or could determine) the stocking plan and schedule, it was possible to work out an optimal "circuit" of lakes that would produce catches of the oldest and largest fish in a given year. If you judged wrongly, you brought home only the scrawny.

So it was that Beartooth Lakes came to bear a human imprint, but their own geology limited the effects. This was not the only human impression. As recently as the 1960s domestic sheep grazed on some plateaus, and some trails they formed, with the guidance of their shepherds, still remain. Although many are now discovered only with difficulty, others persist as visible, useful, and passable routes to and through the high country. Miners searched the Beartooths for gold, and some places in the wilderness still bear names that testify to their efforts, like Placer Basin in the East Boulder Valley of the Gallatin National Forest ("placer" mining refers to removing gold or precious stones using moving water flowing over a stream's sediment deposits – the simplest form is "panning"). I remember seeing Placer Basin for the first time on a mid-June day. The mountain valley spreading out from each side of the East Boulder River (here little more than a modest stream) was just emerging from the remnants of winter snow. To say that the valley on that day was "green" would be like calling the Mona Lisa a "portrait." Even words like "verdant" and "dazzling" fall far short. The valley's color seemed alive and throbbing with vitality so strong it felt as if the landscape could barely contain it. And seeing it, I remember thinking, "If in this world there be any place more beautiful than this, no human eye could bear the sight of it." The Beartooths were not untouched by humans, but the land had reached back and graced the coarseness of the human touch with a beauty of its own, far more than we should have had a right to expect, much less deserve. We merit from the Earth only what we give to nurture it, but sometimes receive by grace more in return than we warrant, once we begin to make even the smallest efforts toward its care.

Seasons of the Wilderness

The Beartooths mark time by seasons. In every year summer fades into fall and fall into winter. A Labor Day snowstorm often covers the high meadows. In a few days this melts, but the harbinger is clear, and the elk know its meaning. In the weeks ahead they begin their descent, no longer in the large cow-calf herds of summer, but in smaller tribes of females and young, the males in still smaller groups, and some alone. In one year at this time the elk of the Line Creek Plateau began moving to lower elevations to the east later than usual, using stands of Douglas fir at the heads of streams and ravines with names like Gold Creek, Ruby Creek, Mill Draw, and Robertson Draw, the last drainage memorable for having the remnants of an old makeshift corral at its head. The elk were more selective now, valuing cover as much as forage. Hunters were searching for them. Documenting range use in fall was as important as summer, especially because elk were using some stands that might have future timber value. I spent the fall locating and measuring the stands and slopes bearing the greatest use by every herd. October turned to November. Snow now covered the high country and was working its way down the slopes.

Working the fall and winter range of the Line Creek elk, I often used a primitive lodging provided by the FS known as the Line Creek Shack. No one ever told me how long it had been there. It never occurred to me to ask. The first time I decided to spend the night I left Old Blue at the Shack, planning to return that evening. From the Shack I was better placed to measure the growth of forage in the burned and unburned areas of North Line Creek, which surrounded it, the next day. But today was devoted to studying characteristics of timber stands on the plateau's east face. I worked my way north, focusing on stand measurements and elk use. Hannah was with me. Zeal for knowledge and pleasure in its discovery are a great joy, but can sometimes be a dangerous combination and a distraction from other elements of your surroundings, especially in November on the East Face. As I made my last measurements in the Gold Creek drainage, I realized I had finished later than intended. The sun was already below the Plateau's rim, and the Shack was miles south. I packed up, and Hannah and I headed out.

The snow had begun falling lightly in the afternoon, but was finding more strength as our hike progressed, following a route called the Face-of-the-Mountain Trail. There had been some snow on it as we had come north that morning. It was getting deeper now. The wind and snow gathered together as if encouraging one another and the white flakes began to fall in diagonal lines, reducing visibility. As the sky darkened, the trail grew more difficult to discern, and the snow slowed our pace. At the southeast extension of the plateau we turned the Trail's "corner" at Robertson Draw and, hiking west, reached the South Face as the last light was failing. Here, west of the old corral, many ravines converged. One led to the Line Creek Shack. The others were unnamed, nondescript draws and gulches, identical in appearance, eventually spreading out into flats of sagebrush. The sun had set, and the ravines looked even more alike in the lengthening shadows. I like solitude, but, if you experience it often enough, you begin to talk to your dog.

"Think it's this one?" I asked Hannah, pointing to one gulley among many. She looked up hopefully, but without reply, prepared to trust my judgment. I led on. As you descend any one ravine from the South Face, the ridges on each side enforce your choice. Once you sink into the belly of the gulch, you have no other ravines in sight, only slopes of sagebrush on either side. After 20 minutes of such descent, now with only dark shapes and shadows about us, I was starting to feel some apprehension. The temperature was falling, not yet causing discomfort, but beginning to get my attention. I wondered if my choice had been the right one. If only Hannah could talk!

At that moment we saw the outline of the Shack, rising out of the sagebrush, Old Blue beside it. The light was enough to finish the journey, barely. I reached the door, drew out my keyring, and found the nondescript, round-headed key. I thrust the narrow end into the bottom of an old bolt lock that bound a latch across the Shack's only door and turned the key. The lock sprang open. I exhaled a deep and grateful breath.

That evening, the Shack offered us, at a very reasonable rate, a single 8- by 10-foot room, a four-legged iron stove (in which we composed an excellent cooking fire), and some beef stew (vintage Dinty Moore) which some FS worker had stored in the cupboard (who knows how long ago), but now bubbling in a pot. I added some dog food for Hannah that I had remembered to bring, all this enjoyed by lantern light, the handle hung from a beam in the roof. After dinner, Hannah, following a relieving foray into the snow outside, returned to recline on the floor. I spread out my sleeping bag, retrieved the little "camp pillow" from my backpack, and lay both on a thin, well-worn mattress atop a black bedspring frame possessing dimensions perhaps best described as "economy-sized twin." Lying down, I discovered that the mattress was past its prime, and permitted a felt connection between each one of the bedsprings and every muscle in my back. But sound sleep is aided by strenuous effort and came swiftly enough, with no time to consider any discomfort.

Since that night, and others that I spent at the Shack, in all weathers, I have had the privilege of staying in some exquisite lodgings – the Old Faithful Inn in Yellowstone, the Grand Hotel on Michigan's Mackinaw Island, the Centro Bristol Hotel in Germany, when I was a guest speaker at the University of Bonn. But even with these accommodations for comparison, I realize now that I have known no lodging more pleasant, or more deeply appreciated, than this humble Shack on that cold and snowy night. And with that perception, I have come to understand that the desirability of a place should not be measured by the amount of ease we gain from it, but by the level of interest we have in it, and the reward of that interest in the depth of need met by it. The Shack was that place, and it belonged to the Beartooths.

The Final Summer

In my last summer in the Beartooths, I took a day of comp time (accrued from many work weeks in excess of 40 hours) to hike to an isolated place – Deep Lake, in northern Wyoming. It is about a three-hour walk, perhaps two-thirds with the luxury of a trail. The limit for cutthroat was lower in Wyoming, only six, so it took less time to catch the fish than to reach them, but there remained satisfaction in success. As I reached the end of my return that afternoon at the juncture of a paved road, the scene had changed from the morning. Where there had been an open green meadow there were now three boldly colored tent pavilions. Under each canopy were tables and chairs, and small but numerous groups of people standing about. As I looked more attentively, I realized that this gathering, which resembled a small circus, was an enterprise of a wilderness outfitter. Under one canopy the "dudes" were registering and paying for their soon-to-begin trip. In the middle pavilion they were having wine and cheese at cloth-covered tables. Under the third tent stood a picket line of saddled horses, awaiting the privilege of bearing their clients.

I should be the last person to find fault with anyone for riding a horse in a wilderness, much less doing so to enjoy the beauty of its scenery. And likewise, I should find no fault with payment for services rendered. The state of Montana paid me for mine. The dudes paid the outfitter for his. Yet I could not help feeling there was something wrong with what I was seeing. There are many pleasures to be enjoyed in a wilderness. Some are intrinsic to its nature, and to our response to that nature. These are the pleasures, and privileges, of toil and (sometimes) danger, of discoveries of your own self-reliance and resourcefulness, of the privileges of human dependence on others, and the provision of divine aid. But there are also the privileges of companionship with and relationship to a landscape and its creatures, met on their own terms and in their own places, not arranged for human comfort and spectacle, but with the courtesy we are right to extend to their unique and distinct lives, that we might respect those lives, without boorish intrusion. These are privileges bestowed upon us from another, a community of land and creatures where nature is not an assembled artifact of human contrivance, but a gracious host whose hospitality we do not deserve but are nonetheless privileged to receive. I realize now these were the privileges that swept over me as I watched the mountain goats on the Silver Run. I did not sense that these privileges would be provided for the tourists in their ride ahead, regardless of the comforts they might enjoy. In our interaction with nature, our most important lessons are not the skills we gain in "penetrating" her secret places, or the advantages we purchase from others to do so, but rather the restraints we come to value, and master in ourselves, that are expressions of our respect toward her life.

The Beartooth Wilderness has boundaries. I knew some of them well, and many times crossed others. Most days I came and left the wilderness through them. Nearly every week I flew over their limits, far above the expanse of land and the creatures within it, a passenger in the FWP's trusty Supercub, aloft with my pilot, Keith. But to know the extent of the Beartooths is not to know the life of the Beartooths. For here I found, after many days and nights in this wilderness, that a man could spend his whole life within its boundaries and never know the height and depth and breadth of all that lived within them, of all its places or its secrets, nor walk its borders all the way round and think this was the same as "seeing" the Beartooths. Scoffers might say I just didn't walk far enough (though they might have said differently if they had walked with me). I once read of a man who, in his affection for Starbucks coffee shops, made it his goal to visit every Starbucks on Earth. His quest was in progress at the time, and I do not know if he ever completed it. But to this day I consider his effort a fool's journey, one in which every visit became the means of "capturing" each Starbucks, instead of enjoying the partaking of their substance, the coffee that had once attracted him. The beauties and joys of the Beartooths cannot be captured or held within its demarcations, but they can be received in the partaking of their substance. That substance is the land itself, and the life within it. That land held pleasures, as well as hardships and dangers, which welcomed me each day. But to be known, enjoyed, and (sometimes) overcome, they had to be each received within each day, one privilege at a time, with full attention to the thing presented, not to its place in some "collection" of conquests or experiences.

As Thoreau said, "We should come home from far, from adventures, and perils, and discoveries every day, with new experience and character."[17] This is what I experienced in the Beartooths. I relate these events, and their effects upon me, as confessions, not arguments. I was not the first to set foot in the Beartooths. I will not be the last. They have no voice, but their communication of their own life is eloquent. If you can bear their company, they will reveal all. Their narrative is their own, and it makes an impression on other lives forever, if you permit it to speak. These impressions change us, if we are open to receive them. No wonder the Beartooths were revered by the Apsáalooke.

That reverence was not a romantic delusion contrived by human artifact. It was a right response to the land that met them, and they greeted it with welcome. So nature, in all its verities and places, should be so to us. The designation of "wilderness" now applied to the Absaroka Beartooths acknowledges its unity as a system and prevents it from being dismembered into component parts representative of "resources" to be added to the current economic enterprise. Such acknowledgment is a correct response to what the Beartooths really are. Such protection is a correct action to ensure what they should continue to be so, and to prevent them from being changed into something they are not. As Harvey Locke has said, "to those of us who have experienced such primeval places, claims of the non-existence of wilderness are absurd and offensive."[18]

Perhaps "fully" natural ecosystems no longer exist, but near-natural ecosystems are very much with us. The designation of the Beartooths as "wilderness" is a human construct, but a construct of correct perception and material effect, directing that perception to the reality of nature, and channeling that effect by changing our intentions toward it and our interactions with it. It is the patterns of human interaction with nature that determine the fate of "natural" nature, not the amount of nature that we choose to "make natural" or "conserve." This book is offered as a voice in the pursuit of those changes in intention and interaction, not in "wilderness" only, but in all places where nature bears a life of its own.

References

1. Thoreau, H. D. 1984. Journal. Volume 2: 1842–1848. Page 158 *in* R. Sattlemeyer, editor. *Henry David Thoreau, Journal, Volume 2: 1842–1848 (The Writings of Henry D. Thoreau)*. Princeton University Press, Princeton, New Jersey, USA.
2. Berry, W. 2003. *The art of the commonplace: The agrarian essays of Wendell Berry*. Edited and introduced by N. Wirzba. Counterpoint, Washington, DC, USA, 24.
3. National Museum of the American Indian. 2023. Homelands – Crow Nation. Smithsonian Institute, Washington, DC. <https://americanindian.si.edu/nk360/plains-belonging-homelands/crow-nation>. Accessed 21 September 2023.
4. Van Dyke, F., W. C. Klein, and S. T. Stewart. 1998. Long-term range fidelity in Rocky Mountain elk. *Journal of Wildlife Management* 61:1020–1036.
5. Van Dyke, F. 2007. Colonization of non-traditional range in dispersing elk, *Cervus elaphus nelsoni*, populations. *Canadian Field-Naturalist* 121:133–141.
6. Van Dyke, F. G., B. L. Probert, and J. J. Rozema. 1994. Vegetation characteristics of elk summer range in south-central Montana. Pages 285–300 *in* D. G. Despain, editor. *Plants and their environments: Proceedings of the first biennial scientific conference on the Greater Yellowstone Ecosystem*. US Department of the Interior National Park Service Technical Report NPS/NRY-ELL/NRTR, 93/XX.
7. Van Dyke, F., and J. A. Darragh. 2007. Responses of elk to changes in plant production and nutrition following prescribed burning. *Journal of Wildlife Management* 71:23–29.
8. Van Dyke, F., and J. A. Darragh. 2006. Short- and long-term changes in elk use and forage production following prescribed burning. *Biodiversity and Conservation* 15:4375–4398.
9. Van Dyke, F., and J. A. Darragh. 2006. Short- and longer-term effects of fire and herbivory on sagebrush communities in south-central Montana. *Environmental Management* 38:365–376.
10. Van Dyke, F. G., J. P. DiBenedetto, and S. C. Thomas. 1991. Vegetation and elk response to prescribed burning in south-central Montana. Pages 163–179 *in* R. B. Keiter and M. S. Boyce, editors. *The Greater Yellowstone Ecosystem: Redefining America's wilderness heritage*. Yale University Press, New Haven, Connecticut, USA.
11. Van Dyke, F. G., and W. C. Klein. 1996. Response of elk to installation of oil wells. *Journal of Mammalogy* 77:1028–1041.

12. Van Dyke, F. 1995. Microhabitat characteristics of moose winter activity sites in south-central Montana. *Alces* 31:27–33.
13. Van Dyke, F., B. L. Probert, and G. M. Van Beek. 1995. Moose home range fidelity and core area characteristics in south-central Montana. *Alces* 31:93–104.
14. Van Dyke, F., B. L. Probert, and G. M. Van Beek. 1995. Seasonal habitat use characteristics of moose in south-central Montana. *Alces* 31:15–26.
15. Berry, W. 2003. *The art of the commonplace: The agrarian essays of Wendell Berry*. Edited and introduced by N. Wirzba. Counterpoint, Washington, DC, USA, 28.
16. Schweber, N., and M. Matsuda. 2023. Trout refugium: The waters of the Beartooth Mountains. Atlas of the Absaroka-Beartooth Wilderness. <https://abatlas.org/from-nutcrackers-to-cutthroat/story-stocking-trout-high-lakes>. Accessed 21 September 2023.
17. Thoreau, H. D. 1946. *Walden: Or life in the woods*. Dodd, Mead, and Company, New York, USA, 182.
18. Locke, H. 2014. Green postmodernism and the hijacking of conservation. Pages 146–161 *in* G. Wuerthner, E. Crist, and T. Butler, editors. *Keeping the wild against the domestication of the Earth*. Island Press, Washington, DC, USA, 157.

2 This Present Age

The Onset of the Anthropocene

Ashes from the Sky

It was an August day. I was working in the upper East Boulder Valley on the Gallatin National Forest in Montana. On most mornings I would have believed that no change in weather could ever detract from the beauty of this place beneath the summer sun. But this day was different. There were no distinct clouds, but the sky was a smoky gray, the sun an obscured orange. The air breathed with a disquieting heaviness. And black ashes were falling from the sky with every breath.

Those ashes were coming from the south. Their source was Yellowstone National Park, its northern border perhaps 20 miles away. The year was 1988 – the year Yellowstone burned (see Figure 2.1). A warm, dry winter, high winds, heavy fuel accumulations, low nighttime humidity, and a hot summer. Thirty-nine fires were ignited by lightning, six others escaped from human control after being started by people. Nearly 800,000 acres of the Park's land area of 2.2 million acres (approximately 36%) burned – to varying degrees. Firefighters labored heroically, but it was September rain and snow that put an end to the burning.

Figure 2.1 Fire at the Norris Geyser Basin in Yellowstone National Park (USA) in Yellowstone National Park's historic fires of 1988. US National Park Service. Public Domain.

DOI: 10.4324/9781003404415-2

I have been a forest firefighter – a member of a Helitack (helicopter-dispatched initial attack) squad based in the Blue Mountains of Oregon. As a firefighter, I looked forward to forest fires, not only because of the sense of excitement they provided (which was more interesting than pitching horseshoes next to the helicopter pad waiting for the next dispatch) but because my salary went up. "Hazard pay" (time and a half) was the compensation when we were on the fire line. Unfortunately, it was a wet year, and our greatest challenge was to reach the fires before they went out. Hazard pay was minimal. The year was 1976, the last year in my lifetime that the Earth's average annual temperature was below the global mean. Humans have been using fire, and changing landscapes with it, for at least the last 400,000 years. It was the first great tool for expanding the human niche. But the ashes in the East Boulder Valley were not evidence of that expansion. They were the harbingers of processes humans now affect, but in no way control.

First Perceptions

In 1864, George Perkins Marsh published *Man and Nature*,[1] what many now consider to be the first book on what we today call "environmental science." Marsh had been a career diplomat for the United States, especially to nations of the Mediterranean. A careful, lifelong student of their culture, language, and landscapes, he had increasingly seen the lands of these nations irrevocably degraded by human activities and processes, including agriculture, grazing, mining, and forestry. Marsh perceived that human beings had changed, and were continuing to change, the face of the Earth, and had become by far the most powerful and disruptive forces shaping its future. "Man," wrote Marsh, "is everywhere a disturbing agent. Wherever he plants his foot, the harmonies of nature are turned to discords."[2] Given that Marsh's words were written over 160 years ago, they must be deservedly declared prophetic. Marsh was one of the first to recognize that human activity on Earth was becoming more powerful than natural forces. He was also perhaps the first to recognize – with comprehensive breadth of vision – the conditions, both present and developing, that we today call "the Anthropocene."

The Anthropocene acts as an "umbrella" term,[3] a name many have given to our present age, the age of humans – the name for a new relationship between humans and nature.[4] It is an attention-getting word because it uses a suffix ("-ocene") that geologists have traditionally used to separate different ages of the Earth distinguished by profound biophysical changes – changes which formerly have not involved humans at all. The name is an assertion that the growing physical and biological changes on the Earth are caused by human activities which have begun to move critical functions of the entire Earth System beyond any historical precedents in ecology, climate, and geology. It is a time scientists describe as being "without analogue." In the words of Frank Oldfield et al., it is an age in which human activities "play a major, integral and ever-increasing role in the functioning of the Earth System."[5]

The *Holocene*, the name given to the present geologic epoch, was first proposed by Sir Charles Lyell in 1833, and adopted by the International Geological Congress in Bologna (Italy) in 1885.[6] The perception we have entered a new geological epoch proceeds from the conviction among many that humanity has become a planetary force of greater or equal effect to the geological and climatic changes used to define different phases of Earth's history, proposing the human impact on Earth will now leave a lasting geological signature.[7] In stratigraphy, the geological community has now officially decreed, by a 12 to 4 vote of the members of the Subcommission on Quaternary Stratigraphy in

March 2024, that the present geological epoch will not be renamed.[8] "The decision is definitive," said geologist Philip Gibbard, secretary-general of the International Commission on Stratigraphy (ICS), the body governing the geologic timescale. "There are no outstanding issues to be resolved. Case closed."[9]

The second, and broader, study of the science of the Anthropocene is *global change science*, first arising through the International Geosphere–Biosphere Programme, started in 1987 – an investigation directing its efforts to measure and understand human-induced changes in the biosphere. Research to investigate the Anthropocene as a geologic stratum was not initiated by geologic events or by new geologic discoveries. Rather, it was provoked by the work of Global Change Science to investigate (and, perhaps, substantiate) the hypothesis that human effects on Earth had reached the point of beginning to produce a recognizable stratal unit. In other words, was there geologic evidence that global, planetary effects of human activities were real – so real they could be recognized stratigraphically? Geology was not the source of discovering the Anthropocene, nor did geological investigations expect to discover evidence of the same kind used to define earlier geologic periods. Rather, the presumed transition to the Anthropocene was the reason for geological inquiry. As Oldfield et al. put it, "The justification for the Anthropocene ... does not rest on the issue of exact equivalence to past epochs in a formal sense, but on the dramatic physical and biological changes caused by human activities."[10]

Where Did the Name Come From?

The first use of the term "Anthropocene" in modern scientific literature appears in the writings of the geologist Aleksei Pavlov, who described the present time as part of an "Anthropogenic system (period) or Anthropocene."[11] The name did not find wider uses until the 2000s, when Paul Crutzen, a climatologist, and Eugene Stoermer, an expert in lake ecology, published a seminal paper, "The 'Anthropocene,'" in the *International Geosphere–Biosphere Programme* newsletter. Stoermer noted he had been using the term since the 1980s, "but never formalized it until Paul [Crutzen] contacted me."[12] Crutzen had first objected to using the name "Holocene" at an earlier conference of Earth System scientists in Mexico in 2000. At one point, frustrated with the use of "Holocene," he blurted out, "Stop using the word Holocene. We're not in the Holocene anymore. We are in the ... the ... the ... the Anthropocene!"[12] But Crutzen would shortly make his most declaratory statement in his paper "The Geology of Mankind," published in *Nature* in 2002. "It seems appropriate," wrote Crutzen, "to assign the term 'Anthropocene' to the present, in many ways human-dominated, geological epoch, supplementing the Holocene."[12]

The name soon escaped from its birthplace in geology and the planetary sciences and became, in Paul Voosen's apt phrase, "an argument wrapped in a word."[13] It was an argument that soon captivated philosophers, ethicists, writers, poets, artists, activists, historians, economists, political scientists, sociologists, and, of course, conservationists. Three new scientific journals emerged: *The Anthropocene* (Elsevier), *Elementa: Science of the Anthropocene* (University of California Press), and *The Anthropocene Review* (SAGE Publishing). A periodical, *Anthropocene*, aimed at a more general audience, was launched in 2015.[3] The "Anthropocene," in both name and concept, spread to these fields and others because, in the eyes of many, the Anthropocene was a declaration of "the end of nature" or, from a human point of view, the claim that "nature is us."[14] In fact, the assertion humans had entered an anthropogenic age implied that an entirely new relationship between humans and nature was now in effect. But what was that relationship? Geographer Noel Castree notes that entry into this broader debate requires the recognition of four meanings of the word "nature" as they have been used in Western thought: 1) *External nature* – the non-human world, particularly those portions least affected by humans (as in, "I left the city to get back to *nature*"); 2) *Universal nature* – the entire physical world, including humans as biological entities and products of evolutionary history (as in, "All of *nature* was affected by the impact of the asteroid"); 3) *Intrinsic nature* – the defining property or essential quality of something (as in, "It was the *nature* of the mother bear to defend her cubs"); 4) *Superordinate nature* – the power or force governing all living things (as in, "The flood's effects were a force of *nature*").[15]

The common thread in each of these is the absence of human agency. Nature is a condition, force, or process external to and independent of human effect, but one which affects humans. "The Anthropocene" changes the meaning of all four. The first, external nature, has been (or is in the process of being) obliterated by human action. The second, universal nature, now consists of human actions in nature and independent of nature. The third, intrinsic nature, will increasingly be affected, if not decided, by what humans define and make nature to be, as we determine the distribution of species, the composition of ecosystems, and the genetics of organisms. The fourth, subordinate nature, is best expressed simply as "nature is us." Human action will now act upon and direct natural forces – their kind, placement, frequency, and intensity.

Is the Anthropocene Something New Under the Sun?

Throughout their history, humans have employed different modes of interacting with nature to gain needed substances and energy from the world around them, as well as different modes of perceiving and valuing nature for itself and its own intrinsic qualities. They have interacted and continue to interact with nature as hunters and gatherers, farmers and shepherds, and industrialists (using fossil fuels as sources of energy). In each of these categories they have changed the world around them.

Since the late Pleistocene (approximately 130,000 to 12,000 years ago), human hunter-gatherer societies have made changes in the environment. The disappearance of many large mammal species has often been historically coincident with human arrival.[16, 17] Lewis and Maslin express it more precisely as the "Megafauna Extinction" event that occurred roughly between 50,000 and 10,000 years ago, noting that

> about half of all large-bodied mammals worldwide, equivalent to 4% of all mammal species, were lost. The losses were not evenly distributed: Africa lost 18%, Eurasia lost 36%, North America lost 72%, South America lost 83%, and Australia lost 88% of their large-bodied mammalian genera.[18]

In fact, within a few thousand years of human arrival, 23 of 24 Australian mammals weighing 100 pounds or more became extinct. A large number of smaller species also disappeared.[19]

Humans affected more than species persistence. Archaeological, fossil, and biochemical evidence (from soil depositions), evaluated in world models of past human land use, suggests that hunter-gatherer societies increased the frequencies of fires in many landscapes, setting the burns or clearing the forests to create habitats favoring animals they hunted for food (and conditions for hunting them), as well as to create soil conditions favorable to the plants they cultivated. Such anthropogenic soils (anthrosols – soils with characteristics driven by human action) include the manure-enriched "plaggen" soils of

Northwestern Europe, the "terra preta" soils of the Amazon basin, soils enriched with charcoal dating back as far as 500 BC, and a wide range of globally distributed anthrosols which have been altered by sustained tillage, irrigation, manuring, and other land-use practices.[17] And humans may have even changed the climate while they were hunter-gatherers or still in a stage of early agricultural development. Eco-geographic modelers Erle Ellis and his colleagues note that

> mid-Holocene increases in CO_2 [carbon dioxide] and CH_4 [methane], [both gases with high capacities to absorb heat reflected from the Earth's surface] resulted from early land clearing and other agricultural practices and that these unprecedented interglacial trends in atmospheric composition set global climate on a trajectory toward warmer conditions long before human use of fossil fuels.[20]

Hunter-gatherer societies employed fire as a tool of "habitat management," or perhaps better as one of "niche expansion" – their own. Many hunter-gatherer societies were mobile, but others formed seasonal or annual settlements, especially where resources were abundant.

Archaeological evidence indicates that by AD 1100 the settlement of Cahokia, in what is now the US state of Illinois, had a population of 10,000 to 20,000 people (see Figure 2.2).[21,22] The Hopewell Culture centered in the US state of Ohio created trade networks (the "Hopewell Interaction Sphere") spanning thousands of miles throughout eastern and southern North America.[23] These communities and others like them throughout the world practiced a combination of hunting, foraging, horticulture, and agriculture as a means of subsistence, but without any identifiable "state" or specialized government structures.

Figure 2.2 Cahokia, near the Mississippi River in the central United States, was by AD 1250 a community of 10,000 to 20,000 people, making it the largest human settlement north of Mexico and one of the largest pre-Columbian settlements in the Americas, and larger than London, England at the same time. Artist rendering by William Iseminger. Photo courtesy of Cahokia Mounds State Historic Site.

Some of these settlements transitioned, over a period of about 3,000 years, to exclusively agricultural- and pastoral-based economies, while others developed what we would call "states" of distinctive government organizations which regulated their life together.[24] Compared to hunter-gatherer societies, early settlements associated with agricultural communities did not improve the lives of their residents in terms of nutrition. Rather, a varied diet of many kinds of plant and animal products was reduced to one of few, or even one, plant crop. Neither did they improve overall health, as such concentrations of people were marked by epidemics of previously unknown diseases, as well as chronic deformities accruing to the back, pelvis, knees, and toes from the daily strains of lifting, carrying, and kneeling which accompanied the work of cultivation.[25] Such early settlements did, however, permit the growth of higher population densities. In the words of historian Yuval Harari, they provided ways to "keep more people alive under worse conditions."[26]

Perhaps the oldest of these early urban areas might have been Çatalhöyük in southern Anatolia (the Asian portion of Turkey), which existed approximately 7,400 to 5,200 years ago.[26] Many other early cities, such as Eridu, Uruk, and Ur, were located in Mesopotamia (modern Iraq) in the historic "fertile crescent" of the Tigris–Euphrates River system, all apparently beginning at about the same time.[27] Both forager communities and agriculturalist–pastoralist settlements changed the world around them, although often differently in both kind and degree, sometimes subtracting from and other times adding to their surrounding biodiversity. Still greater changes would come with the development of industrialization, almost all of it at nature's expense.

When Did Human Effects Begin to Dominate Natural Processes?

The Broad Strokes

Determining when the name "Anthropocene" came into use is easier than determining when the processes characteristic of the Anthropocene got started, but these are tied to the beginnings of industrialization – characteristic of the increasing use of, first, charcoal and, then, coal as sources of energy, and of machine-based manufacturing of specialized commodities. In hunter-gather and agriculturalist–pastoralist societies, production of goods and subsistence are tied to current and ongoing photosynthetic processes (fruits and grains harvested from growing plants, animals eating grass). In industrial societies, production depends on the products of past photosynthetic processes (coal, oil, natural gas). As these changes occurred, social and economic forces began to separate people from nature by removing people from land. Laws prohibited rural dwellers and peasants from hunting and collecting wood in both England and northern Europe from the late seventeenth century onwards, separating rural people from their traditional means of subsistence. Such displacements were early manifestations of the capitalist conception of property – property which must be not only private but exclusive. And exclusion entailed enclosure. Such enclosure was not simply a physical fencing of land, but the extinction of formerly common and customary use rights on which many people depended for their livelihood.

Such changes intensified with industrialization. Coal had been mined from surface deposits in China since 3490 BC, although its use was confined primarily to individual households.[28] Coal began to be used as a household heating fuel in the 1500s in the Netherlands, along with peat. These were examples of energy extraction from fossil fuels, but such use was limited and not an everyday energy subsidy.[29] Up to 1500, "industry" was barely distinguishable from homemade handicrafts, invariably employing little machinery hardly distinct from trade and barter. At this time, coal, if procured at all, was used for subsistence living (for example, heating your home), but such use nevertheless began to produce an increase in affluence for its users. It was not an increase in direct material gain, but, as Fischer-Kowalski et al. put it, "an increase in affluence in the sociometabolic sense in which we use this term: affluence in energy."[30] Such gains were one of the most distinctive signatures of industrial society. It was this "increase in affluence," manifested in the burning of coal, and later, of other fossil fuels, that began to inflate the amount of carbon in the Earth's atmosphere.

The New Organization of Humans and Nature

With the increased use of fossil fuels, other changes began and were soon accelerating, altering not only patterns of land use but human community structures and social relations. The organization of work, which had been shaped by a combination of personal choice and community tradition, came to be dictated by the pace of the machine. The clock, a device all but unknown before the early stages of industrialization, became the worker's master, as emphasis shifted from *production* for personal need (with highly flexible and choice-determined timeframes) to *productivity* (specific outputs of specialized goods or commodities per unit of time).[31]

The textile (weaving) industry exemplified these trends as it expanded in Britain through increased mechanization, concentrating production in factories in urban areas, and provoking mass migrations of workers from rural areas, where weaving had been a "cottage industry," to cities where it became the work of machine operators, with considerable redistribution of income, often resulting in lost wages for formerly more independent rural workers. Protests against such changes spawned a new political party, the Luddites – a powerful protest movement in Britain by the early 1800s – primarily composed of textile workers and artisans who believed that industrialization was leaving them unemployed.[32] The name *Luddite* came from one Ned Ludd, a legendary anti-industrial hero in the growing city of Leicester, England, who reportedly smashed two mechanized knitting frames in a textile workshop in what was described as a fit of passion. The act came to symbolize the frustration of weavers and other industrial workers with various forms of machine technology replacing their traditional skills in favor of faster and greater production. By 1811, such sentiments had become an organized social protest, often expressed with violence. Mills and factory machinery were burned. There were full-scale battles with the British Army, and one mill owner was assassinated. But, eventually, advanced knitting technology prevailed, resulting in more factory production, higher outputs, migration to cities, and lower-paid workers.[32] Industrial methods and devices replaced human labor and skill. Such labor and skill helped build rural communities and gave them independence and self-sufficiency. Industrial methods and devices, and the migrations they required, destroyed them. And one of the most terrible costs of this exchange was the destruction of nature.

Some changes occurring during the Industrial Revolution in Britain were coincident, and connected, to the growing need for coal, which motivated not only increased amounts of coal mining, but better technology for moving coal from the mines to production or use sites. Originally withdrawn from below the surface by digging pits to reach coal deposits, extraction was hindered in deeper pits by striking subsurface groundwater, resulting in water seeping into the pit and, if not removed, causing the pit to flood. Pumps powered by horses were the first approach to removing the water. Steam engines (with water heated by burning coal in a boiler) improved the rate of water removal but ironically required large amounts of coal to operate.[33] James Watt, a Scottish inventor, was asked to repair one of these engines. Watt completed his repair, and then put the engine to the test to see how it would perform. It worked, but in the process, Watt noticed that over 25% of the steam created was lost as latent heat (heat required to simply convert the water to steam) without contributing to operating the engine used to pump the water out of the pit. After a year of work and study, Watt succeeded in solving the problem, producing a far more efficient steam engine and improving the rate and amount of coal that could be mined. However, the coal, often mined in rural areas, had to be transported to other places, especially urban centers, to be available for industrial production. But by 1800, the steam engine had been adapted to power railroad locomotives, which not only made transportation of the coal easier, but spurred the development of railroad transportation.[34]

The development of the railroad industry in Britain, not to transport people but to move coal, fostered urbanization and effectively created the first commuters. To this point, coach traffic (people transported in carriages drawn by horses) carried 14 or 15 passengers per week. Railroad traffic, however, could carry many times more, and it was this commuter traffic, not the transport of coal, that soon became the principal source of railway income. As a result, the network of railroads, originally envisioned as a small number of lines connecting coal fields with neighboring cities, spread throughout all of England to take people from one place to another.[33] In this way, work, which had formerly been tied to residence in agricultural and subsistence living and thus was a key component of community stability, was now separated from residence. The energy derived from coal, which had been perceived as the energy of power to do work, now became the energy of movement to transport both people and goods.

These developments were not unanimously applauded. Britain's most perceptive social critic John Ruskin wrote, with scathing sarcasm,

> There was a rocky valley between Buxton and Bakewell, once upon a time, divine as the Vale of Tempe … You cared … but for cash (which you did not know the way to get): you thought you could get it by what the Times [a London newspaper] calls "Railroad Enterprise." You enterprised a railroad through the valley—you blasted its rocks away, heaped thousands of tons of shale into its lovely stream. The valley is gone, and the Gods with it; and now every fool in Buxton can be at Bakewell in half-an-hour; and every fool in Bakewell at Buxton; which you think a lucrative process of exchange—you Fools Everywhere.[35]

The Causes of the Anthropocene

The "New" World – The Frontier of the Anthropocene

European exploration and colonization of North and South America, initiated in the late fifteenth century, opened vast and (in the eyes of Europeans) inexhaustible supplies of resources which could fuel industrial growth, both in Europe and the Americas. It also offered vast landscapes seemingly waiting to be transformed to more productive uses than Indigenous communities made of them. The changes that followed affected all aspects of human society as well as planetary systems. In 1700, nearly half of all ice-free terrestrial lands were uninhabited by humans, and 45% had only limited human use and small settlements ("semi-natural"). By 2000, less than 25% of the same lands were uninhabited, and less than 20% experienced only limited use and settlement.[36] More than half of all existing terrestrial ecosystems had been transformed by humans by the mid- to late-twentieth century, so much so that biogeographers like Erle Ellis consider older, "classical" biome considerations no longer accurate. Classical biomes reflect a concept developed to describe large areas of regionally similar vegetation occurring in predictable patterns of latitude and climate (see Figure 2.3). But with human influence on landscapes, Ellis and others have proposed a new system of global ecological classification – one based on anthropogenic (human-dominated or human-influenced) biomes or *anthromes*.[36] In contrast to biomes, anthromes are the product of land use and other human interactions with ecosystems. Based on their own modeling studies, Ellis et al. estimate that about half of the landscape transformations occurring during these three centuries have been driven by human-induced changes to lands formerly uninhabited ("wildlands"), and the other half driven by the intensification of land use in semi-wild lands.[37] However, it was not merely the land that would be affected by human agency, but the fundamental systems of the Earth.

Consequences Made Manifest

By the early twenty-first century, CO_2 levels were 35% higher than pre-industrial levels, and higher than any levels documented in the past 800,000 years. In the last 200 years

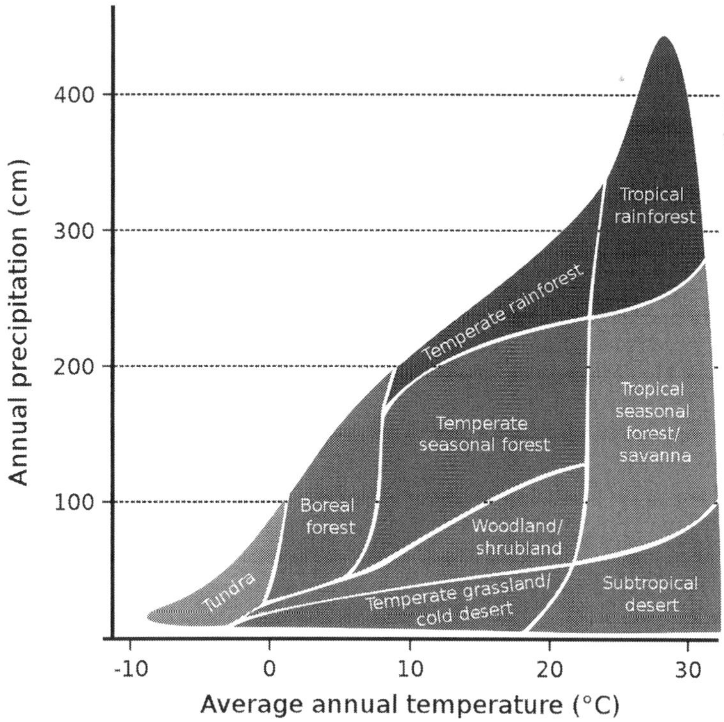

Figure 2.3 Biomes have long been the basic units of classifications of world ecoregions based on combinations of temperature and moisture. Many scientists are now advocating a classification system based on *anthromes* – regions whose biological characteristics are determined by human presence in and use of such ecoregions. Reprinted under CCO 1.0 Universal Public Domain.

the human population has grown from 1 billion to 6 billion. Energy use has increased 40-fold and economic production 50-fold. The amount of land surface devoted to human activity has risen from 10% to 30%[36] and the amount of pastureland has also increased, with accompanying expansion of croplands into what were formerly wild grassland, shrublands, and woodlands.[38]

After 1945 and the end of World War II, the same trends continued but accelerated. The pace of human population growth, atmospheric CO_2 levels, and land transformation per capita have risen dramatically, concurrent with rapidly increasing levels of energy consumption and economic activity. By 2007, the world human population had become more than 50% urban, and more than 80% urban in Europe and North America.[39] The world's human population is now taking 20–40% of global net plant primary productivity for its own use. Today 43% of the Earth's terrestrial ecosystems have undergone wholesale transformation, and 50% of the Earth's terrestrial lands are predicted to have undergone "state shifts" (transition from one kind of ecosystem to another, or to a human-dominated environment in which natural processes are mostly absent) by the time the human population has reached 8.2 billion, a target projected for 2025.[38] This 75-year period (1945–2020) has been named the "Great Acceleration" in recognition of its unprecedented and multiple expansions in human activity in, and human effects on, all of the Earth's planetary biological and climatological systems.[40]

Distinguishing Causes from Consequences

Increased carbon emissions, ecosystem transformation, intensification of land use, increased urbanization, and radioactive sediments are all consequences of the growing human domination of Earth's natural systems. Consequences are *effects*, the results of actions humans have taken in their modern engagement with nature. But consequences are not *causes*. This is the fundamental difference between understanding "how" something happens and "why" something happens. As Yuval Harari explains,

> To describe "how" means to reconstruct the series of specific events that led from one point to another. To explain "why" means to find causal connections that account for the occurrence of this particular series of events to the exclusion of all others.[41]

Consequences can be measured in "pollutions units." Causes are about relationships. The motivations for the Anthropocene are rooted in an economic system that defines nature's value, a method for organizing that value, and a mission to harness that value. As the social scientist Jason Moore has put it, such consequences reflect a

> fundamental transformation of the relations of power, knowledge, and capital that have made the modern world. Shut down a coal plant, and you can slow global warming for a day, shut down the relations that made the coal plant, and you can stop it for good.[42]

To understand the *causes* of the Anthropocene, we must begin by examining its name, *the age of humans*. That name asserts that the Anthropocene is a development characteristic of humanity at a global scale, an inevitable expression of the human presence and its interaction with nature. This is not the case. The Anthropocene is a product of global influences, but it is not a product of "humanity." It reflects a specific set of intentions toward nature which in turn define and direct human interaction with nature, an array of intentions and interactions that is not characteristic of all times, places, and cultures.

The Mission of Science – To Dominate Nature

At the height of his political career from 1613 to 1621, Francis Bacon was Lord Chancellor of England under King James I, and bore the additional titles of Counsel Learned Extraordinary to His Majesty, King's Solicitor-General, His Majesty's Attorney General, Counselor of Estate, and Keeper of the Great Seal of England. At an early age, Bacon began to consider what his life mission should be.

> I therefore asked myself what could most advantage mankind, and for the performance of what tasks I seemed to be shaped by nature. But when I searched, I found no work so meritorious as the discovery and development of the arts and inventions that tend to civilize the life of man…, it seemed to me that such a discoverer would be called the true Extender of the Kingdom of Man over the universe, the Champion of human liberty, and the Exterminatory of the necessities that now keep men in bondage.[43]

No one can fault Bacon for lack of ambition. He believed in what scientists today refer to as the *Inductive Method*, deriving conclusions about the whole (the "system") by finding out the particulars of its parts. Accordingly, Bacon asserted that knowledge should be acquired "not by arguing but by trying."

> There are and can be only two ways of searching into and discovering truth. The one flies from the senses and particulars to the most general axioms, and from these principles, the truth of which it takes for settled and immovable, proceeds to judgment and to the discovery of middle axioms [deductive reasoning, from the whole to a part]. And this way is now in fashion. The other derives axioms from the senses and particulars, rising by a gradual and unbroken ascent, so that it arrives at the most general axioms last of all. This is the true way, but as yet untried.[44]

Bacon was an advocate of progress, and the handmaiden of progress was technology.

> It is by instruments and helps that the work is done, which are as much wanted for the understanding as for the hand. And, as the instruments of the hand either give motion or guide it, so the instruments of the mind supply either suggestions for the understanding or cautions.[45]

As technology was the instrument of progress, so application was the instrument of technology. Knowledge was a resource to be used for specific and purposeful ends – and that end was the subjugation of nature. "Human knowledge and human power," claimed Bacon, "meet in one; for where the cause is not known the effect cannot be produced."[46] The methods and machines of the Industrial Revolution found their highest purpose in the Baconian attitude that nature was something to be controlled by man. "Only let the human race recover the right over nature which belongs to it by divine bequest," he wrote, "and power be given it; the exercise thereof will be government by sound reason and true religion."[47] Bacon profoundly misunderstood the concept of "dominion," which he supposed he was invoking from the Judeo-Christian tradition. In fact, the biblical understanding of dominion was rooted in its sense of "ownership" as a foundation of responsibility for the good of the other, which is why, in the Bible's account of creation, God's declaration to "let them [humans] rule" is followed almost immediately by the injunction to "serve and protect" the creation at hand.[48] Bacon's writing on the use of induction as a new method of knowing in science was his most original philosophical contribution – a subjugation of nature to humanity, by humanity, and through technology. With a misplaced understanding of human dominion, it would become one of the most powerful weapons employed against the natural world.

Capitalism – A New Relation between Humans and Nature

The fifteenth century began with advancements in many fields widely scattered worldwide. In the Middle East, Muslim scholars were making important contributions in mathematics.[49] In Asia, China was entering the ascendancy of the Ming Dynasty. Its seven "Treasure Voyages" from 1405 to 1432 boasted armadas of up to 300 ships and a combined crew of nearly 30,000 sailors. These fleets enforced and extended Chinese economic and political dominance throughout Southeast Asia, India, and East Africa.[50] In the Americas, the Inca Empire governed over 10 million people speaking 30 languages across what is now Peru, western and south central Bolivia, southwest Ecuador and Colombia, and a large portion of modern-day Chile.[51] The capital of the Aztec Empire, Tenochtitlan, in what is now Mexico, had a population of between 200,000 and 400,000 people, making it one of the largest cities on Earth.[52]

The new century did not begin with any momentous events in Western Europe. Yet, by the end of this same century and in the years that followed, Western Europe became master of the "outer world" beyond its borders, especially in the Americas, even though nations such as China had greater naval power and more effective government administration. Rather, Europe conquered the Americas because the Asiatic powers, although stronger, showed no interest in them.[50] The Ming Treasure Voyages accomplished their intention of extending Chinese economic and political influence and, having succeeded, ceased. In contrast, Western Europe, experiencing a merger of capitalism and technology-empowered science, ironically expanded its influence through the recognition of its own ignorance in science and of the known world. Western science, abetted by technology, drove the West to overcome such ignorance. Western exploration, driven to know an unknown world, led it to conquer territories.[53] Both were aided by the development of an economic system that changed the human relationship to nature and financed the colonization of the world. That system was capitalism.

The fifteenth century in Europe began with social and economic life organized in a system known as *feudalism*. Labor was oriented to agriculture and performed by peasants or "serfs." Serfs resided in communities on the land where they performed their work, giving much of the harvest to the landowner (the "manor lord"). The serfs were not wage earners but subsistence farmers. They lived on the food they grew and the animals they raised, and made most of the things they needed, or were needed by the Manor (the lands on which their manor lord resided), through their own craftsmanship. Industry meant handicrafts. "Investment" was most often devoted, by the manor lords, to the land itself, through improvements like irrigation or reclamation of land that had become unproductive for crops. Trade was minimal, not only because there were limited networks of it, but because it was often actively discouraged. As one lord expressed it, "all things necessary ought to be made on the manor so that it would not be necessary to buy or borrow them outside."[54] This admonition did not forbid the possibility of getting things from "outside" the manor, but stressed that it was better to avoid other sources (what today we would call "markets"). As the historian Carlo Cipolla put it, "The market was unreliable and gave no assurance of providing necessities regularly… It was inconceivable that a community could live on commercial activity without engaging directly in agriculture for the daily necessities of life."[55]

The manor lord's "ownership" of the land was not the kind of ownership we associate with modern land possession. The serfs retained rights to move about the lands of the manor more-or-less independently and could make use of its resources (such as game and fish), although some restrictions might apply. They could build their modest homes and communities on the land and not be subject to eviction if they continued to be part of the working community. There were other classes in feudal society. The *clergy* were devoted to the Church, the *chivalry* (knights) to "valor," and then there were the *merchants*. Most wealth was held by the manor lords, the clergy, and the knights. In contrast, the merchants were an itinerant, vagabond class, regarded with disdain and suspicion. They had no homes or lands, and no ties (and therefore no loyalty) to any community. The Church damned them because they were devoted entirely to material profit, engaged in loaning money to others at high rates of interest (the biblical definition of usury), they bought and sold many disreputable goods, and even traded in other human beings as slaves. In feudal society, esteem and rewards belonged to those who pursued noble ends (the lords, the clergy, and the knights), not to those whose production produced the means that supported such ends (the serfs). And the merchants did neither.[56]

Technology, Money, and Markets

Feudal society had little technology, few markets, and a philosophy of "society in the land" (or, we might say now, "society in nature"). But the first to change was technology. Improvements in agriculture, such as the replacement of the "scratch" plow with the "heavy" plow, permitted the cultivation of soils that had previously been too difficult to penetrate.[57] As available agricultural land increased, surplus agricultural production followed. Such surpluses permitted the developments of "towns," small settlements by today's standards, but increasingly the abode of the doctors, lawyers, judges, pharmacists, and merchants. Still a minority element of society, professions that could be practiced in towns, and the merchant class itself, were the one open door for a person to gain material enrichment and social standing at a relatively rapid rate, independent of the established categories of class.[58]

Over time the residents of towns became more specialized in the services they offered as well as the things they produced. Traditional direct barter exchanges were inefficient, and the more specialized professionals began to develop *bills-of-exchange* which specified the particular value of different goods in exchange for other goods. By the mid-fifteenth century, bills-of-exchange had begun to take on many of the characteristics of modern "paper money." Precious metals also began to assume the same function. Gold, for example, was not valued for what one could do with it (which, for most people, was very little), but because it became an increasingly trusted medium of exchange, especially as such exchanges began to involve different countries (which might be using different forms of and exchange rates for paper money).[59]

As craftsmanship changed toward industry and agricultural production produced greater and more diverse surplus, both products became specialized as "commodities." In order to be traded with money, commodities required "markets," systems of exchange in which buyers and sellers negotiated the prices for each commodity. Likewise, as money replaced the direct exchange of goods, labor productivity (the amount of a commodity produced per unit of time) replaced subsistence labor (making things solely for one's own needs and at one's own pace). A parallel transition was occurring in human relations to land. In agriculture, land now became the means of production, and therefore a commodity in itself. Thus, "land" became "property," and the privileges of its use and resources were formalized as "property rights."[60]

The traditional products of labor had been direct use or exchange. As land as property became a commodity, so the products of labor productivity became wages (money paid per unit of time labored), creating a market for labor itself. Workers no longer worked to produce a particular item. They worked for the wages paid according to the time needed to produce it. The measurement of time became essential for determining labor productivity. The outcome was time-based wages (for example, an "hourly" wage) and the institution of wage labor.

Advances in technology increased the capacity to produce individual commodities in large numbers (mass production), requiring the concentration of labor at specific work sites (factories) associated with laborers-in-residence (cities). Operating such technologies required ever-increasing quantities of energy, which, by the mid-1500s, came from coal. These changes made markets, not community relations or different individual needs, the bases of exchange, and drew populations from rural areas and land-based occupations into urban areas and production-based (factory) occupations. Through the growing dominance of markets as the basis for setting rates of exchange, markets became not only the means to determine how much something was valued (its assigned worth) but what *would be valued*.[61] Relations between managers and laborers, which in past ages had been based on land tenure, historic family relationships, or personal loyalties, were now organized on market principles in which labor was but another good to be bought and sold. Because the labor market determined not only *how much* labor was valued but *what kind* of labor was valued, more skilled and more productive workers increased in personal wealth. Less skilled and less productive workers were less marketable and therefore often unemployed. The cumulative effect was part of the reason, even with the increasing productivity of the nascent Industrial Revolution, for an increase in the number of poor at a time of growing national wealth.[62]

Capitalism also contributed directly and indirectly to changes in the human relationship with nature. As people in rural areas left (or were forcibly expelled from) their homes and their livelihoods in domestic agriculture to reside in the human-dominated environments of cities to enter or be forced into becoming labor power for industry, capitalism provided a material condition for making a separation between man and nature "real." Nature was now visibly perceived as something external to human society. Older structures of value in feudalism, as well as the classical philosophy of the Greeks, supported an image of nature at once beneficent and instructive ("Mother" Earth). The Earth was depicted as alive and sensitive, and a source of wisdom for human life. In this view, destructive acts of violence against nature could be considered, and often were condemned, as a breach of ethical behavior. With nature now viewed as external and the capacity to master it the highest and best aspiration of technology, such acts were no longer seen as "destructive" and certainly not unethical. With its use of markets and private property, capitalism no longer perceived "nature" as a home for human communities and a system of intertwined relationships in which they lived and interacted, but something that could be transformed into individual, dismembered parts abstracted as "resources," with each assigned a market price based on its value to commodity production.

Global Influence – Local Effects

Empowered by machine technology and driven by its "highest" purpose – the domination of nature – Bacon's inductive method was employed in the advancement of a belief structure that emphasized progress, mechanization, and applied scientific investigation, all characteristics of the early Industrial Revolution. This kind of transformation required that nature be remade into a "marketable good," ideally one that could be exclusively controlled by a supplier who would gain from its sale or trade to another. Such a view was perceived as morally reprehensible by some, and provoked intense criticism. Writer and poet Alain de Lille drew a picture of such abuse in his work *The Complaint of Nature* (*De Planctu Naturae*). A traveler encounters Mother Earth, personified as a woman – beautiful, yet now attired in clothing, once lovely, that has been violently torn and tattered. Their dialogue proceeds this way:

> I marvel, then I [the author] said, wherefore certain parts of thy tunic ... suffer division in that part of their texture where the fancies of art give the image of man. ...she [nature] answered, "thou canst deduce what the figured gap and rent mystically show. For since ... many men have taken arms against their mother in evil and violence, they thereupon, in fixing between them and her a vast gulf of dissension, lay on me the hands of outrage, and they themselves tear apart my garments piece by piece, and, as far as in them lies, force me, stripped of dress, whom they ought to clothe with reverential honor, to come to shame, like a harlot. This tunic, then, is made with this rent, since by the unlawful assaults of man alone the garments of my modesty suffer disgrace and division.[63]

In the world being created by the Industrial Revolution, this "rent of unlawful assaults" upon nature reflected the human relationship with nature being changed from a complex set of *multiple interactions* to only *two functions* – tap and sink. As a "tap" (source), nature provided natural resources and energy that would supply or power industrial production. With land now a commodity easily bought and sold in a "market," the key action was to locate such resources ("resource-rich property"), acquire the property through cash transaction, and then create exclusive use of the property by the owner through the enforcement of the owner's "property rights." To maximize gain in the transaction, the goal was to extract the resource using methods that minimized costs. In addition to minimizing wages for workers to the lowest level that workers would accept, it also meant using methods of extraction that incurred low costs for maintenance and left the used property in whatever state the extraction changed (reduced) it to. Nature became a "sink" – a receptacle of wastes created in extractive and industrial processes and discarded after use, with little or no further costs.[61] These perceptions were the root causes of the changes in human behavior that would produce "the Anthropocene."

By extracting resources from nature using this method, the system could create an over-supply ("glut") of the resource on the market, not only lowering the cost of extraction but lowering the overall price of the commodity. In northern Michigan, where I reside today, this was the strategy followed during Michigan's "Logging Era" (1860–1920), which resulted in the removal of all marketable stands of red pine (*Pinus resinosa*) and white pine (*P. strobus*) trees in the state. Large plots of land containing extensive stands of these trees were purchased by wealthy capitalists ("timber barons") with the objective of transforming trees into lumber (a marketable commodity) for the purpose of increasing personal wealth (their own). To accomplish this, they converted forested lands into private property, producing a superabundance of the natural resource (trees) removed at minimal costs by employing recent immigrants – especially Irish, Norwegians, and Swedes – at low wages. They sold the timber in markets, both national and international, that provided the highest price for their lumber, with the resource (formerly trees) transformed into money, which accumulated at much faster rates than growing more trees.

In the Logging Era, the relationships between humans and nature were framed by a view of humans and nature as separate entities in which products (resources) from nature provided the supply (tap) to the capitalist institution of markets which transformed the resource into money, with the objective to gain material wealth for the resource (property) owner. Nature was to be sequentially 1) explored, 2) mapped, 3) surveyed (a practical assessment of the landscape and its traits), 4) inventoried, 5) calculated, and 6) guarded (as private property).[61] This sequence of interactions with nature fit the pattern of lumber barons in Michigan, particularly exemplified in one, David Ward. One of nine children in a farming family, Ward aspired to wealth, and perceived the vast timber resources of his state to be the means to it. At an early age he learned the skills of surveying from his father and became an in-demand surveyor for the leading lumber companies. But, very astutely, Ward demanded that he should not be paid in wages, but in the receipt of one-quarter of each parcel of land he surveyed for each company, thereby acquiring control (property rights) over the marketable commodity itself (timber). Gaining this compensation and knowing exactly where the most valuable trees were located, Ward soon amassed thousands of acres of pine timber and then masterfully organized its removal. In following this pattern, Ward was emulating what was already an established practice in the United States, although usually one reserved for young men who were rich, not those from humble beginnings like Ward. Surveying was one of the first skills that a young aristocrat like George Washington, the first president of the United States, learned in his early years in Virginia when it was still a colony of the British Empire. Washington wrote, "The greatest estates we have in this Colony" were made "by taking up and purchasing at very low rates the rich back Lands which were thought nothing of in those days, but are now the most valuable lands we possess."[64]

The *rate* of the conversion of trees to lumber in Michigan's Logging Era was imposed by technology (how fast the trees could be cut and moved), which was initially limited. In the early years of the Logging Era, trees could be felled and transported only in winter, when "roads" could be constructed to the stands, then covered with water, quickly forming a layer of ice. Horse-drawn sleds could then pull the timber on the ice, which was impossible for the horses to move on an ice-free surface. Two technological innovations changed the *rate* of removal. The first was the invention of the "Big Wheel," two enormous wheels (usually about 10 feet high) fixed to a long wooden "tongue" harnessed to a team of horses. Using chains suspended between the wheels, trees cut to shorter lengths (logs) could be hoisted off the ground, reducing the friction present when trying to drag them over a dry surface. Using this device, logging began to proceed in all seasons, and timber removal became a 12-month operation.

Supplanting the Big Wheel was the narrow-gauge railroad, a technique that used lumber itself (even two-by-four planks cut from the felled trees) to fashion railroad lines and ties on-site, such that a small locomotive could be driven directly to the area where the trees were being felled and timber removal could proceed at an even faster pace. The Big Wheel and the narrow-gauge railroad were not the *agents* of deforestation, but rather devices that increased the *rate* of such deforestation in pursuit of material gain (wealth). With the trees gone, the wealth generated by lumber barons and brokers moved into the burgeoning automobile industry, with their sons and other beneficiaries using these profits in founding or financing the beginnings of auto makers like General Motors[65] and Ford.[66]

The removal of the large pine stands in Michigan was over by 1920. Nature (in this case, forests) as a "sink" was exemplified by the cut stands now abandoned without reforestation, covered with the limbs and branches (slash) removed in tree felling and preparation for milling. This landscape created conditions conducive to forest fires. In October of 1871, the Peshtigo Fire, the deadliest wildfire in US history, along with other "smaller" fires combined to burn an estimated 2.5 million acres across the state of Michigan (for comparison, Yellowstone National Park, larger than some eastern US states, is approximately 2.2 million acres).[67] In 1881, the Thumb Fire burned another million acres, killing at least 280 people and sending a plume of smoke all the way to the East Coast.[67] In 1908, the Metz Fire burned between 300,000 and 2 million acres (estimates from the time vary widely).[67]

Many of the laborers ("lumberjacks") who had felled the forests turned to agriculture after the Logging Era, hoping to establish productive farms on the cutover lands. But the soils of most of such areas are predominantly sand (Michigan's "state soil" is "Kalkaska Sand," named after my current county of residence). Using a handheld soil tube, I have shown many students in the field in this area that the topsoil (the "A layer" of soil) is usually only 1–2 inches deep and rapidly exhausted by conventional agriculture. Unable to pay the property taxes to maintain the ownership and property rights on the commodity (land) they had purchased, the former lumberjacks abandoned their "farms," which passed into the hands of the state of Michigan ("receivership"). So it was that men like David Ward and other timber barons created wealth and destroyed communities.

It was these failed farmlands that, over time, formed the bulk of the holdings of what is today the Michigan State Forests, now including nearly four million acres among the over 20 million acres of forests in Michigan, making Michigan the sixth most forested state in the US. Less than one-half of 1% of this forest now bears old growth pine (see Figure 2.4). To these circumstances and conditions, the essayist Wendell Berry speaks truth without comfort.

> This country, as we have made it by the pretense that we could do without it as soon as we have completed its metamorphosis into cash, no longer holds even the possibility of such forests, for the topsoil that they made and stood upon, like children piling up and trampling underfoot the fallen leaves, is no longer here.[68]

The Michigan Logging Era is but one example of the way in which capitalist economic enterprises organize nature. With these economic relations established, supported by a philosophical universe that separated "people" (human society) from "nature" (nature without people), the dismemberment of natural systems into market commodity resources and the transformation of natural systems into "land" that could be bought, sold, and exclusively used as private property, relations between humans and nature were now in place that followed their course to its logical conclusion. Nature could be used most cheaply as a "tap" where resources were disproportionately abundant, like timber in northern Michigan forests. Thus, all over the world, capitalist patterns of resource use followed the trail of natural "commodities." As sociologist and economist Jason Moore explains, "commodity frontiers materialize at sites where ecological abundance is reorganized around commodity production. This ensures a 'yield honeymoon.'"[69] For European colonizers of North and South America, Africa, Asia, and Australia, the vast new tap of resources in land, energy, timber, soil, marketable animals, and other commodities permitted a wave of wealth accumulation. But no honeymoon lasts forever.

It is these relations between humans and nature that both preceded and then fostered the Industrial Revolution in Europe. The *consequences* of the Anthropocene began in 1800. The *causes* of the Anthropocene began centuries earlier. Such relations, aided by a

Figure 2.4 Hartwick Pines State Park in the US state of Michigan contains one of the last remaining old growth stands of red pine (*Pinus resinosa*) and white pine (*P. strobus*) trees in the US Upper Great Lakes States. Such stands covered millions of acres prior to the Logging Era of 1860–1920. Courtesy of Michigan Department of Natural Resources.

technology whose mission was to subjugate nature for human use and channeled through an economic system that dismembered nature into resource commodities, represent the root causes of the Anthropocene.

Human beings behave differently toward nature than any other species. As the ethicist Holmes Rolston III puts it,

> The amoral fauna and flora are checked in their possessive impulses by the limitations of their ecosystems—which provide a satisfactory place, a niche, for each specific form of life, but limit each species to its appropriate sector, where it has adapted fit. The human species is not so checked, but tempted by the fearful power of hand and mind to possess the whole.[70]

What's in a Name – Should We Call this Time "The Anthropocene?"

The causes of the Anthropocene are not a product of all humanity nor are they the inevitable outcomes of human interaction with nature. Rather, the Anthropocene has been, for Western culture, a quest, in Thoreau's memorable words, "of desperate haste to succeed in desperate enterprises."[71] As Jason Moore notes, calling the present time the Anthropocene "tells us that behind the current, disastrous state of world affairs is the Anthropos. It's a trick as old as modernity – the rich and powerful create problems for all of us, then tell us we're all to blame."[72] The name also reveals human hubris – that the present age should be named in representation of human actions, reflecting the relationship between humans and nature that has caused these problems rather than giving it a name that might provide a solution.

Henry David Thoreau lived in a cabin he had constructed at Walden Pond, in the US state of Massachusetts, for two years, the better to understand and define his own interactions with the world around him. But he also traveled about the local countryside and saw other kinds of interactions with nature. One of these was on a nearby farm, owned by the Flint family, which included a pond they had named "Flint's Pond." Thoreau was outraged at both the name of the pond and the treatment of the land around it, and seems to have well understood that both had been desecrated by their conversion to market commodities and values.

> What right had the unclean and stupid farmer, whose farm abutted on this sky water, whose shores he has ruthlessly laid bare, to give his name to it... I go not there to see him or to hear of him; who never saw it, who never bathed in it, who never loved it, who never protected it, who never spoke a good word for it, nor thanked God that He had made it...[73]

Thoreau also thought of Flint's motives that inspired the treatment of his pond and farm:

> – him who thought only of its money value; ... who exhausted the land around it, ... – and would have drained and sold it for the mud at its bottom ... who would carry his God to market, if he could get anything for him; ... on whose farm nothing grows free, whose fields bear no crops, whose meadows no flowers, whose trees no fruits, but dollars...[74]

It was Flint's relation to the pond that bore his name (better, the pond to which Flint *applied* his name) that so angered Thoreau, for that act was symbolic of all that was wrong with Flint's interaction with his own land – with nature. So, the name "Anthropocene" is itself profoundly dishonest. It treats "humanity" as a single entity that somehow dominates the entire planet, while at the same time treating the Earth as if it were a lifeless commodity.

The name is also one that obscures and depoliticizes profound differences between people and cultures with vastly different impacts and claims. The philosopher Thomas Berry has proposed that instead of naming the present age "the Anthropocene", it ought to be christened the *Ecozoic*, "a period when humans would dwell upon the Earth in a mutually enhancing manner."[75] Jason Moore has declared that the present time should be called "the Capitalocene," asserting that this is a better epithet so that the name, and the blame, "is laid squarely at the door of a system: capitalism or the global economic system."[76] And that is indeed where the responsibility lies. Humans cannot overcome the "age of humans," but we can – we *must* – overcome the "age of capital." The present condition cannot continue to be our ongoing relationship with nature.

"The Anthropocene" is the name that is with us. It will remain embedded in these times for the foreseeable future, and so will be the name I use in this book. But it is not the best name, and it does not express the best sentiments, relations, or hopes for a future interaction between humans and nature that could be different – and better – than the one we have today.

References

1. Marsh, G. P. 1864. *Man and nature: Or, physical geography as modified by human action.* Scribner, New York, USA.
2. Marsh, G. P. 1864. *Man and nature: Or, physical geography as modified by human action.* Scribner, New York, USA, 36.

3. Malhi, Y. 2017. The concept of the Anthropocene. *Annual Review of Environment and Resources* 42:77–104.
4. Crutzen, P. J., and E. F. Stoermer. 2000. The Anthropocene. *Global Change Newsletter* 41:17–18.
5. Oldfield, F., A. D. Barnosky, J. Dearing, et al. 2014. The Anthropocene Review: Its significance, implications and the rationale for a new transdisciplinary journal. *The Anthropocene Review* 1:3–7. DOI: 10.1177/2053019613500445.
6. Nordblad, J. 2021. On the difference between Anthropocene and climate change temporalities. *Critical Inquiry* 47:328–348.
7. Steffen, W. 2013. The Anthropocene. Page 486 *in* L. Robin, S. Sörlin, and P. Warde, editors. *The future of nature: Documents of global change*. Yale University Press, New Haven, USA.
8. Zhong, R. 2024. Are we in the "Anthropocene," the Human Age? Nope, scientists say. New York Times. 5 March 2024. <https://www.nytimes.com/2024/03/05/climate/anthropocene-epoch-vote-rejected.html>.
9. Voosen, P. 2024. The Anthropocene is dead. Long live the Anthropocene. *Science* 383:6686. <https://www.science.org/content/article/anthropocene-dead-long-live-anthropocene>.
10. Oldfield, F., A. D. Barnosky, J. Dearing, et al. 2014. The Anthropocene Review: Its significance, implications and the rationale for a new transdisciplinary journal. *The Anthropocene Review* 1:4. DOI: 10.1177/2053019613500445.
11. Shantser, E. V. 1973. The Anthropogenic System (Period). The Great Soviet Encyclopedia. Volume 2:140.
12. Crutzen, P. J. 2002. Geology of mankind. *Nature* 415:23.
13. Voosen, P. 2012. Geologists drive golden spike toward Anthropocene's base. Greenwire, 17 September 2012. <https://subscriber.politicopro.com/article/eenews/2012/09/17/geologists-drive-golden-spike-toward-anthropocenes-base-141533>. Accessed 7 March 2024.
14. McKibben, B. 1989. *The end of nature*. Random House, New York, USA.
15. Castree, N. 2019. The "Anthropocene" in global change science: Expertise, the Earth, and the future of humanity. Pages 25–49 *in* F. Biermann and E. Lovbrand, editors. *Anthropocene encounters: New directions in green political thinking*. 1st edition. Cambridge University Press, Cambridge, UK.
16. Redman, C. L. 1999. *Human impact on ancient environments*. University of Arizona Press, Tucson, Arizona, USA.
17. Kirch, P. V. 2005. Archaeology and global change: The Holocene record. *Annual Review of Environment and Resources* 30:409–440.
18. Lewis, S. L., and M. A. Maslin. 2015. Defining the Anthropocene. *Nature* 519:171–180. DOI: 10.1038/nature14258, 173–174.
19. Harari, Y. N. 2015. *Sapiens: A brief history of mankind*. Harper Perennial, New York, USA, 64.
20. Ellis, E. C., J. O. Kaplan, D. Q. Fuller, et al. 2013. Used planet: A global history. *Proceedings of the National Academy of Sciences* 110(20):7978–7985, 7983.
21. Scott, J. C. 2017. *Against the grain: A deep history of the earliest states*. Yale University Press, New Haven, Connecticut, USA, 468–469.
22. State Historic Preservation Division. 2024. Cahokia Mounds. Illinois Department of Natural Resources, Springfield, Illinois, USA. <https://dnrhistoric.illinois.gov/experience/sites/site.cahokia-mounds>. Accessed 7 March 2024.
23. Scott, J. C. 2017. *Against the grain: A deep history of the earliest states*. Yale University Press, New Haven, Connecticut, USA, 457–458.
24. Scott, J. C. 2017. *Against the grain: A deep history of the earliest states*. Yale University Press, New Haven, Connecticut, USA.
25. Harari, Y. N. 2015. *Sapiens: A brief history of humankind*. Harper Perennial, New York, USA, 83.
26. Renfrew, C. 2006. Inception of agriculture and rearing in the Middle East. *Human Paleontology and Prehistory* 5:395–404. DOI: 10.1016/j.crpv.2005.10.012.
27. Scott, J. C. 2017. *Against the grain: A deep history of the earliest states*. Yale University Press, New Haven, Connecticut, USA, 119.

28. Dodson, J., X. Li, N. Sun, et al. 2014. Use of coal in the Bronze Age in China. *The Holocene* 24:525–530. <https://doi.org/10.1177/0959683614523155>.
29. Fischer-Kowalski, M., F. Krausmann, and I. Pallua. 2014. A sociometabolic reading of the Anthropocene: Modes of subsistence, population size and human impact on Earth. *The Anthropocene Review* 1:8–33.
30. Fischer-Kowalski, M., F. Krausmann, and I. Pallua. 2014. A sociometabolic reading of the Anthropocene: Modes of subsistence, population size and human impact on Earth. *The Anthropocene Review* 1:15.
31. Landes, D. S. 1969. *The unbound Prometheus: Technological change and industrial development in Western Europe from 1750 to the present.* Cambridge University Press, London, UK.
32. Linton, D. 1992. The Luddites: How did they get that bad reputation? *Labor History* 33:529–537. DOI: 10.1080/00236569200890281.
33. Lilley, S. 1976. *Technological progress and the Industrial Revolution, 1700–1914.* Fontana Press, London, UK.
34. Ramanathan, T. 2023. Watt steam engine. Encyclopedia Britannica, 30 June 2023. <https://www.britannica.com/technology/Watt-steam-engine>. Accessed 27 February 2024.
35. Rosenberg, J. D., editor. 1963. *The genius of John Ruskin.* George Braziller, New York, USA, 368–369.
36. Ellis, E. C., K. K. Goldewijk, S. Siebert, et al. 2010. Anthropogenic transformation of the biomes, 1700 to 2000. *Global Ecology and Biogeography* 19:589–606.
37. Ellis, E. C., and N. Ramankutty. 2008. Putting people in the map: Anthropogenic biomes of the world. *Frontiers in Ecology and the Environment* 6:439–447.
38. Steffen, W., J. Grinevald, P. Crutzen, et al. 2011. The Anthropocene: Conceptual and historical perspectives. *Philosophical Transactions of the Royal Society A* 369:842–867. DOI: 10.1098/rsta.2010.0327.
39. McMichael, A. J. 2014. Population health in the Anthropocene: Gains, losses and emerging trends. *The Anthropocene Review* 1:44–56. DOI: 10.1177/2053019613514035.
40. Barnosky, A. D., E. A. Hadly, J. Bascompte, et al. 2012. Approaching a state shift in Earth's biosphere. *Nature* 486:52–59. DOI: 10.1038/nature11018.
41. Harari, Y. N. 2015. *Sapiens: A brief history of humankind.* Harper Perennial, New York, USA, 238.
42. Moore, J. W. 2017. The Capitalocene, Part I: On the nature and origins of our ecological crisis. *The Journal of Peasant Studies* 44(3):594–630. <http://dx.doi.org/10.1080/03066150.2016.1235036>.
43. Bacon, F. 1928. Bacon selections. Page ix *in* M. T. McClure, editor. *Bacon selections.* Charles Scribner's Sons, New York, USA.
44. Bacon, F. 1928. Bacon selections. Page 283 *in* M. T. McClure, editor. *Bacon selections.* Charles Scribner's Sons, New York, USA.
45. Bacon, F. 1928. Bacon selections. Page 279 *in* M. T. McClure, editor. *Bacon selections.* Charles Scribner's Sons, New York, USA.
46. Bacon, F. 1928. Bacon selections. Page 280 *in* M. T. McClure, editor. *Bacon selections.* Charles Scribner's Sons, New York, USA.
47. Bacon, F. 1928. Bacon selections. Page 372 *in* M. T. McClure, editor. *Bacon selections.* Charles Scribner's Sons, New York, USA.
48. Van Dyke, F. 2010. *Between heaven and earth: The history and practice of environmental stewardship in the Judeo-Christian tradition.* Praeger Publishers, Santa Barbara, California, USA.
49. Al-Daffa, A. A. 2020. *The Muslim contribution to mathematics.* Routledge, London, UK.
50. Dooley, H. J. 2012. The Great Leap Outward: China's maritime Renaissance. *Journal of East Asian Affairs* 26:53–76.
51. Morris, C., and A. von Hagen. 2011. *The Incas.* Thames and Hudson, London, UK.
52. Cartwright, M. 2013. Tenochtitlan. World History Encyclopedia. ttps://www.worldhistory.org/Tenochtitlan/>.
53. Harari, Y. N. 2015. *Sapiens: A brief history of humankind.* Harper Perennial, New York, USA, 280.

54. Cipolla, C. M. 1976. The origins. Pages 11–23 *in* C. M. Cipolla, editor. *The Fontana economic history of Europe. Volume I. The Middle Ages.* Harvester Press, Sussex, UK, 11.
55. Cipolla, C. M. 1976. The origins. Pages 211–223 *in* C. M. Cipolla, editor. *The Fontana economic history of Europe. Volume I. The Middle Ages.* Harvester Press, Sussex, UK, 11.
56. Cipolla, C. M. 1976. The origins. Pages 1–23 *in* C. M. Cipolla, editor. *The Fontana economic history of Europe. Volume I. The Middle Ages.* Harvester Press, Sussex, UK.
57. White Jr., L. 1976. The expansion of technology 500–1500. Pages 143–174 *in* C. M. Cipolla, editor. *The Fontana economic history of Europe. Volume I. The Middle Ages.* Harvester Press, Sussex, UK.
58. Bernard, J. 1976. Trade and finance in the Middle Ages 900–1500. Pages 274–338 *in* C. M. Cipolla, editor. *The Fontana economic history of Europe. Volume I. The Middle Ages.* Harvester Press, Sussex, UK.
59. Bernard, J. 1976. Trade and finance in the Middle Ages 900–1500. Page 322 *in* C. M. Cipolla, editor. *The Fontana economic history of Europe. Volume I. The Middle Ages.* Harvester Press, Sussex, UK.
60. Purdy, J. 2015. *After nature: A politics for the Anthropocene.* Harvard University Press, Cambridge, Massachusetts, USA.
61. Moore, J. W. 2015. Cheap food and bad climate: From surplus value to negative value in the capitalist world-ecology. *Critical Historical Studies* 2326–4462/2015/0201-000158.
62. Moore, J. W. 2016. The rise of cheap nature. Pages 78–115 *in* J. W. Moore, editor. *Anthropocene or Capitalocene? Nature, history, and the crisis of capitalism.* PM Press, Oakland, California, USA.
63. de Lille, A. 1908. The complaint of nature. Page 41 *in* A. S. Cook, editor. *Yale Studies in English.* Translated from Latin by D. M. Moffat. Holt and Company, New York, USA.
64. Cleland, H. 1955. *George Washington in the Ohio Valley.* University of Pittsburgh Press, Pittsburgh, Pennsylvania, USA, 237.
65. Detroit Historical Society. 2024. Encyclopedia of Detroit. <https://detroithistorical.org/learn/encyclopedia-of-detroit/durant-william-c>. Accessed 16 May 2024.
66. Bryan, F. R. 2014. The birth of Ford Motor Company. Henry Ford Heritage Association. <https://hfha.org/the-ford-story/the-birth-of-ford-motor-company/>. Accessed 16 May 2024.
67. Sodder, B. 1997. Michigan on fire. *In* D. Weeks, editor. *Michigan on fire.* Thunder Bay Press, San Diego, California, USA.
68. Berry, W. 2002. *The art of the commonplace: The agrarian essays of Wendell Berry.* Edited and introduced by N. Wirzba. Counterpoint, Washington, DC, USA, 26.
69. Moore, J. W. 2010. The end of the road? Agricultural revolutions in the capitalist world-ecology, 1450–2010. *Journal of Agrarian Change* 10:389–413, 400.
70. Rolston, H. III. 1994. Does nature need to be redeemed? *Zygon* 29:205–229, 223.
71. Thoreau, H. D. 1946. *Walden: Or life in the woods.* Dodd, Mead, and Company, New York, USA, 285.
72. Moore, J. W. 2017, The Capitalocene, Part I: On the nature and origins of our ecological crisis. *The Journal of Peasant Studies* 44(3):594–630. <https://doi.org/10.1080/03066150.2016.1235036>, 6.
73. Thoreau, H. D. 1946. *Walden: Or life in the woods.* Dodd, Mead, and Company, New York, USA, 169–170.
74. Thoreau, H. D. 1946. *Walden: Or life in the woods.* Dodd, Mead, and Company, New York, USA, 170.
75. Berry, T. 1999. *The great work: Our way into the future.* Bell Tower, New York, USA.
76. Moore, J. W. 2015. *Capitalism and the web of life: Ecology and the accumulation of capital.* Verso, London, UK, 6.

3 The Ecology of the Anthropocene

Humans and Nature – Ecology as Separation

My PhD institution, the State University of New York (SUNY-Syracuse), placed its principal field center, the Cranberry Lake Biological Station (CLBS), in a roadless area of upstate New York that can only be accessed by boat. Deep in the state's 6-million-acre Adirondack Park (the largest protected area (PA) in the continental United States), the CLBS is emblematic of an ecology of place and space (see Figure 3.1). Historically, that is the kind of place where ecologists have gone to study ecology – a site as far as possible from human presence and effect.

Ecology is the branch of biology that deals with the relations of organisms to one another and to their physical surroundings. The word was coined by the German zoologist Ernst Haeckel (who used the word *oekologie*) to express the "relation of the animal both to its organic as well as its inorganic environment."[1] "Ecology" is derived from the Greek *oikos*, meaning "household," "home," or "place to live." Improving on Haeckel,

Figure 3.1 The Cranberry Lake Biological Station (CLBS), located in a roadless area of upstate New York (USA), is so isolated it can only be accessed by boat. Deep in the state's 6-million-acre Adirondack Park, the CLBS is emblematic of the traditional placement of field research stations in remote areas where ecologists could study ecology without the effects of human activity.

DOI: 10.4324/9781003404415-3

we understand now that plants have ecologies too. So do populations, biotic communities, ecosystems, and microbial organisms, among others.

Today the public face of ecology is that of another "value-free science." Yet, it has ever and always been the home of scientists who have been anything but value-free, and whose values – specifically their concern for the welfare of "nature" – often found them acting as advocates for the conservation of the things they studied and came to love. The Wildlife Society, the Society for Conservation Biology, and the Society for Ecological Restoration, all leading scientific societies today with explicitly value-laden missions (wildlife management, biodiversity conservation, and ecological restoration, respectively), had their roots of formation, leadership, and membership in the older Ecological Society of America (ESA). And the ESA itself has become increasingly forthright in its advocacy for the preservation of biodiversity and ecosystem preservation in recent years.

To study ecology, ecologists have invariably sought "undisturbed" environments to determine the "baselines" of ecosystem processes and community composition that ecosystems "should" have if they are otherwise undisturbed by humans. The father of wildlife conservation in the United States, Aldo Leopold, declared that, for a science of "land health" (ecology and conservation), "the ... most perfect norm is wilderness"[2] – no disturbance at all and as far from human influence as you can get. Such an approach is an ecology of place and space. Ecological study and conservation practice depended on finding the most remote *places* and *spaces* where nature was not disturbed, and then, through various means of protection, ensuring that it never would be.

One manifestation of this thinking can be seen in ecological education. For decades (now approaching centuries), ecologists and their supporting institutions (mainly colleges and universities) have established "field stations" in places remote from their campuses to study "the way nature really works." This pattern is vividly displayed by my undergraduate alma mater, the University of Idaho. The university's academic campus is in the still small city of Moscow and surrounded by intensive agriculture. In contrast, the university placed its College of Natural Resources (CNR) field station – The Taylor Wilderness Research Station – in the Frank Church–River of No Return Wilderness over 300 miles away by road (Idaho is a big state). The CNR proudly proclaims that "the station's remote location provides an abundance of unique research opportunities for study."[3] The word "unique" is the critical modifier – "like no other" (and certainly not like anything you could find in Moscow).

This pattern of field station placement is not exclusive to the University of Idaho or SUNY-Syracuse. These and other facilities like them possess sublime beauty, rich history, and magnificent accomplishments in research, education, and inspiration for generations of conservation scientists. But now we must ask important questions about them. Today, in this world of the Anthropocene, humans press their footprint on nature at all biophysical scales and levels of biological organization, including the most remote locations and PAs.[4] Are these settings still the best places to learn about ecology and conservation given the now global prevalence of human-dominated systems?

Today the human appropriation of the Earth's net primary productivity is estimated at 24%,[5] nearly one-quarter of all the Earth's new plant growth. The Earth has lost 5–10% of all native plant species from more than half of its terrestrial ice-free biomes, and more than 10% from a quarter of these same biomes.[6] And biodiversity "hotspots" – which contain disproportionately large numbers and high densities of species – are now more intensively used by humans than ever and have lost more native species than non-hotspot areas.[6]

In a world in which nature is being shaped not only, or even primarily, by "natural" forces, but by the socioeconomic behaviors of human beings, how valuable is an education that has emphasized the study of "pristine" nature? How important is it to conserve "the real thing" in a world now far from pristine, shaped more forcibly by humans than anything else? The acid rain that has fallen on the trees and lakes of the Adirondack Park surrounding Cranberry Lake for the last several decades has made them among the most affected (and damaged) areas in the world, leaving many soils too acidic to support healthy forests and many lakes devoid of fish. In the River of No Return Wilderness, recreationists use jet boats to move upstream on the Salmon River and deep into the wilderness area, making a lie of the river's namesake, defying its legendary rapids and falls that prevented Indigenous Peoples and early explorers from going back against the current (a river of "no return").

Conservationists today, even in the most remote places, the most historic field stations, and the most pristine ecosystems, must learn and practice not simply the management of nature, but the management of human presence in nature and of human effects on nature. Today such effects are as integral to ecological systems in their processes as they are inescapable in their consequences. Because human socioeconomic and cultural behaviors are now as much a part of ecological reality as biological and geophysical processes, they must become just as much a part of ecological theory, study, and practice. And the science of ecology, and all its applications in conservation, must become part of an interdisciplinary suite of disciplines that examines the linkages between social (human) and ecological (natural) components of the world in rigorous detail. Without this unity of vision, there can be no hope for the conservation of a living Earth.

Humans' Ecological Effects – Past and Present

Humans as Architects of Niche

Humans have always had effects on ecology. They are one of the few species with both the capacity and the intention to change their surroundings for their benefit, that is, to intentionally create and expand their own ecological niche. Past human societies changed the plants, animals, biological communities, and physical landscapes around them in pursuit of a bigger "ecological space" for themselves. For thousands of years, human presence (and often, human arrival) has coincided with the extinction of large mammals on every ice-free continent. At one point, humans were targets (prey) of the large predators, and their numbers suffered. Over time, humans became, first, *competitors* of such predators for prey animals, then the *dominant* predator of larger prey, and finally the *predator* of the large predators themselves.[7]

As some human societies transitioned from hunter-gatherers to pastoralists and agriculturalists, they became mutualists with, rather than predators of, animal species. *Mutualism* is an ecological relationship in which two species interact in ways that produce benefits for both. In the math of ecology, mutualism is a plus-plus (win-win) interaction rather than the plus-minus (win-lose) interactions characteristic of predation and parasitism, or the minus-minus (lose-lose) outcome that is sometimes the case in direct competition. It may seem hard at first to perceive a herdsman and his cattle as "mutualists" (after all, the cows may eventually be eaten), but it follows the equation of mutualism. In exchange for domestication, the herdsman protects the cattle from other predators that might kill some of them if they lived in a "wild" state. The herdsman leads the livestock to water and forage and might (if sufficiently knowledgeable and compassionate) take steps to prevent or reduce disease in the herd, remedying their illnesses if infected and treating their injuries if they are harmed. For the animals, these benefits come at a price. From them, the herdsman, his family, and the community may receive milk, meat, and (raw material for) clothing. For the animals, their numbers are greater with the herdsman than they would be in the wild.[8]

As more human societies became supported by such mutualism, their practices led to changes in the surrounding ecology. One change was that more and more land was used for or converted to systems that produced forage for animals (pastures). Second, predators of the domestic animals were killed when they attacked the animals or hunted down if they were simply nearby. Third, plant communities on the pastures were changed by the kinds and numbers of domestic animals that grazed them because different species graze differently (sheep remove plants differently than cattle), both using and removing forage plants differently than wild herbivores. These differences changed ecological systems.

More recently, many human societies have transitioned from population growth driven by high fertility to controlled fertility and reduced numbers of offspring as their sources of subsistence have become more reliable. As medicine and healthcare increase human lifespan and decrease the risk of loss of family members, and as fossil fuels and renewable energy sources concurrently reduce the need for family members to provide direct energy (their own labor) to acquire subsistence needs, childbearing may be delayed and the number of children per family reduced.[8] The human population of approximately 2 billion in 1950 has quadrupled to nearly 8 billion in less than 75 years. But the rate of increase is declining. Current projections suggest that the world human population will cease to increase somewhere between 10 and 12 billion individuals.[8] A number like 12 billion may offer little optimism for preserving healthy ecosystems, but it does signal that increasing demands on resources due to (human) numbers alone might not continue if the human population ceases to grow, and even more so if humans become more economical in their resource use. If, however, rates of per capita consumption increase, the burden of human presence on the Earth's ecological processes will not be abated. It might get worse, not only because of greater resource use by humans, but because humans change processes that shape ecological communities.

Succession under Arrest – Understanding the Process

One of the most fundamental drivers of ecological change is *succession*, the replacement of plant populations on a given site over time. Succession is often initiated following a disturbance event that reduces or removes vegetation that was formerly established on the site. But we must be precise about what we mean by "disturbance." I'll define it here as any event that is relatively discrete in time and space *that disrupts the structure of an ecosystem, community, or population, and changes resource availability or the physical environment.*[9] Thus, succession is primarily a study of *ecology in time and change*, not *ecology in place and space*.

In the past, ecologists viewed succession as an intrinsic, self-willed process of nature. Ecosystems responded to discrete events caused by natural processes such as fires, droughts, and population changes in their animal and plant communities. Such disturbances might naturally be short-lived and transient, but today anthropogenic effects create disturbances that can be both persistent and continuous either through discrete disturbance "events" or from chronic stresses or changes in the system itself.[10] Disturbance events are key drivers in starting and, in many cases, directing succession, regardless of what or who initiates them.

The fact that succession occurs at all is because the plants themselves, once established, modify their environment. Such modifications, over time, can eventually create conditions more favorable for other plant species, and these then begin to colonize the site. For example, in the northern mixed conifer and deciduous forests around my home in northern Michigan (USA), oaks (*Quercus* spp.) and maples (*Acer* spp.) are both common. Oaks can germinate from their seeds (acorns) on unshaded soil (such as might be present on a disturbed site), but not in shade, not even the shade of their own parents. Most maple species, on the other hand, such as sugar maple (*A. saccharum*), can do so. Under the right conditions, sugar maples can germinate under a canopy of oaks and, over time, replace them. Subsequently, their shade tolerance permits the maples to replace themselves in successive generations. It is not surprising, then, that sugar maples are the more abundant tree in these forests.

This is a simple and partial example. Plant succession may involve the sequential colonization and subsequent replacement of many species over varying lengths of time, and cause changes in (and call for adaptations to) a myriad of variables, not only shade, but air temperature, soil temperature, soil nutrients, soil moisture, rates of animal herbivory, and others. Whatever path the successional process might follow on a given site, the response to disturbances and the changes they create is an expression of the *resilience* of biological communities – their ability to persist, albeit with changed composition, on the same site over time despite changing conditions, which they themselves might create, as well as their ability to recover from discrete disturbance events.

Human activity has long been a disturbance agent provoking (both intentionally and unintentionally) multiple changes in plant and animal communities worldwide. But Anthropocene disturbances are different. Today human disturbances can be of such magnitude and severity on a site that the successional process is *arrested*, that is, "stopped," at some early stage of plant population replacement (or does not get started at all) and the disturbed site never returns to its former state or to the state of undisturbed sites around it.[11, 12] In such cases, the disturbance is severe enough to create an "interaction deficit" in which the interactions between different plant species, and between plants and the physical environment, are so reduced in number and so weakened in effect that the "normal" (expected) processes of succession cannot occur. Sometimes the anthropogenic disturbance is so systemic and long-term it may be "invisible" as a disturbance agent, but all the more powerful in its effects.

For example, in tropical forests today, lianas (woody vines that root in the ground but climb and attach to trees), normally a subdominant component of such forests, have become more abundant, in many cases shading and out-competing forest trees for sunlight (see Figure 3.2).

Lianas can process carbon dioxide (CO_2, an essential reactant for plant photosynthesis) in greater amounts and at greater efficiencies than most tropical forest trees and have a further competitive advantage in their lower evapotranspiration (water loss) rates compared to those trees. In a world in which CO_2 levels have increased from 315 ppm (April 1958)[13] to 427 ppm (May 2024)[14] – a 36% increase – along with the warmer and longer dry seasons that result, such changes favor the lianas. These changes set up a positive feedback interaction (the worse it gets, the worse it gets) in which the more abundant and rapidly growing lianas increase the mortality of the trees they infest, and

Figure 3.2 A large liana (woody vine) snakes up a tree in Thane, India. Liana infestation is one means through which normal successional processes in a tropical forest may be stopped, resulting in "arrested succession." Image by Dinesh Valke. CCBY-SA 2.0.

more trees die. Increased tree mortality creates more forest openings, which are favored by lianas. The trees that survive are nevertheless affected because they become more infested with lianas that shade their leaves and reduce their ability to remove CO_2 from the air, lowering the amount of CO_2 stored by the trees, and exacerbating the increase of atmospheric CO_2 levels – more lianas, fewer trees, even more lianas, even fewer trees. Blaise Tymen and his colleagues, having investigated the problem directly, assert that "the liana-infested forest is in an arrested stage of ecological succession" and "such forest formations can persist for decades with no apparent evidence of transition towards a different structural state."[15] Arrested succession is a signature anthropogenic effect on ecological processes.

Succession Altered – Changing the Process

An "arrest" of ecological processes by human disturbance may not always affect the successional process itself, but instead "shift" the system into a new equilibrium, a "different structural state" such as Tymen et al. describe. On the "lower" elevations (5,000 to 6,000 feet above sea level) of the Custer National Forest in south-central Montana some grassland areas were used by cattle and horses as grazing areas for many decades until the 1960s. Palatable grasses, such as Idaho fescue (*Festuca idahoensis*), bluebunch wheatgrass (*Pseudoroegneria spicata*), needle-and-thread grass (*Hesperostipa commata*), and others occurred in these communities along with less palatable shrub vegetation like big sagebrush (*Artemisia tridentata*) and rubber rabbitbrush (*Ericameria nauseosa*). Preferring the more plentiful grasses, the livestock reduced their abundance, allowing the less palatable shrubs, especially sagebrush, to increase. Forest Service (FS) managers perceived the problem and removed the livestock. But the grasses did not recover, and the sagebrush did not diminish. Big sagebrush is a long-lived shrub with multiple traits that enhance its competitive abilities against other plants, and, once established, it wasn't going anywhere. As the biologist of the Montana Department of Fish, Wildlife, and Parks

(FWP) assigned to assist the Custer Forest in developing their wildlife management plan (Chapter 1), I wondered if there might be a way to "free" this arrested successional process on these ranges, which were important winter foraging areas for elk and mule deer (*Odocoileus hemionus*). But that would require another disturbance.

With the Forest Service (FS) managers, I selected a few designated sites to apply prescribed burning, along with adjacent sites that would not be burned, and a monitoring plan for both (see Figure 3.3). The selected sites were intentionally small in size because this was an experiment, and sagebrush creates an ecosystem of its own, supporting unique animal species, like sage grouse (*Centrocercus urophasianus*) and Brewer's sparrow (*Spizella breweri*) that depend on it. The unburned sites were as important as the burned ones. As "controls" or "counterfactuals" (sites without management application), the unburned sites would answer a critically important, but often overlooked, question in conservation management – *what would happen if we did nothing at all?* Changes in four variables would provide the answer: the species of the plants (plant community composition) in burned versus unburned sites; the abundance of each species on these sites; the nutritional content and digestibility of the plants on burned versus unburned sites; and the use by elk and mule deer (the two dominant wild herbivores) on the sites. I was asking whether fire would create interactions between the plants, and between the plants and animals, that would change the plant community. Would the plants on burned sites provide different (and more palatable) forage for the herbivores? And would the herbivores respond?

Figure 3.3 A prescribed burn in habitat dominated by big sagebrush (*Artemisia tridentata*) at Hart Mountain National Wildlife Refuge, Oregon (USA), a habitat similar to those subjected to prescribed burns by the author and his colleagues in the USFS to disrupt the arrested succession of plant communities on elk (*Cervus canadensis*) winter range in south-central Montana (USA). Photo by Scott Shaff, US Geological Survey.

The unburned sites remained a good home for sagebrush. Some of these (based on results from my own experiment as well as older historical records) did not change for 30 years![16, 17] On burned sites, sagebrush, being woody, was a good fire fuel, completely eradicated in the prescribed burns, leaving the sites to be subsequently (and quickly) colonized by grass species. Over time, the grasses retained their abundance on the burned sites, although their diversity decreased (some grasses were better competitors than others). Grasses on the burned sites were also more digestible than grasses that remained on (unburned) sagebrush-dominated sites. In the first year after the burn, grass species on burned sites doubled their levels of protein compared to the same species on unburned sites.[16, 18, 19] Protein matters, because it is a critical nutrient for herbivores on a harsh winter range. Elk and mule deer can get calories from their stored fat, but they cannot gain protein except through the catabolism (destructive metabolism) of their own muscles, a process that weakens the herbivore, makes it more susceptible to other mortality-inducing stresses, and in females can lead to resorption of embryos (impregnated by males in the fall, female elk and deer carry their offspring in utero through the winter, then give birth in spring). Herbivores responded to these changes, with their use increasing up to 500% on some burned sites in the year after the burn.[16] Over the course of 11 years (1989–1999), herbivore use gradually declined, but remained well above use levels on the unburned sagebrush sites.[16, 18, 19, 20]

It is and always will be important to understand how humans alter ecological processes like succession. But in today's human-dominated world, it becomes even more important to learn how humans can *change* these processes and the effects they produce. Today ecological study must increasingly focus on understanding the mechanisms of *outcomes*, not merely the mechanisms of *processes*. Ecological disturbances happen, and humans are causing more and more of them. But we must be ready to cope with their effects. Many disturbance events, like prescribed burning, flooding, or timber removal, are not only *caused* by humans but *initiated* by humans. These initiatives must be made part of the process of conservation. They can, with skill and practice, become beneficent *tools* rather than harmful *injuries* to the world around us, whether, as in this example, we are restoring grasslands to benefit wild herbivores, removing dams to revitalize the hydrology and ecology of formerly free-flowing rivers (this chapter), or replanting forests fragmented by oil and gas development (Chapter 11), as well as many other interventions in which human initiative can repair what has been harmed and nurture the healing process that follows (Chapter 10).

The Extinction of Interactions

The Sixth Extinction Event and the Debt of Extinction

We are now living in and through the Earth's Sixth Extinction Event. Unlike the five that preceded it, present extinctions are occurring at rates far above those documented at any time in Earth's history. Over 42,000 species of both plants and animals are presently threatened with extinction, including 41% of all amphibians, 21% of reptiles, 27% of mammals, 13% of bird species, 36% of reef corals, and 37% of cartilaginous fish species (sharks, skates, and rays). In plants, extinction threatens 69% of cycads (tropical and subtropical plants that are among Earth's most ancient plant families) and 34% of all species of coniferous trees.[21] Past examples of mass extinctions were induced by natural events and proceeded (by human standards of time if not geological ones) slowly. The current extinction event is entirely human-caused and happening much more rapidly. The International Union for Conservation of Nature's (IUCN) Red List tells us the *extent* of the problem, but not the *effects* of it, effects that strike directly at the ecological processes of the Earth itself. And the causes of such effects today are rooted in human behavior toward nature in the Anthropocene.

Ecosystem functions and processes begin to degrade when human disturbance produces multiple extinctions that reduce the number of interactive relationships among species, as well as interactions with their physical environment. Such disturbances change (usually lower) the amount of biomass energy available to species at multiple trophic (feeding) levels. Those disturbances also degrade the processes of matter and energy transfer that maintain habitats in a healthy state. So, the ecosystem first loses its capacity to retain all the species present, and then the extinction of individual species begins to take place. This time lag creates an "extinction debt."[22] Now the "cost" of the less productive ecosystem must be "paid" by the subsequent loss of species it can no longer support. Put another way, we could say that *the certain future extinction of species in a declining system is due to events that have affected the system in the past*. Such debt is the reason that conservationists, trying to protect species in declining environments, sometimes say with resignation, "They (many of the species) are extinct already." Extinction debt is one of the most pervasive, and saddest, ways humans have changed ecological processes. But the loss of the species themselves and interactions among them also play a part.

Beneficent Relationships

Three of the most important categories of species interactions are those of "benefactor" species with other "beneficiary" species, predator species with prey species, and multiple species with "keystone" species.[23] Benefactor species are those that help other species become established in a community through conditions the benefactor species modify and processes which they affect. This relationship is called *facilitation*, an interaction in which one species (the "benefactor") provides a positive benefit to another species (the "beneficiary") and is not harmed in doing so. In many plant communities, for example, one species enhances or "facilitates" the establishment and subsequent growth of another species that could not otherwise become established in that habitat. In Romania, cattle forage in agricultural lands, pastures, and so-called "wood pastures" (pastures bordering surrounding forests). In these environments, trees may germinate from seeds dispersed from nearby woodlands. If enough of these seedlings can persist to older age classes, their survival can transform the pastures into forests and contribute to expanding forest habitats. Cattle, however, often damage or eat the young tree seedlings, preventing them from surviving to the older age groups.[23]

Many species of spiny shrubs, especially the dog rose (*Rosa canina*), hawthorn (*Crataegus monogyna*), and blackthorn (*Prunus spinosa*), grow in these foraging areas, but such shrub species discourage cattle from eating them because of thorns growing on their limbs and trunks. These shrubs can act as benefactors to young oaks (*Quercus* spp.), basswoods (*Tilia* spp.), hornbeam (*Carpinus betulus*), ash (*Fraxinus angustifolia*), European wild pear (*Pyrus pyraster*), and other tree species. By germinating and growing close to the spiny shrubs, the emerging trees are "protected" and suffer less damage from cattle. As more seedlings survive and become older saplings, the saplings themselves become benefactors for newer, younger seedlings germinating in their proximity. In this way, both the shrubs and the saplings create a "regeneration niche" for new germinating trees. Without their shrub benefactors, most seedlings would not survive. In these pastures, researchers have found that 75% of seedlings present were associated with at least one benefactor shrub or an older sapling.[23]

Trees obtain much of their moisture and many of their nutrients through the association of their roots with soil fungi (mycorrhizae) that grow on the roots, using the root structures for support. Ecanteriana Fodor and her colleagues found that, in these pasture systems, the shrubs often provide mycorrhizae like those found on the parent trees in nearby forests, and so could form mycorrhizal associations (ecological "networks") with the seedlings, providing them with more water and nutrients and consequently more rapid growth.[23] None of the benefactor shrub species are presently endangered. But suppose some were? Their rarity could affect the welfare of other species for which they perform beneficial functions.

Predators and Prey

Humans not only affect predator–prey interactions by being predators *of* predators, but also by changing the relationships of predators *to* predators. Large predators are among the most vulnerable species on Earth. Predators are attractive to "trophy" hunters, who see large predators as the top "prizes" that display their hunting prowess. Large predators also engender animus from humans because they may hunt the same species as humans do (both wolves and humans kill deer and elk). Large predators can also pose threats to domestic livestock, which they can kill and eat. The position of large predators at the top of the food chain, coupled with their tendency to defend exclusive territories (the predator's "property rights"), means that predator densities are always low compared to prey densities, and often kept even lower by their self-limiting tendencies, as well as their need for areas large enough to hold sufficient prey. These traits reduce their populations. Among large predator species, 61% are listed as threatened or endangered by the IUCN. In a recent study of 33 predator species with body weights greater than 15 kg (33 pounds), 24 (73%) were in decline.[24]

Despite their low numbers, large predators impose multiple effects on biological communities directly and indirectly. Directly because they regulate numbers of prey and change prey behavior, and indirectly because they often reduce numbers of "mesocarnivores" (middle-sized predators) through "interguild" competition. Large predators often seek out and kill smaller predators that may use the same prey or be in some way "like them." From experience, I can attest that wolves (*Canis lupus*) will not tolerate coyotes (*C. latrans*) in their home ranges. Coyotes take the same attitude toward red fox (*Vulpes vulpes*). These unfriendly relationships in turn affect the amount of predation on smaller prey, which can then affect the abundance of these smaller species. Ranchers and other private property owners in Australia have long constructed exclusion fences, sometimes enclosing enormous areas, to keep out dingoes (*C. lupus dingo*). A recent study evaluating animal populations inside and outside these fenced areas found that the native diversity of small marsupials and reptiles was higher where dingoes were present compared to areas where they were absent.[25] The likely explanation is that where dingoes were in residence, there were fewer foxes, so more small marsupials and reptiles survived. Similarly, researchers in Zion National Park in the US state of Utah found that multiple species of both plants and animals were related to the abundance of mountain lions (*Puma concolor*), the park's apex predator. Such diverse groups as water plants, wildflowers, amphibians, lizards, and butterflies all had reduced abundance in areas where mountain lions were scarce but were more abundant where mountain lions roamed frequently.[26]

I remember standing in a grove of aspen (*Populus tremuloides*) on the northern range of Yellowstone National Park. It was a pleasant October day, my students around me and a Park Service biologist before us explaining the recent efforts to restore wolves to Yellowstone. Behind her was one of the board-and-wire "pens" where wolves, brought to Yellowstone from Canada, had recently been acclimated to the park and prepared for their new lives there via "soft release" (the technique of keeping the animals to be released in an enclosure for an extended period of time, then, after releasing them, continuing to provide food at the pen as they learn how to live in their new world). The pen was empty now, and the wolves weren't coming back anymore. Their new lives in this new place were a first step in the repentance of the Park Service for its role in exterminating resident wolves in Yellowstone after the park was established – the better to increase the number of elk in the park, which were often within easy viewing of park visitors. As the number of elk increased, the vegetation of Yellowstone changed. We learned a lot listening to the biologist that morning, albeit wishing that some wolves might have been within our field of vision to complement the sound of her voice. I noticed, as she was talking, that the aspen trees around us were in various stages of senescence. None had living limbs within 8 feet of the ground. There were no young aspen beneath them. I asked her why this was so. She replied, "The elk never let the young aspen grow very tall."

Were there really enough elk in the park to enforce their own "management policy" on aspen? I knew from my own studies that elk like to eat aspen seedlings, as well as the limbs and branches of older trees as high as they could reach (and some could reach pretty high). And I had attended more than one conference on the Yellowstone Ecosystem where multiple presenters had decried the damage elk were doing to park vegetation, not just to aspen but to streamside willow communities, which the elk had all but exterminated. The research effort on the Yellowstone wolves was in full swing, but the study was young, and it was too early to draw conclusions.

The findings arrived in the years that followed, and they were clear. Wolves killed elk in Yellowstone, reducing elk numbers. Motivated to stay alive, larger herds of elk broke up, and elk started to travel and forage in smaller groups. Aspen and willow communities began to reappear.[27, 28] Researchers have since noted that, after the 1995–1996 wolf reintroduction, riparian willows on Yellowstone's northern range showed significant growth for the first time since the 1920s (the correlation between elk numbers and proportion of available aspen browsed is about 80%, a strong association!).[29] But declines in elk not only reduced the proportion of aspen being browsed; the reduced browsing also affected aspen physiology. Although aspen is capable of reproducing by seed, it often spreads by "suckering," a process in which new upright shoots of aspen ("suckers") grow above ground from the roots of the parent tree. But, with the decline in elk, aspen generated fewer suckers. Instead, the now more abundant seed-germinated plants produced increased stem heights. This shift was driven by increased "apical dominance." In aspen, an intact, growing apical bud (one that has not been bitten off by elk and what botanists call the "apical meristem") produces higher levels of growth hormones. These hormones inhibit the growth and development of buds lower on the stem, and generate more concentrated growth at the tip, resulting in stem elongation.[30] Thus, low levels of apical dominance produce "shrubby" growth forms (which favor herbivores by putting all the new growth within reach) and high levels produce "upright" growth forms that rapidly rise above the herbivores. Elk still browse aspen in Yellowstone, but their reduced numbers have resulted in measurable aspen recovery.

Keystone Species

The Foundation and the Threat – Coral Reefs

Just as effects on ecological processes can be driven by the extinction of multiple species or by the presence or absence of specific species that may function alternatively as benefactors or predators, ecological processes can also be affected when *keystone species* become endangered. Better described as *function-based indicator species*, keystone species may determine many of the functions an ecosystem can provide or perform. The best example may be the hundreds of species of reef corals found in tropical oceans (see Figure 3.4). Reef corals actively construct their reefs in marine environments that provide foraging, reproductive, and rearing habitats for thousands of (non-coral) marine species. The best recent assessments judge that 830,000 (95% credible limits: 550,000–1,330,000) multi-cellular species are associated with and often dependent on coral reefs worldwide, and that number does not include the fungi. This means that although they occupy less than 0.1% of the ocean floor, coral reef ecosystems provide habitat for at least 25% of known marine species, with many species yet to be discovered.[31] Although the bulk of a coral reef is non-living matter, a layer of living creatures – composed mostly of coral polyps – forms its surface. Like their relatives – jellyfish and anemones – the polyps have soft, column-shaped bodies topped with stinging tentacles. Within them live millions of tiny dinoflagellates (*zooxanthellae*) which are photosynthetically active. As the polyps produce carbon dioxide and water through their cellular respiration, these are taken up by the *zooxanthellae* as the needed reactants for their own photosynthetic processes which produce both food (sugars and fats) and oxygen that can be consumed by the polyps. This mutualism creates a tight recycling of products between the polyp cells

Figure 3.4 Corals building a coral reef ecosystem in Biscayne National Park, Florida (USA). In the Anthropocene, as climate change causes oceans to warm, coral reefs everywhere are under threat. Photo courtesy of US National Park Service. Image by Amanda Bourque.

and the *zooxanthellae*, and its outcome is the growth and productivity of coral reefs. It is this mutualistic relationship that has built the coral reefs of the world, the most biologically diverse environments in the ocean.

As the polyps grow, they secrete calcium carbonate ($CaCO_3$) as a metabolic product, which can precipitate and then harden in water to form the more rigid cup-shaped corals that build up on the reef. Such structures attach to one another and thus serve as "homes" for the polyps. Over many generations of coral, these calcium carbonate secretions construct a coral reef, each generation enlarging the reef by building on the bodies of its progenitors.[32] Coral reefs are thus centers of marine biodiversity because they combine elements of structure, nutrients, water quality, and light to create a favorable environment for living things, making corals the *keystone* species for the entire reef system. Physically, the body of the reef provides a substrate and point of attachment for many species, especially more sedentary species like crustaceans and mollusks. Among more active species, the physical characteristics of the reef provide cavities for shelter and breeding.[32] And on this structure, high densities of prey species attract high densities of predators, and predator species in turn influence the density and diversity of prey populations and the community structure of entire reef systems.[33]

Coral reefs are in a global state of decline. The most serious and systemic threat to corals is warming ocean temperatures, a change created by human additions of CO_2 to the atmosphere, which has caused worldwide temperature increases (Chapter 5). The most common response of corals to warmer waters is "bleaching," which results in the *zooxanthellae* being expelled from coral tissues. Bleaching occurs in temperature-stressed corals, and it does not take much temperature increase to stress them. Water temperatures as little as 1–2°C above normal maximums, if sustained for even a few weeks, can cause bleaching.[34] Although some recovery from bleaching is possible, "recovered" corals typically have fewer individuals, less surface area coverage on the reef, slower growth, reduced competitive ability, and greater subsequent susceptibility and mortality to other diseases.

Bleaching was unknown in corals prior to the 1970s. Now it is a worldwide problem. Recent models of world ocean temperature change estimate that all six of the world's major coral reef regions will exceed thresholds for coral bleaching from 2030–2050, such that, under projected temperature regimes, coral bleaching will be an annual event.[35, 36] Across the world, the area that coral reefs occupy has fallen by 50% since 1957.[37, 38] The infamous "Global Bleaching Event" of 1998 alone killed 8% of the world's living coral.[39] Subsequent disturbance events, occurring between 2009 and 2018, killed 14% (about one-seventh of all corals). Projections of increased sea surface temperatures in the future suggest coral reefs will experience further declines in the coming decades.[40] With 10% of the world's living reefs now degraded, current estimates project another 30% to be lost in the next 20 years. The demise of corals threatens thousands of species that depend on them as the keystone species of the coral reef.

Coral reefs are strongly affected by anthropogenic effects and are in the worst possible locations to avoid harm from them. Today the human population density living within 20 km of the ocean is four times as high as that of the rest of the world.[41] At these population densities, direct human impacts at local scales are restructuring these marine communities.[42] But more distant human populations also contribute to reef destruction, and not only through their contributions to climate change. As coral reefs become more popular destinations for "ecotourists," the problems of divers touching or walking on reefs, or physically breaking off pieces of coral from the reef for souvenirs, cause extensive damage. The pet market for some tropical marine species and the food market for others that concentrate on reefs further impoverish the reef ecosystem, and some techniques, such as "blast fishing" or "dynamite fishing" (using explosives to stun and then collect the fish), can destroy the physical structure of the reef itself. In areas of the ocean contested by different nations, some countries further exacerbate the problem by attempting to strengthen their claim of sovereignty by constructing artificial concrete islands, often on top of existing reefs, destroying the living reef in the process. Creating artificial islands is not new. People have been doing it for centuries, but it is increasing in frequency and advancing in technology.[43] Island building is now a major international enterprise by governments like the People's Republic of China[44] as well as an emerging capitalist "growth industry," with an ever-increasing number of private companies specializing in constructing "islands on demand."[45]

Coral Restoration

In the coral ecosystem humans show their power to alter ecological processes on a global scale. Changing their interactions with reef processes is not exclusively a matter of improved scientific understanding (humans are working on that), but also one of intentional intervention to restore the corals humans have degraded, as well as developing integrated social and economic behavior that accounts for the welfare of corals. Today myriads of skilled scientists and managers are engaged in such intervention – multi-pronged efforts to save degraded corals and establish ("plant") new ones. One method intended to restore biological degradation is "coral gardening." Living corals are grown in an aquatic laboratory (a coral "nursery") and then placed on artificially constructed structures near coral reefs or on damaged reefs that have lost their original corals. Living corals are then transplanted from one location to another, and reef building is accelerated by passing an electric current through water near the reef (electrolysis) to increase the deposition of calcium carbonate and enhance the growth of existing or transplanted corals. In the Caribbean Sea and adjoining western portions of the Atlantic Ocean this approach has enjoyed such success that today "coral gardeners" grow and "outplant" tens of thousands of corals onto degraded reefs each year.[46]

Restoration efforts like coral gardening, however well informed and well meant, risk creating the impression that "science" (at least the western version of it) has the "answer" to the "problem" of reef destruction. But, like politics, all ecological effects are ultimately local, and local communities are often the best (and most cost-effective) agents of local solutions. In Indonesia's National Marine Sanctuary of Gili Matra (see Figure 3.5), the locally based Gili Matra Bersama Foundation (GMBF) coordinates reef restoration efforts involving cooperative work and partnerships among local residents, Indonesian universities, government agencies, and private reef conservation organizations. But it is the local residents who are the ones in (and under) the water building artificial reefs, planting the "iron trees" (branched poles) on which new corals can grow, and conducting weekly surveys and monitoring that measures the results of these efforts. As Cakra Adiwijaya, Director of GMBF, put it bluntly, "Once a week, we [local residents] go diving to maintain our projects and conduct surveys. We need to ensure the growth of our coral so that they can enrich the coral ecosystem in Gili Matra."[47] In other words, Adiwijaya would declare that these efforts are "our projects" and these reefs are "our coral."

Academic, government, non-governmental organization (NGO), international, and local collaborations can be fruitful, not only in reef restoration but in the ways in which

Figure 3.5 The Gili Islands of Indonesia, part of Indonesia's Gili Matra National Marine Sanctuary. Here, coral reef restoration is undertaken by a partnership of international, national, and local conservation organizations, Indonesian universities and government agencies, and local residents. <https://17touragency.com/wp-content/uploads/2022/08/Gili-Matra-1-scaled.jpg>.

each party enables and informs the others. Indonesia's Komodo National Park (KNP) in south-central Indonesia contains some of the world's most beautiful and diverse coral reef systems. It is also a place where corals have been devastated by blast fishing, leaving behind "reefs" that are little more than piles of rubble on the ocean floor. Helen Fox, Conservation Science Director of the conservation NGO Coral Reef Alliance, and her colleagues have been working with members of The Nature Conservancy's Coastal and Marine Program in Indonesia alongside local communities in an extended rehabilitation effort to restore damaged reef systems, including an effort to make coral reef restoration affordable for local communities and Majority World countries. To do this, Fox et al. evaluated coral reef restoration using low-cost materials obtained locally by placing 100-square-meter (10 meters x 10 meters) rock rubble piles at sites that might be suitable for new reef construction, and then monitoring their colonization by corals.[48] The results were encouraging. "Recruitment of hard coral and cover," they noted, "increased significantly in the mid-scale studies. The rock piles quickly developed a 'biofilm' and were colonized by coralline algae and other encrusting organisms."[49] Fox and her colleagues then increased the size of their colonization sites, transforming over 6,000 square meters of dead coral rubble into highly structured designs at four locations. On sites of this size, "Scleractinian [hard coral] recruits quickly settled on the rock piles, with considerable recruitment of hard corals after approximately 1 year."[50] Recruitment on structured rock piles was more than 20 times that of untreated rubble, with establishment of corals in as little as 6 months. And this approach was locally affordable. Costs were about US$5/$m^2$, compared to rehabilitation treatments in the Maldives that cost from US$40 to US$160/m^2 or projects in the Florida Keys National Marine Sanctuary that cost from US$550 to US$10,000/m^2.[48]

Nature in a "Built-Out" World: Human Infrastructure as a Driver of Ecologic Processes

Humans not only change ecological interactions through their effects on species presence or absence, but also by their effects on the physical environments in which these species reside. Ecological economist Sophus zu Ermgassen and colleagues, experts in the effects of human infrastructure on biodiversity conservation, noted,

> We are currently experiencing the most rapid expansion of built infrastructure in history... Over US$60 trillion of infrastructure spending is predicted between 2019 and 2040 (estimated for 56 countries, totaling 88% of global gross domestic product [GDP]) ... [and an] additional 1.2 million km^2 of land will be urbanized between 2000 and 2030 (185% increase), and an additional 3–4.7 million km of roads will be added to the global network by 2050 (22%–34% increase).[51]

Increasing infrastructure development reflects the behavior of a growing human population that is now appropriating formerly "wild" lands for human residence and use. In 2019, one-third (9,053 of 27,159) of all species classified by the IUCN's Red List as critically endangered or vulnerable were threatened primarily by the development of human infrastructure, including approximately half of all threatened amphibians (55%) and birds (46%).[52] The concrete and asphalt surfaces produced in the course of infrastructure development, the physical structures they support, and the network of roads that connect them provide few resources for plants and animals, are a source of direct mortality to them, and are often barriers to their movement. These changes are global. As affirmed by zu Ermgassen et al., "Major extractive, transport, and energy-production infrastructure projects are planned within some of the world's most biodiverse and carbon-rich regions, including the Congo Basin, the Amazon, and Borneo."[51]

Many countries have tried to develop regulations to restrict the negative effects of increased infrastructure and urban development, and among the most common mechanisms are "mandatory compensation policies" which stipulate that for any natural landscapes consumed by human infrastructure some other area must be dedicated or created to compensate for the loss.[51] In the United States, one of these was the "no net loss" (NNL) of wetlands policy first adopted as a national goal during the Carter administration in 1977 (Executive Order 11990, the "Protection of Wetlands" order). Although EO 11990 did not provide for purchase or acquisition of wetlands to compensate for those lost to various forms of development, it did direct federal agencies to "take action to minimize the destruction, loss or degradation of wetlands, and to preserve and enhance the natural and beneficial values of wetlands in carrying out the agency's responsibilities." Various states in the United States have developed or adopted similar mandates, and some have gone further, requiring that any wetland land losses within the state be "made up" by the acquisition or creation of wetland areas of equal size.

Wetlands are only one example of habitats and ecosystems that may be targeted for compensatory regulations. Today half of the world's new infrastructure through 2040 is expected to take place in countries with some forms of mandatory compensation policies in place, and this proportion is likely to increase as adoption of compensatory actions for biodiversity spreads globally. Currently, 37 countries (representing 72% of global GDP) have mandatory compensation policies to mitigate the effects of at least some kinds of infrastructure sectors on habitat types. Another 64 provide guidance on compensatory measures or enable offsets as a voluntary practice ("precursor policies"). If all of these 64 countries that currently enable (but do not demand) biodiversity compensation as part of their impact mitigation strategies were to *require* compensation for losses from infrastructure construction, the amount of infrastructure affected (and of lands and systems dedicated to compensate for it) would dramatically increase.[51]

One of the ecological processes that infrastructure most affects is dispersal. I saw the effects of such infrastructure firsthand through my studies of one of the world's least mobile animal groups – amphibians – in one of the world's most "infrastructure rich" environments – suburban Chicago, where forest preserves in DuPage County (adjacent to Cook County and municipal Chicago) were the only "undeveloped" landscapes remaining.[53] Species once found in the preserves like the wood frog (*Lithobates sylvaticus*) and Blanchard's cricket frog (*Acris crepitans blanchardi*) had disappeared within recent decades. Many amphibians in the county were forest-adapted, using permanent and seasonal ("vernal") forest pools as breeding areas. Some, like the eastern newt (*Notophthalmus viridescens*), also depended on the preserves for the terrestrial stage of their life cycle in which the young newt (the "red eft") lives in moist terrestrial forest environments, often under logs and stones, before it matures into a fully aquatic adult. Concentration and arrangement of wetland areas matter in the preserves, as the diversity of amphibian communities is higher where pools are clustered rather than isolated,[53] but infrastructure development is one of the most important drivers of habitat fragmentation, which can include effects like increasing isolation of critical habitat, such as forest pools.

Today these preserves are surrounded by what landscape ecologists refer to as "impervious surfaces" (like roads and parking lots), which reduce amphibian abundance not only because such infrastructure is not usable habitat, but because it directly impedes amphibian dispersal.[54, 55] Connectivity between habitats is reduced because the preserves are separated, not only by densely populated human suburbs, but also by heavily traveled roads, including multiple-lane freeways with constant high-speed traffic. And no Chicago commuter in the midst of a rush hour is going to brake for a frog. Yet, against all odds, some dispersers succeed. I remember my surprise (and joy!) at finding an American toad (*Anaxyrus americanus*) one morning in my native prairie garden in my yard in Winfield, Illinois (in DuPage County). Even though my neighborhood was surrounded by heavily trafficked highways, the toad had somehow managed to move through densely populated suburbs and succeed (repeatedly) in crossing hard-surfaced roads through heavy traffic. I am glad he found a home in my garden.

Altering the Master Variable: The Dynamics of Dam Removal

Dams Change the Lives of Fishes

As difficult as dispersal may have been for that particular toad, some of the most devastating effects human infrastructure can have on animal dispersal take place in aquatic systems. In these, humans alter ecological effects by fragmenting watersheds. The most systemic, global, and destructive mechanism that creates this fragmentation is building dams across channels of moving streams. Today the ecological processes of over 50% of the world's major river systems are affected by such dams. To a river's ecosystem, flow is the "master variable" that drives all others. When flow is changed, everything else changes with it.

The motive for dam construction is rooted in the signature human activity of the Anthropocene – the capture of energy from natural systems. Here the "extraction" process is imposed where flows are cheaply accessible and disproportionately abundant. The power of flowing water (kinetic energy) is extracted from a stream's natural course and diverted to drive turbines that can produce electrical power. Then that power is transmitted to industrial production and human residences. Dams also have been created to form upstream reservoirs (impoundments) intended to provide stable water supplies for drinking and irrigation. However, dams decline in usefulness as the sediment accumulating behind them degrades the quality of water in their reservoirs, reduces the amount of energy generated, and lowers the dam's capacity to control downstream flooding. Newly constructed dams are now expected to decline even faster in their ability to generate electricity due to decreased water flows associated with the increasingly long droughts generated by global climate change.[56]

In Brazil, fisheries scientist Angelo Agostinho and his colleagues examined such effects in hundreds of dams in multiple watersheds. They found that

> Soon after the reservoir is filled, patterns of thermal/chemical stratification intensify progressively in the water column, along with the decay of the flooded organic matter (vegetation, litter, and soil). Consequently, environmental quality may deteriorate in some layers of the column (e.g. thermal stress, low dissolved oxygen, acidification), especially close to the bottom.[57]

In such cases, non-migratory fish species can colonize inner zones of reservoirs because they have simpler demands for completing their life cycles. In contrast, populations of migratory fish decline due to a shortage of flowing water environments and a loss of habitats needed to complete their more complex life cycles.[58]

In Brazil, the upstream reservoirs created extensive open and deepwater areas at the expense of habitat types more common in the original river channel, and these areas had lower levels of productivity, light, and oxygen. As a result, Agostinho et al. reported that "few species colonize the vast pelagic [open water] areas in reservoirs ... [because] Few species are adapted to persist in such an environment."[59] And, as the reservoir aged, species numbers declined further, falling to an average of 20 species per reservoir in reservoirs more than 20 years old, far less than the number of species found in most free-flowing Brazilian rivers. Because the dams blocked the upstream movement of fish, the loss of migratory species was the greatest contributor to this decline. Only 5% of the reservoirs studied had more than three migratory species among their most abundant fish, and more than 50% had no migratory species at all.[57]

Brazil is not unique. Ecologist Karin Limburg and fisheries biologist John Waldman conducted a comprehensive review of studies examining changes in the status of diadromous fish populations in the North Atlantic Basin. Examining 35 long-term studies of diadromous fish populations they found that, in 24 cases, relative abundance in studied species had dropped by 90% or more. *Every species examined* suffered losses of some local populations, and many that were once common became classified as threatened or endangered.[60]

Many dams worldwide have been breached or destroyed to restore the ecology of the stream and its native species, and as many have been removed because they are no longer cost effective compared to other methods of generating electricity or obtaining water for drinking, irrigation, and industrial processes. Ironically, it was often international organizations like the World Bank that in the past advocated dam construction to create hydropower as a "clean" energy source for local communities which was also intended to foster their economic development, such as in the Mekong River in Southeast Asia, which has seen the construction of 129 dams along its watercourse.[61] In the two most heavily dammed sections of the Mekong, the Sesan and Srepok Basins, fish biodiversity declined from 60 to 29 species in the Sesan and from 42 to 25 species in the Srepok in just seven years (2007 to 2014).[62]

Just as humans have degraded aquatic environments by building dams, they can invigorate the same systems by removing them. An increasing number of studies demonstrate that dam removal is a workable strategy to improve watershed connectivity, reduce environmental hazards associated with aging dams, and promote recolonization by fish and other organisms previously prevented from moving upstream.[63, 64] When the Edwards Dam on Maine's Kennebec River (USA) was removed in 1999, diadromous fish species rapidly reoccupied their historical spawning grounds.[65] After the Fort Halifax Dam, also on the Kennebec, was breached in July 2008, over 1 million river herring (*Alosa* spp.) moved upstream just 11 months later (June 2009). Managed river herring commercial harvests were reinstituted for the first time in decades.[65]

Small-Scale Applications – The Boardman River

Impressive as the results of dam removal have been, there has often been more exertion than analysis. As of 2021, only 139 of the more than 1,200 dam removals in the United States had been subject to an assessment of the removal's ecological effects.[63, 64] My colleague, Dave Mahan, his students, and I did such an assessment – a six-year study before and after the removal of the Brown Bridge Dam on the Boardman River in northern Michigan. To say that we studied the "effects" of dam removal needs specificity, because "effects," although a small word, includes an ecological universe of conditions and processes. We chose to examine the changes in communities of bottom-dwelling (benthic) invertebrates. Benthic invertebrates are often used as indicators of the ecological state of a stream because they form the foundation of food chains that support multiple higher trophic levels of other species (like fish). Many benthic invertebrates are sensitive to water chemistry (what is in the water), water flow (how fast the water is moving), and water volume (how much water is moving through the system at any one point). Because such invertebrates are by definition "benthic," they also respond to those things that can end up at the bottom, such as heavy metal pollutants, chemical toxins, and pesticides. And many benthic invertebrates don't get around much. They may remain within a few meters of the same spot in the stream for weeks, months, or even years. Thus, they are excellent indicators not only of *what things are* in the water but of *where things are* in the water.

As part of the planned removal of the Brown Bridge Dam, the US Army Corps of Engineers designed a massive engineering project in the Boardman River to direct the post-dam downstream flow into the Boardman's past (historic) river channel, which had been changed by the alterations in flow that had been caused by the dam. The channel restoration was no small effort. It required the removal of almost 200,000 m^3 of sediment, alignment of the new channel with the relic stream bed, placement of more than 1,800 m (nearly 2 km) of woody debris for bank stabilization and in-stream habitat, downstream sediment traps to capture the initial pulse of sediment discharge when the dam was first breached, and multiple streamside plantings of vegetation that would help stabilize the banks of the new channel. How would the benthic invertebrates respond to this brave new world?

Studies of ecological effects take time. This one took six years (2011–2016), but that was the length needed to assess the benthic invertebrate communities not only above (upstream of) and below (downstream of) the dam, but to determine the composition of these communities both while the dam was in place and after it had been removed, and to determine what would happen to their communities in the "restored" river channel. We found that dam removal had little effect on the presence and abundance of the more generalist invertebrate species (those that could tolerate a wide range of aquatic variables), but strongly affected the abundance of invertebrate groups that were highly specialized and sensitive to a more narrow range of water conditions (especially invertebrates belonging to the insect orders Ephemeroptera, Plecoptera, and Trichoptera, the so-called "EPTs," a group often used as an index of water quality in freshwater streams). In the first seven months after the dam's removal in 2013, EPT abundance declined at downstream sites, almost certainly as a result of the large and relatively rapid discharge of sediment that had built up behind the dam. However, in the years that followed (2014–2016), EPT insects increased in abundance and began to resemble EPT communities found upstream from the dam, both before and after its removal. And after the dam's removal, the invertebrate communities in the restored channel became dominated by the more specialized EPT taxa that were characteristic of high-quality (low sediment, low temperature, faster moving) water conditions.[66]

Will Changed Intentions Change Human Behavior?

The Boardman River study provided insight into some aspects of *how* human effects, like dam installation and removal, can change the ecological processes and characteristics of biological communities. But the Boardman study and others like it worldwide demonstrate that when human intentions change from extracting the *energy* from aquatic ecosystems to *enhancing the productivity* of those systems, the outcomes are different. Native biodiversity, in all sorts of creatures, is not only *conserved* with such a change of intentions, it *thrives*. The physical properties of the stream are changed from slow-moving, low-oxygen, sediment-accumulating waters to faster, free-flowing, well-oxygenated waters that better discharge pollutants and support the kind of plant biodiversity in aquatic and riparian plants that can remove such pollutants. The economic gains that once came from the harvest of the stream's native fish populations, gains which declined or were altogether lost due to changes caused by the dams themselves, can not only be improved, but even restored to historic levels.

What kinds of intentions will characterize the world that human interaction with nature creates? What outcomes will follow nobler intentions? These questions must be answered in the Anthropocene, for all the ecological processes (that we once called "natural") remain part of the life and health of the Earth's biophysical system. But we are now their dominant influencers. We can no longer ignore this question. Today an answer is required of us.

References

1. Haeckel, E. H. P. A. 1866. *Generelle Morphologie der Organismen. Allgemeine Grundzüge derorganischen Formen-Wissenschaft, mechanische Begründet durch die von Charles Darwin-reformirte Descendenz-Theorie. Volume I: Allgemeine Anatomie der Organismen.* 32 + 574

pages; *Volume II: Alllgemeine Entwickelungsgeschichte der Organismen.* 140 + 462 pages. GeorgReimer, Berlin, Germany.

2. Leopold, A. 1949. *A sand county almanac and sketches here and there.* Oxford University Press, New York, USA, 196.
3. University of Idaho (2023) Taylor Wilderness Research Station. <https://www.uidaho.edu/cnr/taylor-wilderness-research-station>. Accessed 20 October 2023.
4. Ellis, E. C. 2015. Ecology in an anthropogenic biosphere. *Ecological Monographs* 85:287–331. <https://doi.org/10.1890/14-2274.1>.
5. Gingrich, S., M. Niedertscheider, T. Kastner, et al. 2015. Exploring long-term trends in land use change and aboveground human appropriation of net primary production in nine European countries. *Land Use Policy* 47:426–438.
6. Ellis, E. C., E. C. Antill, and H. Kreft. 2012. All is not loss: Plant biodiversity in the Anthropocene. *PLOS ONE* 7(1):e30535. DOI: 10.1371/journal.pone.0030535.
7. Brain, C. K. 1981. *The hunters or the hunted? An introduction to African cave taxonomy.* University of Chicago Press, Chicago, USA.
8. Lehmana, C., S. Loberg, M. Wilson, et al. 2021. Ecology of the Anthropocene signals hope for consciously managing the planetary ecosystem. *Proceedings of the National Academy of Sciences* 118(28):e2024150118. <https://doi.org/10.1073/pnas.2024150118>.
9. White, P. S., and S. T. A. Pickett. 1985. Natural disturbance and patch dynamics: An introduction. Pages 3–13 *in* S. T. A. Pickett and P. S. White, editors. *The ecology of natural disturbance and patch dynamics.* Academic Press, New York, USA. DOI: 10.1016/b978-0-12–554520-4.50006-x.
10. Newman, E. A. 2019. Disturbance ecology in the Anthropocene. *Frontiers in Ecology and Evolution* 7:147. DOI: 10.3389/fevo.2019.00147.
11. Sarmiento, F. O. 1997. Arrested succession in pastures hinders regeneration of Tropandean forests and shreds mountain landscapes. *Environmental Conservation* 24:14–23. <https://doi.org/10.1017/S0376892997000052>.
12. Goldsmith, G. R., L. S. Comita, and S. C. Chua. 2011. Evidence for arrested succession within a tropical forest fragment in Singapore. *Journal of Tropical Ecology* 27:323–326. <https://doi.org/10.1017/S0266467411000010>.
13. Keeling, C. D. 1960. The concentration and isotopic abundances of carbon dioxide in the atmosphere. *Tellus* 12:200–203.
14. CO_2 Earth. 2024. Daily CO_2. <www.co2.earth>. Accessed 12 June 2024.
15. Tymen, B., M. Réjou-Méchain, J. W. Dalling, et al. 2105. Evidence for arrested succession in a liana-infested Amazonian forest. *Journal of Ecology* 104:149–159. <https://doi.org/10.1111/1365-2745.12504>, 156–157.
16. Van Dyke, F., and J. A. Darragh. 2006. Short- and longer-term effects of fire and herbivory on sagebrush communities in south-central Montana. *Journal of Environmental Management* 38:365–376.
17. Watts, M. H., and C. L. Wambolt. 1996. Long-term recovery of Wyoming big sagebrush after four treatments. *Journal of Environmental Management* 56:95–102.
18. Van Dyke, F., and J. A. Darragh. 2007. Responses of elk to changes in plant production and nutrition following prescribed burning. *Journal of Wildlife Management* 71:23–29.
19. Van Dyke, F., and J. A. Darragh. 2006. Short- and long-term changes in elk use and forage production following prescribed burning. *Biodiversity and Conservation* 15:4375–4398.
20. Van Dyke, F. G., J. P. DiBenedetto, and S. C. Thomas. 1991. Vegetation and elk response to prescribed burning in south-central Montana. Pages 163–179 *in* R. B. Keiter and M. S. Boyce, editors. *The greater Yellowstone ecosystem: Redefining America's wilderness heritage.* Yale University Press, New Haven, Connecticut, USA.
21. International Union for the Conservation of Nature. 2024. The IUCN Red List of threatened species Version 2023-1. <https://www.iucnredlist.org>. Accessed 12 March 2024.

22. Valiente-Banuet, A., M. A. Aizen, J. M. Alcantara, et al. 2015. Beyond species loss: The extinction of ecological interactions in a changing world. *Functional Ecology* 29:299–307. DOI: 10.1111/1365–2435.12356.
23. Fodor, E., O. Hâruța, and S. Dorog. 2018. Nurse plants and the regeneration niche of tree seedlings in wood-pastures from Western and North-Western Romania. *Reforesta* 6:41–59. <https://dx.doi.org/10.21750/REFOR.6.04.57>.
24. Ripple, W. J., J. A. Estes, R. L. Beschta, et al. 2014. Status and ecological effects of the world's largest carnivores. *Science* 343(6167). DOI: 10.1126/science.1241484.
25. Letnic, M., F. Koch, C. Gordon, et al. 2009. Keystone effects of an alien top-predator stem extinctions of native mammals. *Proceedings of the Royal Society B* 276:3249–3256. DOI: 10.1098/rspb.2009.0574.
26. Ripple, W. J., and R. L. Beschta. 2006. Linking a cougar decline, trophic cascade, and catastrophic regime shift in Zion National Park. *Biological Conservation* 133:397–408.
27. Ripple, W. J., E. J. Larsen, R. A. Renkin, et al. 2001. Trophic cascades among wolves, elk and aspen on Yellowstone National Park's northern range. *Biological Conservation* 102:227–234.
28. Painter, L. E., and M. T. Tercek. 2020. Tall willow thickets return to northern Yellowstone. *Ecosphere* 11(5) DOI: 10.1002/ecs2.3115.
29. Klaptosky, J. 2016. The plight of aspen: Emerging as a beneficiary of wolf restoration on Yellowstone's Northern Range. *Yellowstone Science* 24(1):65–71.
30. Farmer, Jr., R. E. 1962. Aspen root sucker formation and apical dominance. *Forest Science* 8:403–410.
31. Fisher, R., R. A. O'Leary, S. Low-Choy, et al. 2015. Species richness on coral reefs and the pursuit of convergent global estimates. *Current Biology* 25:500–505. DOI: 10.1016/j.cub.2014.12.022.
32. Hinrichsen, D. 1997. Coral reefs in crisis: An overview of these vanishing ecosystems, the problems that plague them, and the means for saving them. *BioScience* 47:554–558. <https://doi.org/10.2307/1313160>.
33. Boaden, A. E., and M. J. Kingsford. 2015. Predators drive community structure in coral reef fish assemblages. *Ecosphere* 6(4):46. <http://dx.doi.org/10.1890/ES14-00292.1>.
34. Spalding, M. 2004. *A guide to the coral reefs of the Caribbean*. University of California Press, Berkeley, California, USA.
35. Hoegh-Guldberg, O. 2005. Climate change and marine ecosystems. Pages 256–273 *in* T. E. Lovejoy and L. Hannah, editors. *Climate Change and Biodiversity*. Yale University Press, New Haven, Connecticut, USA.
36. Hoegh-Guldberg, O., E. S. Poloczanska, W. Skirving, et al. 2017. Coral reef ecosystems under climate change and ocean acidification. *Frontiers in Marine Science* 4:158. DOI: 10.3389/fmars.2017.00158.
37. Ashworth, J. 2021. Over half of coral reef cover across the world has been lost since 1950. Natural History Museum, London, UK. <https://www.nhm.ac.uk/discover/news/2021/september/over-half-of-coral-reef-cover-lost-since-1950.html>. Accessed 13 March 2024.
38. Eddy, T. D., V. W.Y. Lam, G. Reygondeau, et al. 2021. Global decline in capacity of coral reefs to provide ecosystem services. *One Earth* 4:1278–1285. <https://doi.org/10.1016/j.oneear.2021.08.016>.
39. Souter, D., S. Planes, J. Wicquart, et al, editors. 2020. Status of coral reefs of the world: 2020 executive summary. *United Nations Environmental Programme*, Nairobi, Kenya.
40. Williams, G. J., N. A. J. Graham, J.-B. Jouffray, et al. 2018. Coral reef ecology in the Anthropocene. *Functional Ecology* 33:1014–1022. DOI: 10.1111/1365–2435.13290.
41. Kummu, M., H. de Moel, G. Salvucci, et al. 2016. Over the hills and further away from coast: Global geospatial patterns of human and environment over the 20th–21st centuries. *Environmental Research Letters* 11:034010.
42. He, Q., and B. R. Silliman. 2019. Climate change, human impacts, and coastal ecosystems in the Anthropocene. *Current Biology* 29:R1021–R1035.

43. Zhang, J., and F. Zwao. 2020. Applications of light steel and light concrete structure system in island building. *Journal of Coastal Research* 107(SI):73–76. <https://doi.org/10.2112/JCR-SI107-019.1>.
44. Saunders, I. 2019. Artificial islands and territory in international law. *Vanderbilt Journal of Transnational Law* 52:643–684.
45. Islands Building Company. 2023. We build artificial islands worldwide. <https://www.islandsbuildingcompany.com/>. Accessed 10 May 2024.
46. Lirman, D., and S. Schopmeyer. 2016. Ecological solutions to reef degradation: Optimizing coral reef restoration in the Caribbean and Western Atlantic. *PeerJ* 4:e2597.
47. Maqoma, R. I. 2023. How local communities in Indonesia's Gili islands are restoring coral ecosystems amid rising sea temperatures. *The Conversation*. <https://theconversation.com/how-local-communities-in-indonesias-gili-islands-are-restoring-coral-ecosystems-amid-rising-sea-temperatures-216390>. Accessed 10 May 2024.
48. Fox, H. E. P., J. Mous, J. S. Pet, et al. 2005. Experimental assessment of coral reef rehabilitation following blast fishing. *Conservation Biology* 19:98–107.
49. Fox, H. E. P., J. Mous, J. S. Pet, et al. 2005. Experimental assessment of coral reef rehabilitation following blast fishing. *Conservation Biology* 19: 98–107, 104.
50. Fox, H. E. P., J. Mous, J. S. Pet, et al. 2005. Experimental assessment of coral reef rehabilitation following blast fishing. *Conservation Biology* 19: 98–107, 105.
51. zu Ermgassen, S., P. Utamiputri, L. Bennun, et al. 2019. The role of "no net loss" policies in conserving biodiversity threatened by the global infrastructure boom. *One Earth* 1:305–315, 305. <https://doi.org/10.1016/j.oneear.2019.10.019>.
52. International Union for Conservation of Nature and Natural Resources. 2019. The IUCN Red List of threatened species. Version 2019-1. <http://www.iucnredlist.org>.
53. Van Dyke, F., A. J. Engel, S. M. Harju, et al. 2017. Amphibians in forest pools: Does habitat clustering affect community diversity and dynamics? *Ecosphere* 8(2):e01671. 10.1002/ecs2.1671.
54. Houlahan, J. E., and C. S. Findlay. 2003. The effects of adjacent land use on wetland amphibian species richness and community composition. *Canadian Journal of Fisheries and Aquatic Sciences* 60:1078–1094.
55. Eigenbrod, F., S. J. Hecnar, and L. Fahrig. 2008. The relative effects of road traffic and forest cover on anuran populations. *Biological Conservation* 141:35–46.
56. Leslie, J. 2021. As warming and drought increase, a new case for ending big dams. YaleEnvironment360. <https://e360.yale.edu/features/as-warming-and-drought-increase-a-new-case-for-ending-big-dams>. Accessed 10 May 2024.
57. Agostinho, A. A., F. M. Pelicice, and L. C. Gomes. 2008. Dams and the fish fauna of the Neotropical region: Impacts and management related to diversity of fisheries. *Brazilian Journal of Biology* 68(4, Supplemental):1119–1132, 1122.
58. Agostinho, A. A., F. M. Pelicice, and L. C. Gomes. 2008. Dams and the fish fauna of the Neotropical region: Impacts and management related to diversity of fisheries. *Brazilian Journal of Biology* 68(4, Supplemental):1119–1132, 1122.
59. Agostinho, A. A., F. M. Pelicice, and L. C. Gomes. 2008. Dams and the fish fauna of the Neotropical region: Impacts and management related to diversity of fisheries. *Brazilian Journal of Biology* 68(4, Supplemental):1119–1132, 1124.
60. Limburg, K. E., and J. R. Waldman. 2009. Dramatic declines in North Atlantic diadromous fishes. *BioScience* 59:955–965.
61. International Rivers. 2015. The World Bank and dams – Part 1: Lessons not learned. International Rivers. <https://www.internationalrivers.org/wp-content/uploads/sites/86/2020/07/world_bank_and_dams_fact_sheet_web.pdf>. Accessed 10 May 2024.
62. Liu, Z., S. Fagherazzi, Q. He, et al. 2024. A global meta-analysis on the drivers of salt marsh planting success and implications for ecosystem services. *Nature Communications* 15:3643. <https://doi.org/10.1038/s41467-024-47769-5>.

63. Foley, M. M., F. J. Magilligan, C. E. Torgensen, et al. 2017. Landscape context and the biophysical response of rivers to dam removal in the United States. *PLOS ONE* 12:e180107.
64. Bellmore, J. R., J. J. Duda, L. S. Craig, et al. 2017. Status and trends of dam removal research in the United States. *WIREs Water* 4:e1164. <https://doi.org/10.1002/wat2.1164>.
65. Natural Resources Council of Maine. 2018. Edwards Dam and Kennebec River restoration. <https://www.nrcm.org/programs/waters/kennebec-restoration/>. Accessed 15 July 2018.
66. Mahan, D. A., J. E. Betts, E. C. Nord, et al. 2021. Response of benthic macroinvertebrates to dam removal in the restoration of the Boardman River, Michigan, USA. *PLOS ONE* 16(5):e0245030. <https://doi.org/10.1371/journal.pone.0245030>.

4 Geology in the Anthropocene

Three Letters – Three Numbers

N P K 12-12-12 – three letters and three numbers on bags of chemical plant fertilizer familiar to millions of farmers and gardeners worldwide. The letters are the chemical symbols designating Nitrogen (N), Phosphorous (P), and Potassium (K) – three soil nutrients that plants need and often lack in "natural" soil. Historically the distribution and abundance of these nutrients were determined by natural biogeochemical cycles. Those cycles remain in force today, but they are no longer "natural." Human intrusion, exemplified in agriculture but not limited to it, has altered these processes in the Anthropocene. Humans are today the most powerful factor affecting planetary geology. Our impacts now exceed the effects of natural geological processes of erosion, transport, and deposition of sediment.[1, 2] Humankind has been changing geological processes for millennia, but our advancing technologies have transformed these actions from local, site-specific practices of the past into the global change agents of the present. When we look at landscapes today, even those in which human presence seems minimal, we are seeing, in the words of biologist Jon Harding and his colleagues, "the ghosts of land use past."[3] What, exactly, are these ghosts? What is being affected, and how are those effects being revealed? The question of our present time is, *what are the Anthropocene's manifestations (that is, human manifestations) as a geological force?*

The Nitrogen Cycle

Nitrogen (N) is intimate with us in the air we breathe, and, as a biogeochemical cycle, it passes through living biological organisms and non-living elements of the ecosystems around us. As a gas, N forms paired (diatomic) molecules (N_2) that make up nearly 80% of the Earth's atmosphere. But, unlike oxygen, which is both a critical reactant for photosynthesis and an essential element for our own respiration, nitrogen gas does not have an immediate, useful pathway into most life processes. Animals do not need it to breathe. Plants do not need it for photosynthesis, nor can they use it, in its gaseous form, as a nutrient. Nitrogen is in fact essential for plants, but its molecular bonds are extremely stable and difficult to break and recombine with other elements. Nitrogen's packaging must be changed.

One of the most important roles of N is its structural contribution to proteins. Proteins, an essential dietary component and structural constituent for both plants and animals, are composed of long chains of chemical compounds called *amino acids*. Although there are many different kinds of amino acids, they share a signature structure, *the amino*

DOI: 10.4324/9781003404415-4

group – a cluster of a single N atom joined to two hydrogen (H) atoms (NH_2). Without amino acids, you cannot build proteins.

When lightning strikes the ground, it can spontaneously convert nitrogen gas into nitrate (NO_3^-), but it's a risky strategy to base your protein metabolism on being struck by lightning. The nitrogen gas in the atmosphere is more abundant and closer at hand, but plants cannot "breathe" N_2 and reassemble it into the amino acids and proteins they need. However, some organisms can start the process. Bacteria in the genus *Rhizobium* can use plant roots as structures for their own attachment and support (the bacteria colonies often appear as small white "balls" on the plant's roots, see Figure 4.1) and the bacteria benefit from the plant's own production and secretion of organic acids from their roots, which the *Rhizobium* can absorb. But there is mutual benefit from this plant–bacteria association (symbiosis). For their part, *Rhizobium* can take in nitrogen gas, split it, and attach hydrogen to the nitrogen atom, constructing ammonia (NH_3). Although *Rhizobium* can assimilate the ammonia in their own metabolism, they make more than they can use, secreting the excess into the surrounding soil. The plant can employ the ammonia as a nitrogen source and gain nitrogen from other compounds that can be derived from the ammonia.

Two of the compounds so derived are *nitrate* (NO_3^-) and *nitrite* (NO_2^-), which form from ammonia and other nitrogen-bearing compounds in the presence of water (such as is present in soil) and other kinds of bacteria ("nitrifying" bacteria) (see Figure 4.2), as well as gaining nitrogen from deposits of animal feces which release *ammonium*, which is ammonia with an additional hydrogen atom attached (NH_4^+). The "cycle" is completed by another group of bacteria, the "denitrifying" bacteria, which remove oxygen from nitrate and release the product, nitrogen gas (N_2), into the atmosphere from the soil. Both ammonia and nitrate form naturally and nourish plants, but their quantities in soil are low, making them a scarce nutrient. Not all plants live in association with *Rhizobium*, but many that take part in this mutualism are *legumes* – beans, peas, and lentils – which

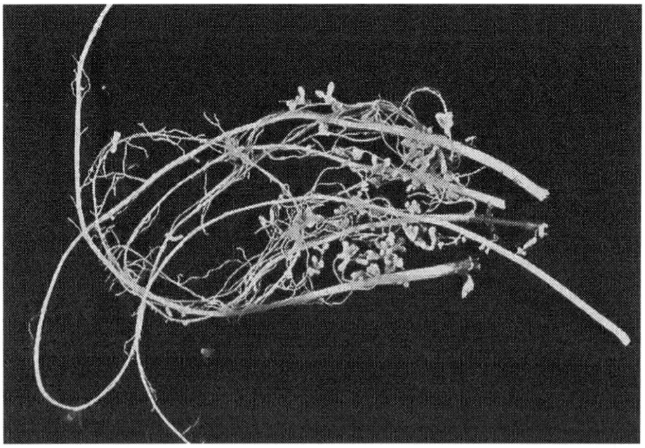

Figure 4.1 On some plants, colonies of *Rhizobium* bacteria (the globular nodes) can live in symbiotic association with the plant's roots, as is the case on the roots of this soybean (*Glycine max*). The bacteria can convert (fix) nitrogen gas into ammonia by combining it with hydrogen, which can then be used as a nitrogen-rich nutrient for the plant or can be converted to other compounds from which the plant can also derive nitrogen. Photo courtesy of USDA.

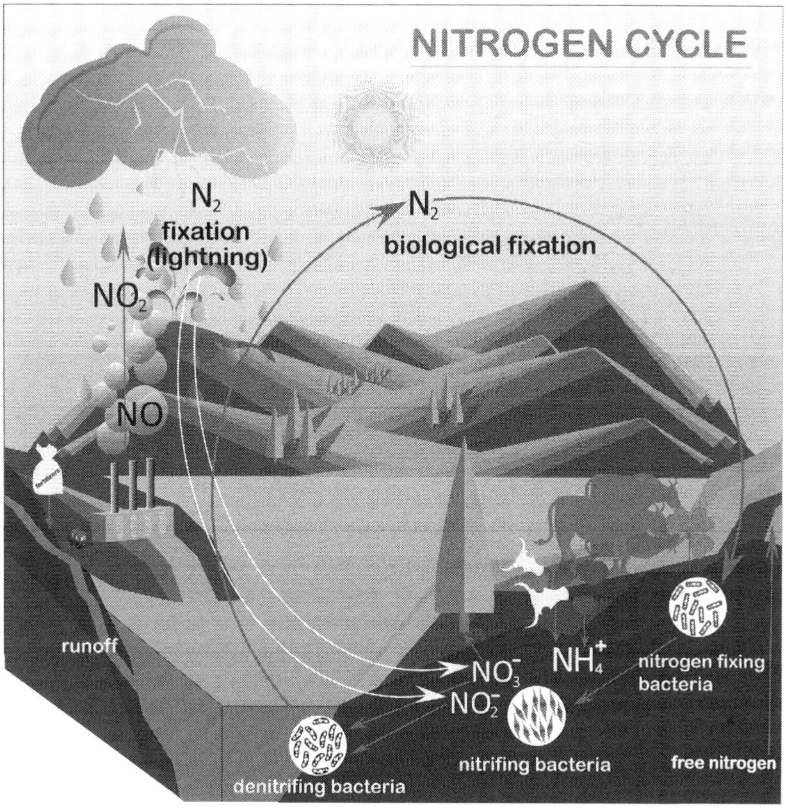

Figure 4.2 A simplified illustration of the nitrogen cycle. <danylyukkl/Shutterstock.com>.

can thus grow in N-poor soils and can add nitrate (NO_3^-) and nitrite (NO_2^-) to those soils. Although not all plants have *Rhizobium*, they all need NH_3.

In 1840, the German botanist Carl Sprengel declared his "theorem of the minimum," based on his studies of plant physiology and growth. He argued that plant growth was not determined by the total nutrient resources at hand, but by the scarcest available resource.[4] His younger contemporary, Justus von Liebig, popularized Sprengel's "theorem of the minimum" as the "law of the minimum" – that plant yield was limited by the nutrient present in the least quantity relative to its demands for plant growth.[5] A reflective reader might be forgiven for thinking that this statement seems less than profound. It sounds suspiciously like saying, "Plants are limited by what limits them." But the law of the minimum was a radical departure from traditional thinking about plant nutrition and growth. Its premise directed scientists to a previously unexplored question: what are essential nutrients for plants and in what amounts? The law also pointed to a new goal for crop management: identify the limiting nutrient and add that to the soil. In natural plant communities, plants that die and decay return nitrogen to the soil. Plants that are eaten have some of their nitrogen returned to the same destination by animal feces and the deaths and decomposition of the animals themselves. When this is the primary process that makes nitrogen available in natural ecosystems, nitrogen is a scarce nutrient. Such scarcity has contributed to the evolution of plant communities in both terrestrial and aquatic environments that are highly diverse in their strategies of nitrogen needs, uptake, and use, and nearly all of their adaptations are means of coping with nitrogen scarcity. But plants cultivated as agricultural crops are notoriously "nitrogen needy." Their high demand for nitrogen (more specifically, for ammonia and nitrate) makes it a limiting nutrient that constrains their yields. But an understanding of the critical effects of limiting nutrients, as articulated in the law of the minimum, provoked plant scientists to increasingly experiment with additions of chemical nitrogen-based fertilizers to soils, and the results were greater crop yields.

Building on more refined studies of the physiology of crop plants and even greater demands for chemical fertilizer that followed this new understanding, industrial chemist Fritz Haber and chemical engineer Carl Bosch perfected a process (which came to be called the Haber–Bosch process) of combining hydrogen with nitrogen gas to produce ammonia that could be used for fertilizer. They initially developed the process from 1908 to 1910, and by 1913 the Haber–Bosch process was being used on an industrial scale.[6] Enabled by a method that could generate ammonia in large quantities on demand, agriculture, first in Europe and then worldwide, increased the global use of nitrogen and its abundance in soil. In the past 50 years, world agricultural production has grown three-fold,[7] and increased application of nitrogen fertilizer – not only ammonia, but other forms of nitrogen that plants can assimilate such as nitrate (NO_3^-) – has been the most important element of that growth. However, the use of nitrogen fertilizer has risen even more than the growth in crop yields. Nitrogen-based fertilizer is now applied at more than *ten times* the rate at which it was used in the 1950s. As a result, humans have more than doubled the rate at which biologically available nitrogen enters the terrestrial biosphere compared to preindustrial levels.[8] And increased nitrogen fertilization has not only increased world food production in traditional agricultural areas, but has also enabled agriculture to expand into lands that were once considered non-arable.

Nitrogen application is not distributed equitably – 10% of world croplands today are found in developed countries, but that 10% accounts for 32% of the Earth's current global nitrogen surplus[9] because agriculture in developed countries has the resources to acquire and then add large amounts of nitrogen-based fertilizer to increase yields. But such practices create an ongoing need for more nitrogen inputs. In a crop production system, most of the nitrogen-containing plant biomass is removed from the site through harvesting. And crop soils, unlike natural soil systems, are regularly disturbed by cultivation, so nitrogen-fixing bacterial populations are also reduced (or even eliminated by chemical agents). The available nitrogen that remains in the ground is volatilized into the air, leaving the soil in a nitrogen-depleted state. Now the nitrogen losses created by harvesting, denitrification, leaching, utilization, and soil erosion must be compensated by applying synthetic N fertilizers and animal manure.[7] But nitrogen is a "leaky" nutrient. Only about half of anthropogenic N inputs are taken up by the crops[10] while the remainder is lost to the environment at a much higher rate than in natural ecosystems.[11] This is what creates a "nitrogen surplus" and why much of the surplus does not remain in the soil. It travels to other ecosystems, creating excess nitrogen within them.

When nitrogen increases in natural systems, its abundance favors a few native species with high nitrogen demands, as well as many non-native species that can become invasive. The outcome is that high-diversity natural plant communities with few non-native plants are transformed into low-diversity plant communities with high infestations of non-native species. In North America, experimental nitrogen additions ("nitrogen loading") in grasslands dramatically decreased plant diversity.[12] In the UK, experimental additions of nitrogen to grassland ecosystems led to increased dominance by a few species of nitrogen-demanding grasses, but suppression of many other plant species.[13] In old growth tropical forests, plant species' richness, density, and cover decreased by 80% in treatments receiving nitrogen inputs relative to control treatments which received no nitrogen additions.[14] But nitrogen is not the only cycle that humans affect.

The Phosphorus Cycle

Compared to nitrogen, phosphorus (P) is a "slow" cycle in natural ecosystems. Here, most reservoirs of P exist in soil or rock. The phosphorus in these systems reaches plant and animal species only through weathering and associated erosion. P cycling is also "tight," with the quantity of phosphorus actually transferred between biota and soil being much larger than its relatively small gains or losses into other systems.[15] Humans have "sped up" and "loosened" the global P cycle through their own uses and applications of it, especially by mining phosphorus-containing (phosphate, PO_4^{3-}) rock, applying commercial P fertilizers, and changing the pathways of P losses (see Figure 4.3). Rising phosphorus availability in croplands increases phosphorus inputs into terrestrial ecosystems as well as losses from land which move phosphorus into aquatic ecosystems in freshwater and coastal zones. These effects last. Excessive applications of phosphorus by humans create accumulation in upland soils where the (long-lived) P deposited in them can have "legacy" impacts through slow, chronic phosphorus release over many years.[15]

Phosphorus is a valuable plant nutrient because it is essential for forming carbohydrate polymers, proteins, and nucleic acids (essential components of every organism's

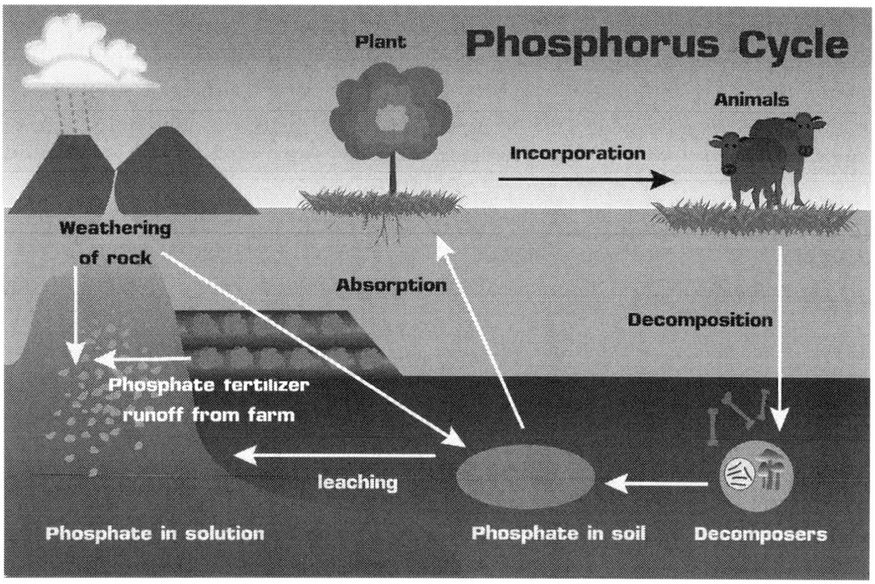

Figure 4.3 A simplified representation of the global phosphorus cycle. A "slow" cycle in natural ecosystems, most reservoirs of phosphorus exist in soil or rock and reach plant and animal species only through weathering and associated erosion. Human additions of phosphorus have sped up this cycle and greatly increased the amount of phosphorus in natural systems. <CharunnesHn/Shutterstock.com>.

DNA). The energy needed for synthesizing all complex molecules of life is supplied by breaking the phosphorous bond that reversibly moves between two important metabolic compounds – adenosine diphosphate (ADP) and adenosine triphosphate (ATP). I can still remember accounting for all the ATP produced and utilized in the reactions of the infamous "Krebs Cycle" of metabolism in my first college biology course. By the time the final exam was over, I had gained an indelible appreciation for the importance of phosphorus in the metabolism of living creatures.

In natural systems, phosphorus is not frequently encountered in its elemental form, but more often as phosphate (PO_4^{3-}), which may be concentrated in deposits of sedimentary rock. Even in older, more traditional human agricultural systems, crop farming relied on P in soil and local P from recycled organic wastes (such as manure), practices which had continued for thousands of years. But by the mid-twentieth century humans had discovered that, where phosphate rock was present in sufficient concentrations, it could be "mined." Mining of phosphate rock and its subsequent transformation and use as fertilizer, detergent, animal feed supplement, and others has more than doubled P inputs to the environment over natural background P from weathering. The application of phosphorus as an agricultural fertilizer has the greatest effect on the phosphorus cycle by intensifying P releases from the soil into both proximate and distant ecosystems.[15] Globally, humans are estimated to have increased annual cropland applications of phosphate 13-fold compared to pre-industrial levels.[16] And the increase has been accelerating. Phosphorus extraction from phosphate rocks grew slowly to over 2 megatonnes (MT) (about 2,200 pounds) of phosphorus per year by 1940, but then experienced a ten-fold increase from more intensive mining for fertilizers from the 1940s to the 1980s. Since 2000, growing fertilizer demand, especially in Majority World countries, has produced another surge of phosphate rock extraction, with an annual increase of 4.2%.[15] Today, when the use of livestock slurry and manure is included, the total world phosphorus application in agriculture amounts to 22–26 MT per year.[8] But this level of phosphate application is not needed for maximum crop yields.

Globally it is estimated that 70% of lands receiving phosphate are having it deposited at surplus levels. As was the case with nitrogen, the surplus applications are not distributed equitably.[15] In developed countries, 10% of croplands account for 40% of the phosphorus surplus.[8]

My wife, Linda, is from the town of Soda Springs, near the southeastern corner of the US state of Idaho. For as long as most residents of Soda Springs can remember, the town's chief employer has been the phosphate processing plants, as well as the phosphate mines throughout the surrounding countryside that provided material to the plants. Removing phosphate from the rocks in which it is embedded requires subjecting the rocks to high heat, and rock that remains after the phosphate has been removed is discarded as waste or "slag." Such slag, more properly referred to as *phosphogypsum*, contains a variety of non-phosphate products that are removed in the refining process, including small amounts of naturally occurring radionuclides, mostly uranium and radium.[17] I remember my first visit to Soda Springs and seeing, with something like shock, the towers of slag – all covering acres of land surface and some more than 100 feet high – surrounding the Monsanto plant, the town's single largest employer. The topmost slag, only recently discarded from the furnaces, was still hot and smoldering. Against the night sky the piles glowed a dull red, strands of smoke rising from them. It was a visual image of Tolkien's Mordor.

Some of the mines surrounding Soda Springs began their operations in the first decade of the twentieth century. Today many of these mines, and others that followed, have been abandoned and shut down. Among those permanently closed is the Ballard Mine, 13 miles northeast of the town. The site is owned by P4 Production, a subsidiary of Monsanto, and the mine operated from 1951 to 1969. Today it is designated by the US Environmental Protection Agency (EPA) as a toxic waste "Superfund" site.[18] Superfund is more formally the Comprehensive Environmental Response, Compensation, and Liability Act (CERCLA) of 1980.[19] CERCLA provides a Federal "Superfund" supplied with the money necessary to clean up uncontrolled or abandoned hazardous-waste sites. Through CERCLA, the EPA was given the power to seek out parties responsible for releasing toxic substances and assure their cooperation (financial and otherwise) in the cleanup. The Ballard Mine site is among these. It consists of 534 acres of mining-disturbed land that includes widespread contaminated soil, six waste rock dumps, and six open pits. The waste rock dumps, the pits, and other land on the site contain about 19 million cubic yards of waste rock bearing elevated levels of selenium, arsenic, naturally occurring uranium, and radioactive material derived from decaying uranium, so-called "uranium-daughter products" such as the radioactive isotopes radium-226 and radon-222[18] (see Figure 4.4).

The Mine stands in stark contrast to the lands around it. The lower elevation valleys are home to large farms and sprawling cattle ranches brimming with spectacular landscapes. I have had the privilege of seeing many of these up close and even being a guest of the ranch owners on some. I could almost envy a cow's life on this range. It is a land that deserves its own Western movie. The mountain slopes bordering the ranches are part of the Caribou National Forest, a portion of the Greater Yellowstone Ecosystem. The Blackfoot River, a pristine Blue Ribbon trout stream, and the main "character" of the novel *A River Runs*

Figure 4.4 One of the "pit" areas of the now inoperative Ballard Mine in southeast Idaho, USA. Currently a toxic waste site designated by the US Environmental Protection Agency (EPA), the mine's contaminated soil, waste rock dumps, and open pits eventually accumulated about 19 million cubic yards of waste rock bearing elevated levels of toxic materials including selenium, arsenic, uranium, and radioactive material derived from decaying uranium. Photo courtesy of US EPA.

Through It, is the principal drainage. The Ballard Mine itself acknowledges the stream as part of its own identity. Its address is on Blackfoot River Road. The Indigenous Peoples of the Shoshone–Bannock Indian Nations hold treaty rights on lands nearby.

The toxic wastes produced by the Ballard Mine have not all remained on site. By the movement of wind, water, and the mine's own rock and sediment transport, such wastes have traveled throughout Caribou County, the town of Soda Springs, and the Shoshone–Bannock lands, and they have been there for a long time. The EPA issued its final cleanup plan (officially its "Record of Decision") in October 2019, 50 years after the mine's closure. As part of the cleanup strategy, the state of Idaho's Department of Environmental Quality and the Shoshone–Bannock Nations signed the associated Consent Decree for Remedial Design and Remedial Action, which began in 2021. Cleanup is still in progress.[18] P4 Production proclaimed in an official statement:

> We fully support this plan and will continue to work collaboratively and transparently throughout this process. As members of the Soda Springs community, we have great respect for the people and the environment in this historic and beautiful area of Southeast Idaho.[20]

P4 admitted no fault for the contamination, but will pay hundreds of thousands of dollars to the EPA, the tribes, and other agencies to cover the costs of removing and decontaminating the toxic waste. P4 must also make an additional $89 million available to the EPA to guarantee that the cleanup work will be completed.[20] This is an expensive remittance when you do not admit that you have done anything wrong.

Eutrophication and Annoxiation

If terrestrial plant communities are nutrient limited, aquatic systems are arguably more so. In fact, nutrients are so important in lake ecosystems that lakes are categorized by their nutrient levels. Nutrient-poor or *oligotrophic* lakes are characterized by clear, often colder waters with high oxygen levels. "Middle-nutrient" or *mesotrophic* lakes have intermediate levels of soluble nutrients and less water clarity. Nutrient-rich or *eutrophic* lakes have high levels of nutrients, extensive plant and bacterial growth, low levels of oxygen, and minimal water clarity. N and P losses from cropland ("nonpoint" sources of pollution) have been recognized as dominant contributors to the nutrient enrichment of streams and lakes through erosion and runoff. Globally, about 56% of aquatic P inputs are induced by surface runoff from cropland, an increase of 21% since the early 1900s,[15] and such inputs have changed lakes worldwide. Using the records of hypoxia (oxygen depletion) in 365 lakes around the world, Jean-Phillipe Jenny and his colleagues found that ~20% of these lakes had shifted to a hypoxic condition since the middle of the nineteenth century, and the number was closely correlated to total P release.[21]

Eutrophication, the transition of a mesotrophic or oligotrophic lake to a eutrophic (nutrient-enriched) state, occurs when nutrients are released into aquatic systems from upstream or surrounding agricultural areas (in the form of fertilizer runoff) or from towns and cities (in the form of human waste).[22] In oligotrophic ecosystems, biomass or diversity may at first increase with increasing nutrient load, but as levels of reactive nitrogen and phosphorus continue to rise, phytoplankton capable of better assimilating these nutrients are favored over species that are more limited by other factors (such as diatoms, which require silica, or plants growing on the lake bottom, which require light). Blooms of low-diversity algae or cyanobacteria ("blue–green algae") populations can result, leading to the depletion of oxygen in surface waters and the release of toxic compounds. This in turn affects organisms at higher feeding levels, such as invertebrates and fish. Sedimentation and decomposition of biomass from phytoplankton blooms deplete oxygen in bottom waters and surface sediments, especially in ecosystems with low water turnover rates. This oxygen depletion further shifts the bottom-dwelling community toward a less diverse assemblage of fewer species able to tolerate such changes.[22]

As the higher levels of nitrogen and phosphorus trigger increases in primary producers (formerly limited by a scarcity of these same nutrients), periphytic (attached) algae and submersed macrophytes initially increase, then decline as phytoplankton and cyanobacteria become more abundant and reduce the amount of light that penetrates the water, causing the algae and macrophytes to die. Dead organisms accumulate as sediment, and bacteria that remove minerals from decaying organic matter (decomposers) consume large amounts of oxygen, which they extract from the water. The loss of some species and high mortality rates among those that remain ("fish kills") follow as oxygen is depleted, but cyprinid fishes (Family *Cyprinidae*, carps and minnows) increase because they can survive in poorly oxygenated waters and are efficient predators of zooplankton – organisms that increase during eutrophication. As a result, populations of grazing zooplankton decrease. Levels of phytoplankton, the prey of zooplankton, then increase, further increasing turbidity.[22] As eutrophication progresses, the lake declines in quality as a source of drinking water, recreation, and food. The final result can be *annoxiation*, the loss of oxygen from the lake's waters, making them uninhabitable to all but anaerobic bacteria.

In the open ocean, some effects of nutrient enrichment are different, but not better. Nutrient additions have caused changes in the composition and functioning of estuarine and nearshore ecosystems and contributed to long-term declines in coastal marine fisheries. Increased N and P input reduce food-web diversity, alter phytoplankton community composition, and increase toxic blooms of phytoplankton. Ocean anoxic events (OAEs) are becoming more common worldwide. One mechanism responsible for past OAEs has been increased phosphorus supply, leading to higher ocean productivity and oxygen demand in subsurface water. These circumstances can be amplified by low oxygen levels because such conditions promote additional release of phosphorus from ocean sediments, leading to a potentially self-sustaining condition of deoxygenation.[23]

The ocean biogeochemical system is sensitive to even small changes in oxygenation or net primary production, changes that can affect the volume of oxygen-deficient water and the area of sediments overlain by oxygen-deficient waters. Much of the global ocean today can in fact be described as being "on the edge of anoxia." If phosphate concentration in the deep sea were to increase significantly (or if changes in ocean circulation were to increase the efficiency of the existing biological utilization of PO_4^{-3}), even more ocean water would be depleted of oxygen. This sensitivity of the low-oxygen regions to relatively small changes in circulation or oxygen supply may be part of the explanation for the observed changes occurring today in an increasing number of "oxygen minimum zones" (OMZs), whose already low oxygen levels are exacerbated by the global warming of ocean waters.[23]

Land Transport, Erosion, and Sedimentation

Human beings not only add nitrogen, phosphorus, and other chemicals to the Earth's soil, they move the soil itself – and the rocks beneath it – in greater amounts and at faster rates than any natural geologic force. By the start of the twenty-first century, humans had become the premier geomorphic agent sculpting landscapes, with exponentially increasing rates of earth-moving.[24] The latest estimates suggest that >50% of Earth's ice-free land area has been directly modified by human actions involving moving earth or changing sediment fluxes.[25] Human additions to sediment erosion and transport lie at the very heart of Paul Crutzen's original definition of the Anthropocene, characterizing the period as a time in which "The yearly impact of mankind equals or exceeds exogenic geological processes of, e.g., erosion, transport and deposition."[26] In "terraforming" cities and building the dams and agricultural land that water and feed them, humans have wrought roughly an order of magnitude change in the long-term rate of erosion and sedimentation.[24, 27] And sediment can be moved indirectly, yet still in massive quantities, simply by changing the surface landscape. In the Piedmont Mountains of the US state of Maryland, geologist Ellen Wohl notes that "sediment yields from forested areas prior to European settlement were less than 36 ton/km^2/y and river channels were stable. With the advent of agriculture after 1700, sediment yields increased to 100–300 ton/km^2/y and river channels degraded."[28]

On a global scale, the deliberate shift of around 57,000 megatons/year of material by human activities is greater than the annual sediment transport by rivers to the oceans.[29] Humans have increased sediment transport by rivers globally through soil erosion by 2.3 billion metric tons/y, yet reduced sediment flux to the oceans (by 1.4 billion metric tons/y) because of sediment storage in reservoirs.[30] Reservoirs worldwide now store >100 billion metric tons of sediment.[31]

Humans do not merely move existing sediment, they add their own. Environmental scientist Erle Ellis describes this in detail:

> More than 170,000 'synthetic mineral-like substances', produced only by human activities, have now been identified ... – compared with 5,000 'natural' minerals. The total volume of Earth now transformed by humans, including the soils altered by agriculture and ocean sediments disturbed by trawling, has also been estimated. The 30 trillion tonne mass of this 'physical technosphere' is ... 100,000 times greater than Earth's living human biomass ... Plastic materials alone now far exceed human biomass, growing from 2 million tonnes produced annually in 1950 to 300 million in 2015. Total historical production, now 5 billion tonnes, is enough to wrap the Earth's entire surface in a thin layer of plastic film.[32]

Mining and Extraction

As a boy growing up in the Hocking Hills of southeastern Ohio (USA), long before I had any quantitative skills or scientific training, I received a colorful but ugly introduction to the effects of mining on the world around me. I remember wondering each spring why the local Hocking River turned orange. I knew instinctively that this was "not right." Even then I surmised the cause – the mines. I discovered them now and again in long walks through the woods – old openings, fallen down, filled in, covered over, all silently communicating neglect. These were the entrances to abandoned shaft mines. I remember seeing long strips of hillsides denuded of vegetation – the legacy of "strip mining," still an active livelihood, even as nearly all the shaft mines had closed, so that today the land's topography is written in its scars. I know now that the color manifested the tailings of iron oxide pollution seeping into streams, and then into the Hocking River, from the abandoned coal mines. I did not realize then that this was more than a local problem. From my Aunt Forrest's house I could walk to the Hocking River. But the color I was seeing stained all of Appalachia, the region I thought of as home. Geologists determined that acid mine drainage in Appalachia polluted over 16,900 km of streams from coal mining operations. The US EPA found that 8,230 km of streams with fisheries were impacted by acid mine drainage in Maryland, Ohio, Pennsylvania, Virginia, and West Virginia, with Pennsylvania alone having 5,212 km of impacted streams.[33] As geologists Robert Turka and Richard Gray put it,

> As AMD [acid mine drainage] is mixed with oxygen and additional water, the pH [acidity of the water] increases and the metals in solution drop out, forming a precipitate that is generally yellow–brown to red from the iron or white from aluminum ... The chemical load on the receiving stream lasts much farther downstream [from the discharge] than the precipitate, frequently destroying all or much of the life in the stream for significant distances ... It is impossible to remediate these problems quickly because so much waste has been discharged.[34]

Like the Ballard Mine in Idaho these changes seemed local, if also personal, to me. But the full extent of "local" mining can change the landscapes of entire regions, even with very primitive technology. In California, the discovery of gold generated such a level of hydraulic (water displacement) mining – much of which was "panning" for gold by individual miners with hand-held pans (placer mining) – that it dislocated a quantity of sediment approximately eight times more than the United States moved in excavating the Panama Canal – over 1 billion cubic meters – between 1853 and 1884.[28] The quest for coal and gold, and the roots of their value, run all the way down to the oldest and most fundamental causes of the Anthropocene. Mining coal was and remains a quest for energy in rapidly convertible and transferrable forms. In Appalachia, the great disparities of capitalism – the affluence of coal barons and the poverty of coal miners, some of whose children and grandchildren I knew and played with at school – were alive and etched in the land itself. The quest for gold runs even deeper. Originally used as a means of exchange that would be accepted in multiple markets (Chapter 2), gold had become its own independent form of wealth, one that could be accumulated without limit. It offered a path to and a promise of getting rich. Many miners in California followed the first and believed the second. Just as astronomer Carl Sagan said, "The visions we offer our children shape the future. It matters what those visions are ... Dreams are maps."[35] I learned in the Beartooths (Chapter 1) that maps tell you the truth about the country and lead you into its realities. But in the Hocking Hills false dreams had imposed themselves on the landscape. Rather than the map conforming to the country, the country had been conformed to the map, a map derived from misdirected dreams. It was the very goals of capitalism "at its best" that motivated the moving of the Earth.

The geologic force of human action comes not merely from transporting sediments across the Earth's surface, or even from removing minerals from beneath it. Some of its most powerful effects act slowly – unresting, unhasting, and silent as night. As minerals or fluids (like water or oil) are removed from subsurface deposits, the consequence is "land subsistence," more literally "land lowering." The mechanism of subsistence can be explained mechanically. Geologist Thomas Holzer and hydrologist Devin Galloway relate the phenomenon this way:

> As fluids are withdrawn from porous media, pore-fluid pressures decrease ... Because deformation of porous media is controlled by effective stress – the difference

> between the total stress and the pore-fluid pressure – a decrease in pore-fluid pressure causes a decrease of power volume. This phenomenon is known to hydrologists as compaction ... When effective stress [of the rock and sediments above] exceeds the yield strength of the granular skeleton of the media [what remains after the fluid has been withdrawn], the compaction is permanent and irreversible.[36]

Holzer and Galloway go further to precisely define its effects:

> Lowering of the land surface of large areas has been a major unintended consequence of groundwater and petroleum withdrawal by humans. Approximately 26,000 km^2 of land in the United States has been permanently lowered. The decrease of land-surface elevation, known as land subsidence, typically occurs at rates measured in centimeters per year. However, the irreversible accumulation of its effects clearly qualifies humans as major geologic agents. Subsidence causes permanent inundation of land, aggravates flooding, changes topographic gradients, ruptures the land surface, and reduces the capacity of aquifers to store water.[37]

In California, approximately 44 km^2 of land adjacent to the San Francisco Bay subsided ~0.5–2.5 m, placing it at risk at high tide. The estimated cost of such subsistence, caused by groundwater withdrawal by humans, was US$131,100,000 in 1979[36] – a value of US$555,358,413 in 2023.

The Mine in Madagascar

Today the United Nations Sustainable Development Goals recognize economic development as an important means of alleviating poverty, but require that such development be reconciled with biodiversity conservation.[38, 39] In a document these are only words on paper, but as a requirement for an organization securing a loan from the UN, or any of its constituent lenders, these requirements matter, and more and more countries are taking them seriously. Such "reconciliation" between economic development and biodiversity conservation often takes the (hopefully) measurable form of "No Net Loss" (NNL) of biodiversity associated with the development activities.[40] Currently, 101 countries mandate some form of biodiversity compensation or support voluntary measures to achieve it.[41] In other countries with less environmental governance, lender requirements, such as the International Finance Corporation performance standards, are important drivers of NNL commitments.[42]

Developers often try to meet such requirements by creating "offsets," areas dedicated to protection or activities that will result in gains in biodiversity that will compensate or "offset" losses to biodiversity caused by the activity where the development takes place (so-called "averted loss" offsets).[43] The work of quantifying gains in biodiversity offsets requires the creation of "counterfactual" sites or environments that can be used to determine the amount of biodiversity loss or gain that would have occurred without any protection or conservation activity at all. To make this comparison "right" requires careful monitoring of both the offset and counterfactual sites as well as careful matching between sites in the two categories.[44]

Madagascar is the world's fourth largest island (587,713 km^2) and the largest in the Indian Ocean. It is a single nation, the Republic of Madagascar, known worldwide for its exceptional biodiversity. Estimated to have been separated from mainland Africa for

80 million years, such isolation has fostered the evolution of plants and animals found nowhere else on Earth. Of its nearly 12,000 plant species, 82% are found only on the island, as are 90% of its animal species. Such unique creatures, long known to the Indigenous Peoples of Madagascar, are only now being described by Western science. From 1999 to 2010, scientists identified 615 new species, including 41 species of mammals and 61 species of reptiles.[45] Rich in biodiversity, Madagascar's 30 million people are poor in material wealth, having one of the world's lowest median incomes and highest poverty rates.[46] Many of Madagascar's plants and animals are now threatened or endangered species, increasingly at risk from an unprecedented rate of deforestation driven by the necessity of meeting economic and subsistence requirements for human households, and exacerbated by current patterns of shifting and expanding human agriculture seeking to feed the growing human population (see Figure 4.5).[47]

In 1960, a large nickel deposit was discovered in Madagascar at a location called Ambatovy, near the town of Moramanga. By 1970, additional studies estimated nickel reserves at these locations to be approximately 40 million tons. The Phelps Dodge Corporation, with the creation of its subsidiary Phelps Dodge Madagascar (PDM), began a feasibility study for the development of a mine and processing plant at the Ambatovy site in 1994.[48] In 1995, PDM gained the rights to the Ambatovy deposits and received permits from the government to conduct further exploration. From 2003 to 2005, PDM created the Dynatec Madagascar Corporation which then acquired rights to the Ambatovy deposit from Phelps Dodge. Construction of the mine began in 2010 and the first commercial shipment of nickel cobalt had reached international markets by 2012, with commercial production in full operation by 2014.[48] Today Ambatovy is one of the world's largest nickel and cobalt mines. Billing itself as a world-leader in sustainable mining committed to ensure NNL of biodiversity, Ambatovy has followed an offset strategy that has served as a pilot program for similar development ventures created by the international

Figure 4.5 The endangered ring-tailed lemur (*Lemur catta*) is one of thousands of endemic species of Madagascar whose habitats may be degraded or lost due to deforestation. Madagascar's Ambatovy Mine, one of the world's largest nickel and cobalt mines, has been following a plan for environmental protection to compensate for forest losses on the mine site with "offset" areas where the mining company has reduced deforestation.

Business and Biodiversity Offset Programme that, informed by the Ambatovy project, has shaped guidelines now widely used in mitigating biodiversity loss from development throughout the world.[49] From the beginning of its actual operations, Ambatovy's strategy has been to compensate for biodiversity loss at the mine site by preventing an equivalent amount of loss in four offset areas, all of which were also facing high deforestation rates due to shifting agriculture.[50] Dynatec, working with partnering conservation NGOs and agencies of the Republic of Madagascar, implemented conservation activities at the four offset sites aimed at slowing forest clearance within them. These efforts included ecological monitoring, fostering the establishment of community forest management associations, and supporting such associations with the monitoring and enforcement of resource-use restrictions, implementing environmental education programs, and promoting alternative income-generating activities in surrounding communities.[50]

An independent scientific team led by Katie Devenish developed a rigorous experimental design to compare trends of biodiversity gain and loss at the mine to that in the four offset sites, using both the offset sites and the counterfactual sites in the same area which had similar habitat and faced similar pressures from shifting agriculture, but where no conservation measures were implemented.[51] Based on carefully matched pairings of site-specific differences, Devenish et al. determined that two of the mine offset sites reduced deforestation relative to the counterfactual sites. One offset showed no difference and the fourth could not be properly assessed. Overall, the mine site destroyed or degraded 2,064 ha of natural forest. In contrast, the conservation measures at two offsets reduced deforestation by an average of 96% and 66%. In total, the offsets avoided the loss of 1,644 ha or 79% of the forest losses incurred at the mine site. Deforestation at the mine site began earlier than conservation at the offset sites, and has since slowed, while reductions in deforestation continue at the same or increasing rates at the offset sites. Given the trajectories now in place at both, Devenish et al. concluded that "Ambatovy is on track to achieve NNL of forest earlier than anticipated," sometime before 2033.[52]

The Ambatovy Mine has employed practices that should be followed by development efforts worldwide. From the beginning, the mine partnered with conservation NGOs, government organizations, local communities, and scientific institutions to design its NNL plan. It not only instituted the plan, but also subjected the evaluation of the outcomes to careful monitoring, sophisticated modeling, and rigorous experimental design, all carried out by independent scientific investigators, not by the mining company. It did not simply purchase non-mine sites and designate them as "protected areas" (PAs). Such action would have displaced Indigenous residents, excluded local communities from the conservation effort, and done nothing to make the resources of these areas available to people in and around them striving to earn a livelihood and escape poverty. The mine did not impose conservation actions or restrictions in the designated offsets, but rather fostered the development of local community forest protection organizations that would administer and take responsibility for conservation activities and the enforcement of required restrictions. Ambatovy supported these organizations with the technical know-how and equipment to encourage the rejuvenation and implementation of forest conservation that would be undergirded by the resident communities' own conservation-positive knowledge and practices for good land and water governance. Together, such combinations of knowledge and technology in conservation education would, over time, better inform the forest associations and their decisions regarding conservation. The mine also struck directly at the root of deforestation – poverty – by promoting alternative income-generating activities, which reduced people's motivation to make a living through deforestation activities. And the mine based its comparisons on "real world" pressures being applied to the offset areas – the actual burden of current agriculture practices in Madagascar. It did not simply create PAs, but permitted the question – can biodiversity conservation be achieved in areas experiencing human population growth and poverty through actual changes in human behavior and economic activities?

The Ambatovy Mine, like Idaho's Ballard Mine, exploits nature as a tap (Chapter 2), but, unlike Ballard, does not treat nature as a cost-free sink. There is a real value to what is lost, and that loss must be accounted for and reclaimed, to be "gained back" in other ways. Such accounting is not only important in terrestrial surface mining. It is a determination that changes the human *interaction with nature* because it rests on a different *intention toward nature* – an intention to save what can be saved, and a conviction that what can be saved is worth saving. Saving nature comes with a price, and that price has been borne from the beginning of the mine's operation. Only by such deliberate changes can the natural world around us be preserved amid potentially overwhelming human technology employed to advance profit-driven development.

Stopping Human Intrusion in Biogeochemical Cycles

The Ambatovy Mine developed a method of "pricing" nature that placed the value of biodiversity conservation within the mine's budget. But mining is still mining. Sediment is moved, substrates are withdrawn, and deforestation occurs. Human effects on geological processes and events continue just as they do in agriculture through human additions of nitrogen, phosphorus, and other chemical fertilizers. A different result requires a different objective.

Among states in the United States, mention "Iowa" and the first things Americans will think of are corn, soybeans, and cows – farming! That's understandable. Iowa is the only state with *all* of its land area fully embedded in the US "corn belt." It is the heart and soul of conventional US agriculture. I lived in Iowa for nine years. During that time, one farmer I became friends with was Paul Mugge. Paul was not your ordinary farmer. Although he had grown up in a traditional farm family, Paul had not planned to be a farmer at all. After completing a university degree in engineering, Paul left home for Seattle (Washington) to work in the aerospace industry. But when his father retired, Paul left his career as an aerospace engineer and returned to Iowa to manage the family farm.

He started with a business-as-usual approach – row crops of corn and soybeans, and raising pigs. However, as time passed, Paul began to question conventional wisdom, and became more troubled by his growing awareness of the collateral effects of chemical fertilizers on soil and water, including his own. Being as thrifty as he was intelligent, Paul experienced increasing frustration with the amount of money he was spending on chemical additives, and how much of that was ending up somewhere other than his crops. He began to experiment.

To better protect his soil, Paul began using a three-year rotation of corn, soybeans, and a small grain crop with an under-seeding of red clover (*Trifolium pratense*). The clover, a legume, developed symbiotic associations with *Rhizobium* bacteria on its roots (recall that *Rhizobium* can take in nitrogen from the air and combine it with hydrogen in the soil, thus producing ammonia in proximity to the roots of the clover). Paul harvested his grain in July and let the clover remain until late autumn. At this point Paul tilled the clover, adding 80–100 pounds of nitrogen back into the soil in the process! Other crop plants can use the nitrogen as a nutrient, so the earth was enriched, and even more improved because, in so doing, nitrogen fixed in the tissues of the clover was returned to the soil. The clover also held the soil in place better than corn or soybeans, thus reducing erosion. Forming a dense cover, the clover further reduced weed infestations and at the same time provided habitat for beneficial insects – some were plant pollinators, others were predators of insect pests.

Paul also began to make changes in his hog-raising operation. This was no small undertaking because Paul raised about 2,500 pigs a year. Changing his practices so that the pigs' manure would not be lost to no effect, Paul collected the manure and employed it as crop fertilizer using a technique known as "windrow composting" – piling the biodegradable waste (pig manure) in long rows ("windrows") on the fields in which he would soon be planting crops. These changes began to pay off in reduced costs for crop fertilizer, as well as in less expense for pesticides to control weeds and harmful insects. To reduce soil loss and conserve beneficial invertebrates in the soil Paul switched to "no till" planting with his soybeans and small grain crop using a device that makes a shallow groove in the ground for seed placement without breaking the soil surface. He then began planting his corn and soybeans later than he had previously been doing under conventional farming practices. The late plantings reduced weed and pest infestations even more. Of course, the weeds did not give up easily. Some still invaded his crops. But now, instead of applying a widespread application of herbicides and spreading or spraying it all over the fields, Paul began using a "flame cultivator" – a device pulled by a tractor that employs one or more flamethrowers (yes, flamethrowers) to kill weeds. The weed is exposed to the flame for a fraction of a second, raising the temperature of the plant's cell walls to 2,000°F. At that temperature the cell walls are ruptured and the weed later wilts and dies. Although killing weeds with direct flame sounds energy intensive, it isn't. The flame cultivator burns about 7 gallons of propane per acre, much less fuel than required to apply conventional herbicides and chemical fertilizers. An added benefit is that the flame cultivator does not till (break the surface of) the soil, and so creates no loss of organic matter from it. The flame cultivator and the late planting strategy were effective. Paul's costs for herbicides and pesticides went down. With his fields vacant longer in spring because of his late planting strategy, Paul began seeding rye (*Secale cereal*), which can be used for both grain and straw, in early spring to further stabilize the soil and reduce weed infestation.

Initially, Paul had not been planning to "go organic." He was simply trying to produce crops and animals in a healthy, efficient way, conserve the soil on his land, and reduce his expenses. Paul's yields did not increase. But his increasing use of biological controls (clover and rye) to reduce both plant and insect pests (and put nitrogen back into the soil), his targeted use of his flame cultivator, his late plantings, and his organic (pig-produced) fertilizer led him to appreciate that not only was he saving money by not using pesticides and chemical fertilizer expenses, *he did not need them at all*. Paul realized he was becoming an organic farmer.

Organic agriculture uses fertilizers of organic origin and emphasizes techniques like crop rotation and companion planting. There are no additions of chemical fertilizers, pesticides, or genetically modified organisms (GMOs) to enhance production. In other words, it consists of all the things that Paul had begun to do. It is one thing to realize that you are an organic farmer. It's another thing to convince other people, especially federal regulators, that you really are one and merit certification. The standards for organic farming are now international, and the certification process is rigorous, detail-oriented, and lengthy – a minimum of three years of external documentation confirming not only that you are not using chemical fertilizers or pesticides, but that *there are none present on your land.* You must sell more than $5,000 in organic products each year and submit to regular and ongoing inspections even after certification so that regulators can ensure your crops remain fertilizer- and pesticide-free. But some of these substances, especially the pesticides, can blow in from neighboring farms. To reduce such contamination, Paul planted shelter belts of natural vegetation around his fields. These plantings not only stopped the incursion of wind-borne chemicals, they also (with some intelligent management by Paul) created habitat for wildlife (including game birds like ring-necked pheasant (*Phasianus colchicus*)), and, to Paul's pleasant surprise, the shelterbelts were beautiful. With certification in hand, Paul now appreciated that he had a distinctively different crop to market – an *organic* crop. Produced without chemical fertilizers or pesticides, the organic crops were healthier and offered less risk to consumers – and therefore merited higher prices. Paul's profits rose at the same time his expenses declined. But not all of Paul's costs decreased. Paul determined that although his expenditures for chemical additions dropped to zero, his costs for labor (mostly his own) increased. Paul was speaking to one of my classes about these changes with the aid of a well-designed pie chart. One student asked, "Paul, aren't you concerned that, with these changes, you're spending a lot more in labor costs?" Without hesitating Paul replied, "No. That labor cost is me doing farm work. I like farming. That's what I want to spend my time doing."

Historically, all farming was organic farming, and today many of the agricultural systems of Indigenous cultures remain so throughout the world. Modern organic farming became identifiable and established as a "movement" in more recent times in response to the growing use of chemical fertilizers and pesticides. By making no additions of nitrogen, phosphorus, and other chemical fertilizers, organic farming adds none of these to natural biogeochemical cycles. Thus, it stops the effects of those additions – like lake eutrophication – and restores natural cycles to its own agricultural system and the ecosystems around it. By 2018, organic farming efforts covered approximately 71,500,000 hectares (nearly 180,000,000 acres) worldwide.[53] An impressive total, yet less than 2% of global farmland area.[54] But even as organic farming has increased in acreage, organic farming is still farming. It may be sustainable, but it is not, in itself, conservation. Can the conservation of biodiversity be practiced as part of farming itself?

Farming for Wildlife

Iowa is not only home to Paul Mugge's farm, but also to the DeSoto National Wildlife Refuge – one of 588 such refuges in the United States managed by the nation's Fish and Wildlife Service (FWS). DeSoto's 8,365 acres make it a small sanctuary by FWS standards. It is also relatively young, established in 1959. Its landscapes are not spectacular. They are flat and wet, set aside primarily to conserve habitat for waterfowl in floodplains and bottomlands adjacent to the Missouri River, especially as a resting and stopover area for the migratory snow goose (*Anser caerulescens*), where up to 550,000 individuals may gather at DeSoto in the spring and fall (see Figure 4.6).

Before there was a refuge, the land was covered in conventional row crop agriculture of corn and soybeans, usually grown in a two-year rotation. When the Refuge was dedicated, some cropland lay within its designated boundaries. The Refuge staff set about converting much of these lands to native prairie communities in upland areas while at the same time restoring wetlands in lowland areas within the Missouri River's historic floodplains. Tallgrass prairies were once the dominant ecosystem of Iowa and eastern Missouri, but

Figure 4.6 DeSoto National Wildlife Refuge, USA, set aside in 1959 as a "stopover" refuge for migrating snow geese (*Anser caerulescens*, shown here in both white and dark morphs). Inheriting conventional row crop farmlands within the Refuge's boundaries, DeSoto staff developed methods of agriculture that used no pesticides or chemical fertilizers, benefited wildlife, and involved local farmers.

were rapidly converted to farmland with white settlement, and adjoining wetlands were drained. These practices in the United States and worldwide have made both grasslands and wetlands among the Earth's most endangered ecosystems. In North America over 99% of original grasslands have been destroyed, most by conversion to agriculture.[55, 56] In Iowa at least 90% of historic wetland areas – imperfectly estimated at 4 to 6 million acres – have been lost. Today Iowa's wetland acreage is approximately 420,000.

The opportunity to study prairie communities of plants and birds in DeSoto's restored grasslands brought me to the Refuge through a grant from the FWS in the 1990s. Using managed and control treatments on the remaining prairies, I determined how plant and bird communities would respond to prescribed treatments of mowing and burning compared to no management treatments at all.[57, 58] But not all of the farmland "inherited" by the DeSoto Refuge had been restored to prairies. Some of it was still being *farmed.* While "hanging out" with the staff at DeSoto, I learned how farming was practiced on the Refuge.

Being dominated by wetlands and dedicated to their preservation and enhancement, DeSoto was no place to add chemical nitrogen- and phosphorus-based fertilizers that would have degraded the very wetlands that supported the unique species the Refuge had been created to protect. Likewise, the Refuge staff were not about to allow the use of chemical insecticides to control crop pests. Instead, they devised, with the help of local farmers, a different kind of agriculture that could benefit wildlife *through farming.* Rather than using the conventional two-year rotation of corn and soybeans, the staff and farmers implemented a three-year rotation of corn, soybeans, and a forage legume (clover or alfalfa, depending on the year and conditions, much like the rotation Paul Mugge used on his farm). The corn and soybeans were harvested in conventional ways (although, by design, the wildlife, mostly white-tailed deer (*Odocoileus virginianus*), were permitted to eat about 50% of the corn crop). The clover or alfalfa was plowed back into the ground to enrich the soil with the ammonia produced by the *Rhizobium* bacteria on the plants' roots, fertilizing the soil for the corn and soybeans in subsequent rotations.

To determine how conventional farming would fare under such conditions, the staff permitted some farmers using the Refuge to plant corn and soybeans in conventional two-year rotations, with chemical fertilizer added on these select plots, and compared production in these to the three-year "wildlife friendly" rotation. Corn yields from the conventional rotation and the DeSoto (three-year) rotation were nearly identical (100.6 to 103.3 bushels per acre, respectively), as were yields of soybeans (35.3 to 37.7 bushels per acre, respectively).[59] Corn outputs in both rotations were lower than those on surrounding farms (deer can eat a lot of corn, and the animals got a good share of the crop in both rotations!), but, on the Refuge, the three-year rotation did as well as the conventional two-year practice. Despite the constraints, local farmers participated enthusiastically in DeSoto's farming program. There was always a waiting list, with more farmers wanting to farm in DeSoto than the staff had allotments to offer. Farmers appreciated gaining additional yields with no property taxes or expenses for fertilizer and pesticides. Many indicated they would use some of the DeSoto practices on their own land.

The DeSoto rotation contained one year in three with no income for farmers, a condition economically unsustainable on their own farms. Since the 1990s, DeSoto has increased its prairie restoration efforts, and in so doing, phased out its farming program. But the program demonstrated that farming could be practiced with community support, participation, and economic benefit in ways that were intentionally beneficial to wildlife conservation. Conservation intention could be part of farming practice. This is a new idea for Western agricultural practices, but long part of the traditional knowledge of the agricultural systems of Indigenous Peoples (Chapter 8).

Humans and Global Geology

Humans alter biogeochemical cycles, move more sediment than all natural processes combined, raise and lower landforms, and change the ecology of streams, lakes, and oceans. In their interactions with nature, humans have sought to maximize wealth, whether in maximum yields from agriculture or maximum extraction of precious metals from mining. Centuries ago, the Roman poet Ovid (43 BC–AD 17) put the problem elegantly in his master work *Metamorphoses*. In his section on *The Creation*, Ovid describes the "development" of humankind and its changed relations to the Earth.

> The rich Earth
> Good giver of all the bounty of the harvest,
> Was asked for more; they dug into her vitals,
> Pried out the wealth a kinder lord had hidden
> In Stygian shadow, all that precious metal,
> The root of evil. They found the guilt of iron,
> And gold, more guilty still. And War came forth.[60]

Indeed, as Wendell Berry has said, "Until we understand what land is, we are at odds with everything we touch."[61] His words are proven by the wars that have been fought, the treaties that have been broken, and the land that has been desecrated in myriad quests for such gain and wealth which Ovid decried centuries ago. We have examples today, although these are still few, in which we see that humans can alter their role in the Earth's geology. Such changes require three reformations: 1) *Reform* the objective of the interaction from maximizing *human wealth* to maximizing *environmental health* – investing in actions that will strengthen the natural processes of the system, in this case the lithosphere and the ecology of the soil, so that the use of the land is both fully productive and fully sustainable. This is the example of organic farming. 2) *Adapt* conservation strategies and actions to fully incorporate them into normal and ordinary human endeavors. This is the lesson from the DeSoto farming program. 3) *Change* what is valued from the commodity alone to include the natural world from which it is removed, and make full accounting for the losses of what is taken and the gains of what can be added. This is the lesson from the Ambatovy mine. Employed together, these are practices that can change the times we live in. They can change the Anthropocene.

References

1. Zalasiewicz, J., M. Williams, A. Smith, et al. 2008. Are we now living in the Anthropocene? *GSA Today* 18:4–9. DOI: 10.1130/GSAT01802A.1.
2. Wilkinson, B. H. 2005. Humans as geologic agents: A deep-time perspective. *Geology* 33:161–164. DOI: 10.1130/G21108.1.
3. Harding, J. S., E. F. Benfield, P. V. Bolstad, et al. 1998. Stream biodiversity: The ghost of land use past. *Proceedings of the National Academy of Sciences USA* 95:14843–14847.
4. Sprengel, C. 1826. Ueber Pflanzenhumus, Humussaure und Humussaure Salze (About plant humus, humic acids and salts of humic acids). *Archiv Fur Die Gesammte Naturlehre* 8:145–220.
5. Liebig, J. V. 1840. *Die organische Chemie in ihrer Anwendung auf Agricultur und Physiologie (Organic chemistry in its applications to agriculture and physiology).* Friedrich Vieweg und Sohn Publishing Company, Braunschweig, Germany.
6. Flavell-While, C. 2010. Fritz Haber and Carl Bosch – Feed the World. Chemical Engineer. 1 March. <https://www.thechemicalengineer.com/features/cewctw-fritz-haber-and-carl-bosch-feed-the-world/>. Accessed 3 November 2023.
7. Bouwman, A. F., A. H. W. Beusen, and G. Billen. 2009. Human alteration of the global nitrogen and phosphorus soil balances for the period 1970–2050. *Global Biogeochemical Cycles* 23:GB0A04. DOI: 10.1029/2009GB003576.
8. Galloway, J. N., F. J. Dentener, D. G. Capone, et al. 2004. Nitrogen cycles: Past, present, and future. *Biogeochemistry* 70:153–226. <https://doi.org/10.1007/s10533-004-0370-0>.
9. Penuelas. J., B. Poulter, J. Sardans, et al. 2013. Human-induced nitrogen–phosphorus imbalances alter natural and managed ecosystems across the globe. *Nature Communications* 2934. <https//doi.org/10.1038/ncomms3934>.
10. Smil, V. 1999. Nitrogen in crop production: An account of global flows. *Global Biogeochemical Cycles* 13:647–662. DOI: 10.1029/1999GB900015.
11. Van Drecht, G., A. F. Bouwman, J. M. Knoop, et al. 2003. Global modeling of the fate of nitrogen from point and nonpoint sources in soils, groundwater and surface water. *Global Biogeochemical Cycles* 17:1115. DOI: 10.1029/2003GB002060.
12. Tilman, D. 1987. Secondary succession and the pattern of plant dominance along experimental nitrogen gradients. *Ecological Monographs* 57:189–214.
13. Vitousek, P. M., J. D. Aber, R. W. Howarth, et al. 1997. Human alteration of the global nitrogen cycle: Sources and consequences. *Ecological Applications* 7:737–750.
14. Lu, X., J. Mo, F. S. Gilliam, et al. 2010. Effects of experimental nitrogen deposition on plant diversity in an old-growth tropical forest. *Global Change Biology* 16:2688–2700.
15. Zengwei, Y., S. Jiang, H. Sheng, et al. 2018. Human perturbation of the global phosphorus cycle: Changes and consequences. *Environmental Science and Technology* 52:2438–2450.
16. Galloway, J. N., W. H. Schlesinger, C. M. Clark, et al. 2014. Biogeochemical cycles. Pages 350–368 *in* J. M. Melillo, T. C. Richmond, and G. W. Yohe, editors. *Climate change impacts in the United States: The third national climate assessment.* US Global Change Research Program. DOI: 10.7930/J0X63JT0.
17. US Environmental Protection Agency. 2024. Radioactive material from fertilizer production. US Environmental Protection Agency. <https://www.epa.gov/radtown/radioactive-material-fertilizer-production>. Accessed 17 June 2024.

18. US Environmental Protection Agency. 2023. Ballard Mine, Soda Springs, ID Cleanup Activities. <https://cumulis.epa.gov/supercpad/SiteProfiles/index.cfm?fuseaction=second.Cleanup&id=1002859#bkground>. Accessed 3 November 2023.
19. US Environmental Protection Agency. 2023. Superfund: CERCLA Overview. <https://www.epa.gov/superfund/superfund-cercla-overview>. Accessed 3 November 2023.
20. Boone, R. 2021. Phosphate mine company will clean up waste to settle lawsuit. Spokesman-Review. 4 March. <https://www.spokesman.com/stories/2021/mar/04/phosphate-mine-company-will-clean-up-waste-to-sett/>. Accessed 3 November 2023.
21. Jenny, J.-P., P. Francus, A. Normandeau, et al. 2015. Global spread of hypoxia in freshwater ecosystems during the last three centuries is caused by rising local human pressure. *Global Change Biology* 22:1481–1489.
22. Brönmark, C., and L. A. Hansson. 1998. *The biology of lakes and ponds.* Oxford University Press, Oxford, UK.
23. Watson, A. J., T. M. Lenton, and B. J. W. Mills. 2017. Ocean deoxygenation, the global phosphorus cycle and the possibility of human-caused large-scale ocean anoxia. *Philosophical Transactions of the Royal Society A* 375: 20160318. <http://dx.doi.org/10.1098/rsta.2016.0318>. Accessed 21 March 2023.
24. Hooke, R. L. B. 2000. On the history of humans as geomorphic agents. *Geology* 28:843–846. <https://doi.org/10.1130/0091-7613(2000)28<843:OTHOHA>2.0.CO;2>.
25. Hooke, R. L. B., J. F. Martín-Duque, and J. Pedraza. 2012. Land transformation by humans: A review. *GSA Today* 22(12):4–10. DOI: 10.1130/GSAT151A.1.
26. Crutzen, P. J. 2002. Geology of mankind. *Nature* 415:23.
27. Wilkinson, B. H. 2005. Humans as geologic agents: A deep-time perspective. *Geology* 33:161–164. DOI: 10.1130/G21108.1.
28. Wohl, E. 2005. Disconnected rivers: Human impacts to rivers in the United States. Pages 19–34 *in* J. Ehlen, W. C. Haneberg, and R. A. Larson, editors. *Humans as geologic agents.* Geological Society of America, Boulder, Colorado, 20–21.
29. Price, S. J., J. R. Ford, A. H. Cooper, et al. 2011. Humans as major geological and geomorphological agents in the Anthroposphere: The significance of artificial ground in Great Britain. *Philosophical Transactions of the Royal Society A* 369:1056–1084.
30. Syvitski, J. P. M., C. J. Vörösmarty, A. J. Kettner, et al. 2005. Impact of humans on the flux of terrestrial sediment to the global coastal ocean. *Science* 308:376–380. DOI: 10.1126/science.1109454.
31. Syvitski, J. P. M., C. N. Waters, J. Day, et al. 2020. Extraordinary human energy consumption and resultant geological impacts beginning around 1950 CE initiated the proposed Anthropocene Epoch. *Communications Earth and Environment*, 1, 32. <https://doi.org/10.1038/s43247-020-00029-y>. Accessed 22 March 2024.
32. Ellis, E. C. 2018. *Anthropocene: A very short introduction.* Oxford University Press, Oxford, UK, 148.
33. Turka, R., and R. E. Gray. 2005. Impacts of coal mining. Pages 79–86 *in* J. Ehlen, W. C. Haneberg, and R. A. Larson, editors. *Humans as geologic agents.* Geological Society of America, Boulder, Colorado, USA.
34. Turka, R., and R. E. Gray. 2005. Impacts of coal mining. Page 84 *in* J. Ehlen, W. C. Haneberg, and R. A. Larson, editors. *Humans as geologic agents.* Geological Society of America, Boulder, Colorado, USA.
35. Sagan, C. 1994. *Pale blue dot: A vision of the human future in space.* Random House, New York, USA.
36. Holzer, T. L., and D. L. Galloway. 2005. Impacts of land subsidence caused by withdrawal of underground fluids in the United States. Pages 87–99 *in* J. Ehlen, W. C. Haneberg, and R. A. Larson, editors. *Humans as geologic agents.* Geological Society of America, Boulder, Colorado, USA, 88.
37. Holzer, T. L., and D. L. Galloway. 2005. Impacts of land subsidence caused by withdrawal of underground fluids in the United States. Pages 87–99 *in* J. Ehlen, W. C. Haneberg, and R. A. Larson, editors. *Humans as geologic agents.* Geological Society of America, Boulder, Colorado, USA, 87.

38. Thacker, S., D. Adshead, M. Fay, et al. 2019. Infrastructure for sustainable development. *Nature Sustainability* 2:324–331.
39. Blicharska, M., R. J. Smithers, P. Rönnbäck, et al. 2019. Biodiversity's contributions to sustainable development. *Nature Sustainability* 2:1083–1093. <https://doi.org/10.1038/s41893-019-0417-9>.
40. zu Ermgassen S. O., S. E. Utamiputri, P. L. Bennun, et al. 2019. The role of "no net loss" policies in conserving biodiversity threatened by the global infrastructure boom. *One Earth* 1:305–315.
41. Maron, M., S. Brownlie, J. W. Bull, et al. 2018. The many meanings of no net loss in environmental policy. *Nature Sustainability* 1:19–27.
42. McKenney, B. A., and J. M. Kiesecker. 2010. Policy development for biodiversity offsets: A review of offset frameworks. *Journal of Environmental Management* 45:165–176.42.
43. Bull, J. W., K. B. Suttle, A. Gordon, et al. 2013. Biodiversity offsets in theory and practice. *Oryx* 47:369–380.
44. Quétier, F., and S. Lavorel. 2011. Assessing ecological equivalence in biodiversity offset schemes: Key issues and solutions. *Biological Conservation* 144:2991–2999.
45. Alexandre, A., R. Smith, and B. Evans. 2022. Madagascar's extraordinary biodiversity: Challenges & opportunities for the future. Kew Botanical Gardens. <https://www.kew.org/read-and-watch/madagascars-extraordinary-biodiversity>. Accessed 5 November 2023.
46. World Bank. 2023. The World Bank in Madagascar. <https://www.worldbank.org/en/country/madagascar/overview#:~:text=The%20situation%20is%20exacerbated%20by,using%20the%20national%20poverty%20line>. Accessed 5 November 2023.
47. Suzzi-Simmons, A. 2023. Status of deforestation of Madagascar. *Global Ecology and Conservation* 42:e02389. <https://doi.org/10.1016/j.gecco.2023.e02389>. Accessed 22 March 2024.
48. Ambatovy. 2022. History of the project. <https://ambatovy.com/en/who-we-are/history/>. Accessed 5 November 2023.
49. von Hase, A., A. Cooke, A. Andrianarimisa, et al. 2014. Working towards NNL of biodiversity and beyond: Ambatovy, Madagascar – a case study. Forest Trends and Ambatovy. <https://www.forest-trends.org/wp-content/uploads/bbop/ambatovy-bbop-nnl-2014-final-pdf.pdf>. Accessed 16 October 2024.
50. Bidaud, C., M. Hrabanski, and P. Meral. 2015. Voluntary biodiversity offset strategies in Madagascar. *Ecosystem Services* 15:181–189.
51. Devenish, K., S. Desbureaux, S. Willcock, et al. 2022. On track to achieve no net loss of forest at Madagascar's biggest mine. *Nature Sustainability* 5:498–508. <https://doi.org/10.1038/s41893-022-00850-7>. Accessed 8 March 2024.
52. Devenish, K., S. Desbureaux, S. Willcock, et al. 2022. On track to achieve no net loss of forest at Madagascar's biggest mine. *Nature Sustainability* 5:498–508. <https://doi.org/10.1038/s41893-022-00850-7>. Accessed 8 March 2024, 501.
53. Food and Agricultural Organization of the United Nations. 2020. The world of organic agriculture – statistics and emerging trends. Food and Agriculture Organization, Rome, Italy.
54. Ritchie, H., and M. Roser. 2019. Land use. OurWorldInData.org. <https://ourworldindata.org/land-use>. Accessed 20 March 2024.
55. Samson, F., and F. Knopf. 1994. Prairie conservation in North America. *BioScience* 44:418–421.
56. Lunt, I. D. 2003. A protocol for integrated management, monitoring, and enhancement of degraded Themeda triandra grasslands based on plantings of indicator species. *Restoration Ecology* 11:223–230.
57. Van Dyke, F., S. E. Van Kley, C. E. Page, et al. 2004. Restoration efforts for plant and bird communities in tallgrass prairies using prescribed burning and mowing. *Restoration Ecology* 12:574–584.
58. Van Dyke, F., J. D. Schmeling, S. Starkenburg, et al. 2007. Responses of plant and bird communities to prescribed burning in tallgrass prairies. *Biodiversity and Conservation* 16:827–839.

59. Gage, G., editor. 1996. DeSoto National Wildlife Refuge Annual Narrative Report. US Fish and Wildlife Service, Missouri Valley, Iowa, USA. <https://data.amerigeoss.org/dataset/444beece-6894-4363-936a-e2ec6bb463af/resource/424944dd-cbc8-451e-86d2-15e7f85b0904>. Accessed 22 March 2024.
60. Ovid. 1957. *Metamorphoses*. R. Humphries, translator. John Calder, London, UK, 7, lines 136–142.
61. Berry, W. 2002. *The art of the commonplace: The agrarian essays of Wendell Berry*. Edited and introduced by N. Wirzba. Counterpoint, Washington, DC, USA, 27.

5 Climate Change

The Signature of the Anthropocene

What is Climate Change?

Our Current Condition

We live in a warming world. That is both a current condition and an ongoing event, not a subject for argument and debate. It is a worldwide phenomenon that is the focus of scientific measurement, but increasingly demanding social and political action. Of all the signatures that humankind has penned in this age of the Anthropocene, none is writ so large as the human stroke on climate. Its effects are so pervasive that the Anthropocene and climate change are often conflated, sometimes to the point of identification. They are intimately related, yet distinct. Before we can understand the effects of climate change, much less imagine our responses, we must define what we mean.

Climate change is *not* variations in weather. *Weather* is the instantaneous state of the atmosphere. It is perhaps the most real and present part of our "environment." No matter how contrived our surroundings of human comfort, we cannot escape it. Weather will find us, and it will affect us. Here in northern Michigan (USA) where I live, weather is so changeable, so forcefully present, and (sometimes) so severe that the local news always leads with the weather report. Weather is the first and most important thing. It is more important than the city council meeting, the new highway construction, or the governor's visit. Weather matters.

The statistical ensemble of weather defines *climate* – a set of weather conditions prevailing over a region over a period of time, generally over three or more decades.[1] You could describe climate as "average weather." Weather is unstable. Climate is (historically) stable. Both can be modeled, but climate is generally more amenable to that effort. As scientist Michael MacCracken, an expert in the biological evaluation of climate change attribution, explains, "climate generally represents the conditions with which a region is associated ... However, scientifically the climate of a region also includes the seasonal, annual, and longer variations and cycling (e.g. El Niño and La Niña)."[1] Weather is a *daily* event. Weather has no "old news." One must keep up. Climate, in contrast, has been around long enough to reveal patterns and tendencies – seasons, temperature changes, moisture patterns, and "events." The event element is less common and predictable – thunderstorms, hurricanes, tornadoes, floods – but, over time, it occurs within identifiable constraints and boundaries. The climate of every region is different, and so are the adaptations of plants, animals, and people to it. But how would we know if climate was changing?

First, we would be careful to determine what *is not* change and identify what *is* normal variability. Short-term annual cycles ("seasons") are changes, but they are changes we

DOI: 10.4324/9781003404415-5

can count on. Ask any farmer. Longer-term, but regular, fluctuations in temperature, like Milankovitch cycles, caused by variations in the Earth's distance to the sun, occur over thousands of years, but they are predictable too. Neither are changes in climate. Climate change is made of sterner stuff. To deserve its name, it must display at least one of three characteristics (see Figure 5.1). These are *discontinuities* or *jumps*, *increasing variability*, or a *unidirectional trend* in a historically up-and-down (oscillating) variable, and these must be changes we can link to changes in the various internal couplings and linkages of the components of the climate system.[2] *Climate discontinuities* and *climate jumps* occur when climate variables, relative to the length of the temporal cycle, show a large change in magnitude over relatively short time spans. After the jump or discontinuity, a new baseline of the variable's average value emerges that is different from the old one. *Unidirectional trends* in climate refer to changes in which the range of variation remains within historical norms, but the movement of the variable's mean, such as temperature, is always in one direction, consistently greater or consistently less than the historic average. *Increasing variability* occurs when the means of climatic variables remain the same but their range of variation increases. In these cases, individual "weather" events become

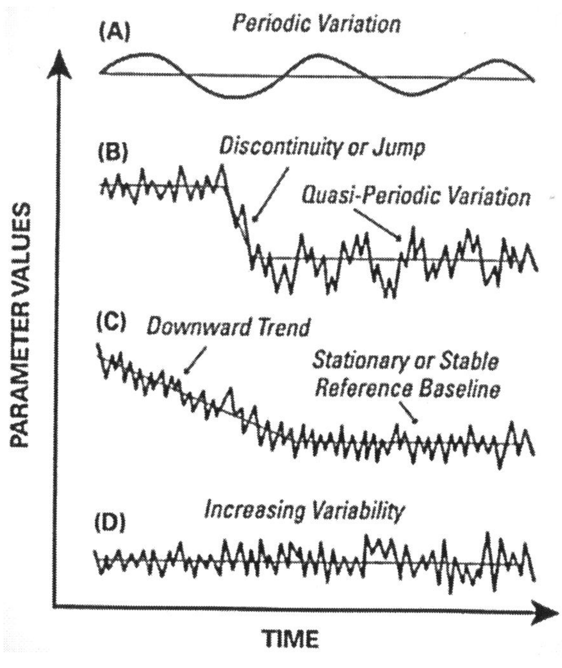

Figure 5.1 (A) Milankovitch cycles are examples of regular *periodic variation* in climate due to longer-term (22,000-year) regular variations in the Earth's distance from the sun. (B) *Discontinuities* or *jumps* can represent different forms of climate change. After the discontinuity or jump, a new baseline of climate variability emerges that is different from the old one. (C) *Unidirectional trends* in climate refer to changes in which the range of variation of climatic variables remains more or less within historical norms, but there is a movement of the mean in a single direction, consistently greater or consistently less than the historic average. (D) *Increasing variability* occurs when the means of climatic variables remain the same, but their range of variation increases. When (B), (C), or (D) occurs, it is appropriate to speak of such variation as climate change. T. R. Karl and K. E. Trenberth, Yale University Press. © 2005 by Yale University.

more unpredictable and less tightly clustered around historic means – and they often become more severe. When any of these sustained directional movements occur, that variation is an expression of *climate change*.

The Earth's Climate System

Five major components interact to create the Earth's climate over time: the atmosphere, the hydrosphere [all the water on the Earth], the cryosphere [the places on Earth where water appears as ice], the lithosphere [the Earth's hard outer layer], and the biosphere [all ecosystems].[3] Because this system is complex, its mechanisms, interactions, and outcomes cannot be perfectly predicted. Therefore, climate can often display erratic, even chaotic behavior. Small changes in any one part of the system can have far-reaching and unexpected effects, and the changes associated with them may be reversible or not reversible.[4]

Over both the land and water surfaces of the Earth, temperature varies spatially, seasonally, and historically. This wide range of temperatures past and present is driven primarily by the absorption of solar radiation from the sun. The net incoming energy from such solar radiation is, as an "Earth average," 340 watts per square meter (W/m^2).[2] A watt is a unit of power or "energy flux" that quantifies the rate of energy transfer from the source of the energy to its destination or "sink." We are most familiar with watts in our measurements of electrical power (a "kilowatt," that unit used to determine our electric bill, is 1,000 watts – a kilowatt hour (kw/hr) is a rate of electrical energy use of 1,000 watts per hour). These "watts" of energy reach the Earth as "light" traveling in waves of different lengths ("wavelengths"). In the light that we can see (the visible spectrum), what we perceive as different "colors" are in fact light traveling at different wavelengths. There are wavelengths we cannot see, such as infrared (IR) radiation and ultraviolet (UV) light, because these travel in lengths that are greater or less than the lengths of waves of light we can see.

About 100 W/m^2 (29%) of these 340 W/m^2 of incoming light are reflected back into space by clouds and atmospheric particles (aerosols). These never reach the Earth's surface. The remaining 240 W/m^2 (71%) is energy available to warm both the Earth and its atmosphere (see Figure 5.2).[5] If that were the end of the story, the Earth would simply heat up without relief and be inhospitable for life. But that is not the case. The Earth's atmosphere is transparent to radiation from the sun in the visible band, but not transparent to invisible (IR) radiation. The atmosphere reflects incoming IR radiation, but absorbs IR radiation reflected back from the surface of the Earth. Thus, the Earth's atmosphere absorbs heat energy but cannot "store it." Instead, the Earth's atmosphere balances heat input by creating heat output – reflecting some of the incoming energy back through the gases in the atmosphere and into space. However, when this happens, the wavelength of the reflected radiation ("light") is changed from shorter wavelengths to longer wavelengths of so-called long-wave IR radiation. Now things start to get interesting.

Although the gases in the Earth's atmosphere are mostly transparent to the shorter wavelength of incoming radiation, some atmospheric gases absorb portions of the reflected, longer-wave IR radiation. The gases that do this best are not the gases that are most abundant in the atmosphere (nitrogen and oxygen), but other gases, in various molecular arrangements, which are less common. These include water vapor (H_2O), methane (CH_4), ozone (O_3), nitrous oxide (N_2O), and carbon dioxide (CO_2). By taking up IR radiation, these gases absorb some of the heat latent in that radiation and create a net

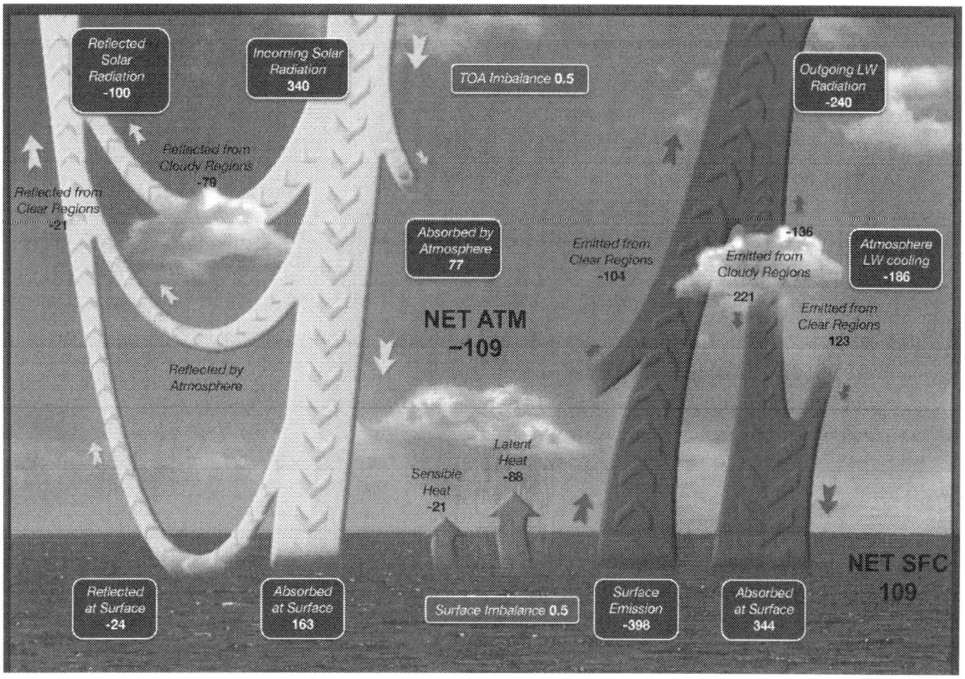

Figure 5.2 The components of the Earth's annual radiation budget (energy balance) in terms of what is received, reflected, and absorbed by different components. At the Earth's surface, an energy balance is achieved through evaporation and convection. At the top of the atmosphere, the balance is achieved by the heat energy emitted into space by the atmosphere and clouds. Reprinted by permission from Elsevier, from Loeb, N. G., and B. A. Wielicki. 2015. Satellite and satellite remote sensing: Earth's radiation budget. Pages 67–76 *in* G. R. North, J. Pyle, and F. Zang, editors. Encyclopedia of Atmospheric Sciences (2nd edition). Academic Press, Oxford, UK. © 2015 Elsevier Ltd.

energy gain (a net influx) of heat in the lower layers of the Earth's atmosphere. As the Irish physicist John Tyndall put it over a century ago,

> ...when the heat is absorbed by the planet, it is so changed in quality that the rays emanating from the planet cannot get with the same freedom back into space. Thus the atmosphere admits of the entrance of solar heat; but checks its exit, and the result is a tendency to accumulate heat at the surface of the planet.[6]

By the fundamental law of the conservation of energy, energy can neither be created nor destroyed, only *converted* from one form to another. So, the energy balance of a system *must be conserved* (the heat gained must in some way be compensated by equivalent heat loss).

To be in heat equilibrium then, the atmosphere must reflect the amount of energy equal to what it absorbs. For this relationship to work, the atmosphere needs to absorb about 80% of the heat radiation emanating from the Earth's surface to maintain the present global average temperature.[7] If the atmosphere absorbs heat energy that would otherwise be radiating out into space by Earth, the resulting surface temperature of the Earth will be warmer than the atmosphere. If the amount of greenhouse gases in the atmosphere increases, the point at which the atmosphere becomes transparent to thermal radiation will move to a higher altitude. But since the temperature at the higher altitude is colder, the thermal radiation emission "power" at this higher altitude will decrease and less radiation will be emitted outward. *This means there will be an imbalance in outgoing radiation versus incoming radiation.* The Earth will heat up and eventually a new temperature equilibrium will be established.[7] Therefore, there must be some change in the non-radiative energy states of the system, including temperature.

The process just described is called "radiative forcing." The surface temperature *must* warm to balance energy input and output because more energy is being absorbed by CO_2 (and other greenhouse gases) and less is escaping to outer space. A consequence of radiative forcing is that the Earth's surface will be warmer than it would otherwise be in the absence of an atmosphere.[7] Without such warming (radiative forcing), the average temperature near the surface of the Earth would be -18°C (in Fahrenheit, about 0°) instead of the actual current average of 14°C (57°F). The troposphere acts as the Earth's "blanket."

Carbon dioxide (CO_2) is one of the most important greenhouse gases that can absorb reflected radiation because, of all the relatively rare greenhouse gases, CO_2 is the most abundant. So, using CO_2 as our example, the difference in emitted radiation due to increased CO_2 concentration corresponds to the additional amount the atmosphere will radiate back down toward Earth. This is what is meant by "radiative forcing." To state this as precisely as possible, *radiative forcing is the change in total radiation power (watts/m^2) emitted by Earth into space as some quantity, such as CO_2 abundance, is changed.*[7] Hence, those previously mentioned gases that have high absorption capacities are commonly known as "greenhouse gases" because, like the panes of glass in a well-constructed greenhouse, they do not permit all incoming sunlight to be reflected back out of the atmosphere. Therefore, on a sunny day, the inside of the greenhouse (where you are growing the plants) is warmer than the outside. But whereas greenhouse glass keeps the inside air warm in large part because it traps the warm air heated by the sun inside of it, on Earth the warm air is free to rise and so warm the lower atmosphere.[7]

The Carbon Cycle

A positive value of radiative forcing means that less energy escapes from the atmosphere into space and, as a result, there is a positive imbalance in incoming solar radiation versus outgoing IR radiation. That imbalance will cause the Earth's surface temperature to rise. Understanding radiative forcing is likewise the key to defining the precise relationship between a change in the amount of a greenhouse gas in the atmosphere, like CO_2, and the degree of change in total radiation power emitted by the Earth into space. This is why units of radiative forcing, not degrees in temperature, are the standard units for quantifying the changing IR absorption in the atmosphere and its effect on changing the energy balance between incoming radiation from the sun and outgoing radiation from the Earth. Understanding the effects of greenhouse gases is *not* just about understanding that there is a relationship between CO_2 concentration in the atmosphere and the temperature (although there is), but that there is a relationship (which can be much more precisely expressed) between CO_2 concentration and radiative forcing.[7]

Like the nitrogen and phosphorus cycles (Chapter 4), carbon is an element that also participates in a biogeochemical cycle involving both a biological component (being stored in and moving through living things) and a geological component (in which carbon is stored in rocks, sediments, water, and the atmosphere). Visualize the carbon cycle as having four parts (compartments) (see Figure 5.3). Start with a non-living element, the atmosphere. Starting here (Step 1), carbon enters the atmosphere as a gas – carbon dioxide. Plants use CO_2 directly (Step 2) in the process of photosynthesis ($6CO_2 + 6H_2O \leftrightarrow C_6H_{12}O_6 + 6O_2$ – the first numbers (coefficients) are included to "balance" the equation so that there are the same quantities of each element on each side). In so doing, CO_2 is absorbed ("taken up") by green plants and other organisms (some kinds of bacteria and some microorganisms) such that it is removed from the atmosphere and stored in their living tissues. Next, animals consume plants and, in so doing, incorporate carbon that was stored in the plants into their bodies and physiological systems (Step 3). At some point, both animals and plants die. Their bodies decompose (Step 4) and, in that decomposition, the carbon in them is released back into the atmosphere.

In the "natural" carbon cycle, the movement of carbon from living things to the atmosphere and back again is slow and involves relatively small quantities of carbon from any individual source. It is equally "natural" in the Earth's carbon cycle for some carbon in living things to be "taken out of circulation." Plant deposits in the form of peat or, over longer time periods, coal, oil, and natural gas eventually remove large amounts of carbon from the cycle and hold ("store") it in subsurface forms, from which little or no carbon escapes. But the rise of capitalism was supported (dare we say "fueled") by the increasing demand for and development of technology that could use large quantities of energy quickly and locally, such as those that could be obtained from burning coal, or, later, oil or natural gas. Through these developments (the Industrial Revolution), humans were

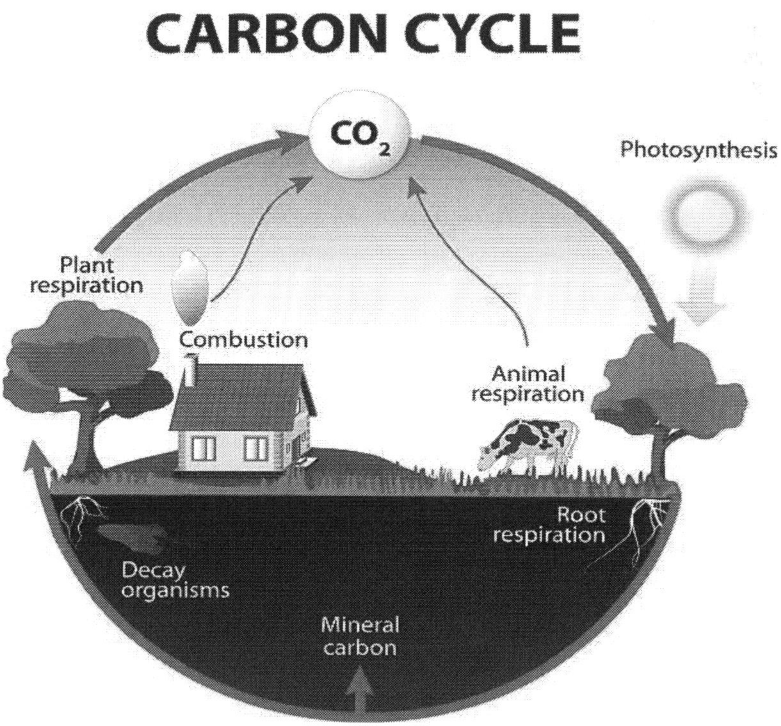

Figure 5.3 The carbon cycle, a fundamental biogeochemical cycle of the Earth by which carbon moves through the atmosphere, the oceans, and living matter.

able to access large quantities of carbon in short periods of time that took thousands or millions of years to form.

From the beginnings of the modern industrial period, humans have labored on their energy-extraction technologies, and through them have increased their capacity to acquire and use energy. In only the past 60 years, human energy use from fossil fuels (carbon formerly "taken out" of the carbon cycle) has resulted in the emission of about 400 Gt (Gigatons) (1 Gigaton is 1 billion metric tons – a metric ton is 2,205 pounds) of carbon to the atmosphere. That is about two-thirds as much CO_2 as previously *existed* in the atmosphere over the past million years,[7] but not all of this CO_2 remained in the atmosphere, as the amount of CO_2 in the atmosphere has increased "only" 30% during this period. This means that about half of the emitted CO_2 has been taken up by surface layers of the ocean and terrestrial soil. By digging deep into the Earth, humans have extracted quantities of carbon that otherwise would have required thousands of years for any previous (natural) carbon cycle to generate.[7]

Climate Change Past

Much has been written about the climate stability of the Holocene, which began about 12,000 years ago. However, the Holocene was not climatically stable.[8] Some of its changes were natural events of a self-willed nature. Others show the fingerprints of human activity. The Holocene did have mostly stable temperatures, but it had large variations in hydrological and ecological systems at regional to subcontinental scales that were often abrupt, especially in subtropical, semiarid, and temperate regions.[8] There were prolonged droughts in North Africa during which grassland ecosystems collapsed and were replaced by what we call today the Sahara Desert.[8] In the US state of Nebraska, the Nebraska Sand Hills, in the state's north-central region, is today dry grassland. It is rich in bird life, especially raptors. I saw that firsthand with my class in Field Ornithology, and I was very glad I had chosen the Sand Hills as one of our field components. Golden eagles, northern harriers, sharp-shinned, Cooper's, Swainson's, red-tails, rough-legs, and ferruginous hawks, kestrels, merlins, prairie falcons, and peregrine falcons – all sailing and soaring over rolling open ridges, and not a human artifact between you and them as far as the eye could see. But the Sand Hills sit atop one of the world's largest underground water reservoirs, the Ogallala Aquifer. In past times of drought, the aquifer has not provided sufficient water to support the grasses. Through repeated cycles of these droughts and (relatively) wet periods the Sand Hills earned their name, alternating between active dunes and stabilized grasslands.[8]

Forests in the eastern Great Plains of the United States changed to prairie in the early Holocene, then reverted to forests in the middle to late Holocene. More recently, and more human-induced, the effects of the Columbian Exchange (1492) were most fully realized by 1610. There had been, by this point, rampant death among Indigenous Peoples in the Americas that followed the arrival of European colonizers to the New World. The most well-informed estimates put the toll at 50 million Indigenous People exterminated, largely by smallpox (Chapter 8). Because many Indigenous Peoples were farmers – especially in Latin America – such declines in the human population meant that many fields were no longer cultivated, and these were reforested by natural successional processes, removing more carbon dioxide from the atmosphere, lowering global CO_2 levels and reducing temperatures worldwide.[9]

Yet more recent was the year 1816, variously known as "The Year Without a Summer," "Poverty Year," and "Eighteen Hundred and Froze to Death." Its abnormally cold weather was provoked by the eruption of a volcano, Indonesia's Mount Tambora, in 1815, an event that triggered a change in the global climate. Atmospheric scientist Andrew Schurer and his colleagues explain that

> The eruption injected a huge amount of sulfur dioxide into the stratosphere, which would have quickly spread across the world, oxidizing to form sulfate aerosols. These volcanic aerosols block large amounts of incoming shortwave radiation (portions of incoming sunlight), which can cause widespread, long lasting surface cooling.[10]

The forcings created by the increased sulfate aerosols in the atmosphere led to a drop in global temperature in the following year of 1–3°C (2–5°F). It was the coldest year in at least the last 250. Each season showed temperature anomalies, but the summer was the most drastic. In that year, the renowned German plant scientist Justus von Liebig (Chapter 2), then just a child, witnessed the failure of crops throughout Europe. Thus, at 13 years of age, Liebig lived through the year without a summer, when the majority of food crops in the entire Northern Hemisphere failed. His native Germany was among the hardest hit. Millions went hungry. Thousands died. It is likely that the experience shaped Liebig's later work, for it left an indelible impression on the young Liebig, and provided a lifelong motivation in his relentless research in plant physiology and crop production to find ways of ensuring reliable and plentiful harvests. Due in part to his own innovations in fertilizers and agriculture which helped prevent future crop failures of this magnitude, the 1816 famine became known as "the last great subsistence crisis in the Western world."[11]

The reason for the cooler temperatures in 1816 was, in principle, the same reason for the warmer temperatures the Earth is experiencing today – it occurred because of a change in the levels of radiative forcing in the atmosphere. With less solar energy reaching the surface, blocked by the sulfates and aerosols spewed from Mount Tambora, the system lost heat from the "external" source (sunlight) that was (had to be) compensated by equivalent heat loss in the "internal" system to maintain equilibrium (energy in equals energy out) – a "forcing" manifested in lower temperatures on the Earth's surface and in its lower atmosphere.

The Present: How Do We Know that Current Climate Change is Us?

The Keeling Curve

The atmospheric scientist Edmond Mathez, who has worked with other scientists at international levels modeling changes in the Earth's climate system, said

> When [climate] models were run with both anthropogenic and natural forcings, the simulations closely resembled the observed warming. But when they were run with only natural forcings ... [like volcanic eruptions and variations in solar irradiance] the simulations displayed a significant and increasing diversion from the observed trend as of around 1960 ... to reproduce the actual late-twentieth-century warming, the models required the input of anthropogenic forcings, specifically those due to greenhouse gases and aerosols.[12]

Humans have been studying climate change for only a short time, and have begun to understand the relationship between CO_2 and temperature for even less. It was only in

March 1958 that the atmospheric scientist Charles David Keeling installed his first IR gas analyzer at Mauna Loa, in Hawaii. On the first day of its use, the analyzer recorded a concentration of 315 parts per million (ppm) of CO_2. This was the first reading in what would become one of the most important continuous terrestrial scientific projects ever carried out. It has given the world its first quantitative assessment of the impact of global industrial activity on the composition of the atmosphere.[7]

Over the first few months of its installation, Keeling noticed a pattern of monthly increase in CO_2 concentrations until May. After May, these concentrations declined until October, and this pattern repeated itself in 1959. It was as if the Earth was breathing in and out once each year. What Keeling was actually beginning to detect was the seasonal effect that plant life was having on the atmosphere in the Northern Hemisphere. Recall one of the biological components of the carbon cycle. In conducting photosynthesis to manufacture sugars like glucose ($C_6H_{12}O_6$), plants take in carbon dioxide from the atmosphere and chemically combine it with water, releasing oxygen (O_2) as a residue. As Keeling put it later, "We were witnessing for the first time nature's withdrawing CO_2 from the air for plant growth during summer and returning it each succeeding winter."[13]

Every day for nearly 70 years, at Mauna Loa CO_2 levels have been recorded using Keeling's gas analyzers. The result has been one of the most famous "plots" in the history of science, the "Keeling Curve" (see Figure 5.4). On May 22, 2024, the CO_2 reading was 428 – an increase of 113 ppm, or nearly 1.7 ppm per year for 66 years. The level of fossil fuel consumption since 1960 has increased at the rate of 1.4 ppm per year – a similar order of magnitude to the increase in carbon dioxide. But global CO_2 emissions have increased by a factor of five since 1960.

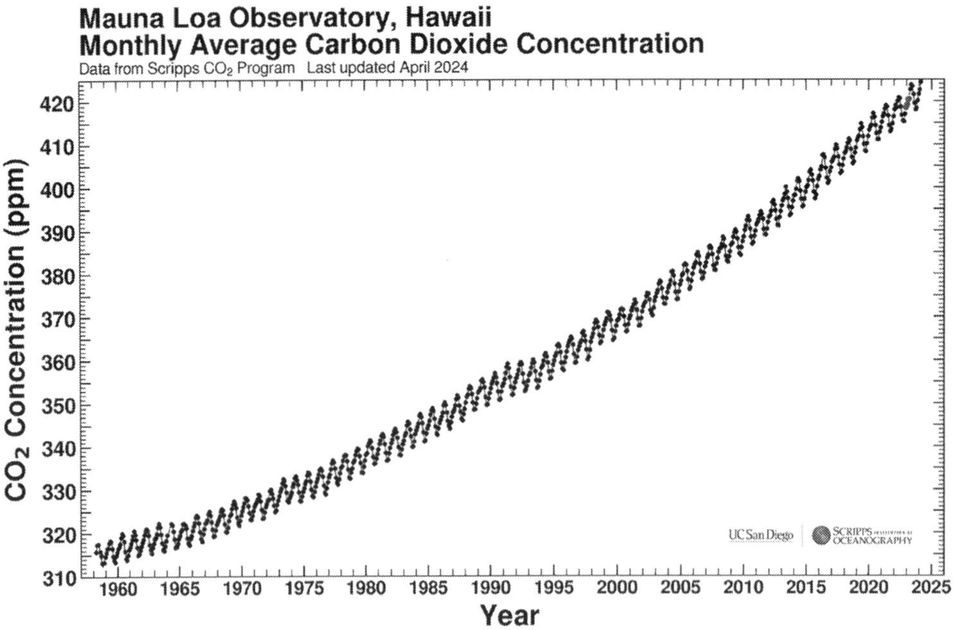

Figure 5.4 The Keeling Curve, one of the most famous illustrations in modern science, shows the fluctuations of CO_2 levels at Mauna Loa, Hawaii, USA since 1958. Scripps Institution of Oceanography at UC San Diego.

Historical Comparisons and Unequal Effects

Compared to the last half million years, our present time is unprecedented in levels of atmospheric CO_2. In every case over this period, higher CO_2 levels have been correlated with warming periods and lower levels of CO_2 with ice ages. Our current increase in CO_2 concentrations began with the modern industrial era (~1800) and the magnitude of CO_2 increase is commensurate with the increased human use of fossil fuels.[6] Today we are seeing the effects of releasing carbon from fossil fuels into the atmosphere. In 2023, the US National Oceanic and Atmospheric Administration (NOAA) reported that: 1) the Earth's temperature has risen by an average of 0.11°F (0.06°C) per decade since 1850, or about 2°F in total; 2) the rate of warming since 1982 has been more than three times as fast – 0.20°C (0.36°F) per decade; 3) 2023 was the warmest year since global records began in 1850; and 4) *the ten warmest years in the historical record have all occurred in the past ten years (2014–2023).*[14]

Earth is a planet of wondrous variety, and so are the effects of climate change. Among the regions already experiencing the clearest effects of global warming on a year-to-year basis are the equatorial and tropical regions. By the end of the twenty-first century, the tropics and subtropics will have seasonal average summer temperatures warmer than the warmest summer so far almost 100% of the time.[7] These regions are home to the Earth's equatorial forests – vegetation communities that have the greatest capacities to remove CO_2 from the atmosphere and mitigate CO_2 additions from other sources. Such a forest not only removes CO_2 from the atmosphere, but its plants, especially trees, also release (transpire) large amounts of water vapor into the atmosphere as part of their own metabolic processes. Such transpiration is part of the moisture-laden environment that contributes to the high levels of rainfall that maintain the forest. But if deforestation (forest loss) reaches a critical level ("tipping point"), the forest will stop increasing the local humidity and rainfall so that these conditions will no longer be self-sustaining, and the forest will degrade into a drier savannah. This change would then release large amounts of CO_2 currently stored in the rainforest trees and vegetation into the atmosphere. About 1 acre in 6, roughly 17%, of the Amazon Rainforest has already been lost. One estimate of the critical tipping point is 20–25% deforestation.[15]

The other landscapes expected to suffer disproportionate effects of global warming are the polar regions. These will experience an increasing loss of sea ice and reduction in the size of polar ice caps, the retreat and shrinkage of glaciers, the dieback of boreal forests, and the thawing of permafrost. The last effect is significant in that permafrost is one long-term storage sink for carbon. With its warming, such stored carbon is volatized from the permafrost and released into the atmosphere, thereby increasing the concentration of atmospheric CO_2, with resultant increased absorption of reflected heat from the Earth's surface.[15] Loss of ice caps and sea ice reduces the Earth's *albedo* – the proportion of incident light or radiation that is reflected by a surface. Snow and ice have a high albedo relative to bare ground or vegetation, so they reflect more light and absorb less heat, a feedback system that contributes to their maintenance. The larger the overall value of the albedo of the Earth's surface, the colder the temperature of the atmosphere. But as snow and ice are lost, albedo diminishes, and polar regions will warm at disproportionately greater rates.[15]

Climate Change in Oceans

Were it not for the Earth's oceans, climate change effects could be worse. The ocean has absorbed 93% of the extra energy from the enhanced greenhouse effect, and approximately

30% of the CO_2 generated by human activities.[16] But the oceans are not a carbon "slush fund" for the continued increase of carbon dioxide. Warming the atmosphere also warms the oceans. Marine biologist Eva Hoegh-Guldberg explains,

> Many early studies concluded that climate change was unlikely to influence the ocean by much given its size and thermal inertia ... The ocean is highly structured, however, and as a consequence, the upper layers of the ocean (a relatively thin layer of several hundred meters) acts somewhat independently of the rest of the ocean. As a consequence, changes in the temperature of the surface layer of the temperature of the ocean are not too far behind that of the surface temperature of the planet as a whole ... While average global temperature has increased by approximately 0.7°C since 1950, average global sea surface temperatures (SSTs) in the Indian, Atlantic, and Pacific Oceans have increased 0.65°C, 0.41°C, and 0.31°C, respectively.[17]

The oceans are affected by climate change in other ways. Global mean sea level rose by 0.11 m between 1901 and 2010, and the rate of rise has been increasing since the 1980s (3.2 mm/y).[18] The temperature and pH of ocean waters, which affect many dimensions of their physiochemical properties and environments, are also changing. Together, these changes have already led to more frequent extreme climate events (tropical storms and hurricanes) and spatiotemporal redistributions of climatic conditions (altered strength and location of upwelling currents and changes in ocean circulation).[19]

The loss of ice from polar ice caps and glaciers will add more water to the world's oceans, contributing to rising sea levels. However, it is not added water from these sources that will make the greatest contribution to this increase – it is the *thermal expansion* of the water in the oceans. When water warms, it increases in volume such that the same mass of water occupies more space. The primary driver of rising sea levels is warmer water, not more water. Even if CO_2 emissions were to *stop now* (and they are not going to), current global climate models estimate that sea levels will rise by approximately 1 m, contributed by these factors:

> Thermal expansion of warmer ocean waters ~ 0.23 +/–0.09 meters
> Input from Greenland Glacier ice melting ~ 0.16 meters
> Input from Antarctica Ice Cap melting ~ 0.12 meters
> Input from other glaciers and ice caps melting ~ 0.37 +/–0.02 meters
> Total ~ 0.88 +/–0.12 meters.[7]

Rising sea levels will have their greatest impacts in coastal areas, where human populations are most concentrated. Of the current 8 billion people on Earth, about 1 billion live less than 10 m above high tide levels along coasts or on islands. Rising sea levels will also affect rivers in the areas in which the rivers deposit sediment – the oceans–delta regions where the river's flow rate slows.[7] The Mekong Delta of Southeast Asia, for example, is the largest rice-producing region on Earth and one of the world's largest freshwater fisheries, but it would take only a small increase in surrounding sea levels to make the freshwaters in its delta brackish. Sand removal (by people living in the Delta) has already reduced the river's elevation by 1.4 m since 2008. A further rise of just 1 m in the surrounding sea level before the end of this century (as current global climate models predict) would place 75% of the total area of the Delta (most of Vietnam) below sea level – an area with a human population of 12.3 million (70% of the people in this region). And both the rice farming and freshwater fishery of the Mekong Delta would collapse.[7]

Climate Change, Ecology, and Conservation

Climate Change Attribution and the Climate Niche

Does climate change matter to biological change? The process of attributing some elements of an observed (detected) biological change to detected trends in climate is what we refer to as *climate change attribution*.[20, 21] To make the question more specific, more practical, and more measurable, we could put it this way – do the non-human species on Earth show changes that can be attributed to climate change? MacCracken says "Yes," and the effects are measurable.

> Of five major meta-analyses assessing changes in distributions and abundances of individual species in recent decades, about half of species for which longer term data exist have shown changes in their distributions over the past 20 to 140 years. These changes are not random but systematically in the direction expected from regional changes in climate [83–90% of changes consistent with changes in regional climate] ... Responses have now been documented in every major taxonomic group for which long-term data exist ... [birds, butterflies, trees, mountain flowers, fish, sea urchins, limpets] across diverse ecosystems...[1]

Every creature on Earth is uniquely adapted to survive within a specific set of conditions. Every creature has a *niche* – that ecological "space" including all of the environmental factors and interspecies relationships that influence the species' behavior and distribution.[22] Some niche factors are physical – structures that may be needed for support, nesting, or reproduction. Some are resource-based – other creatures that provide food or mutual benefit, or aid the species in its own dispersion. But a critical set of niche factors are *climatic* – a range of environmental conditions, such as tolerances of high and low temperatures, moisture, or seasonality (or lack thereof), within which the creature can live, and outside of which it cannot. This is the *climate niche* – the sum of a species' tolerances to ranges of climatic variables. When climate changes, the climate niche is "dislocated." The unique tolerances for which the species is adapted may "move" to a new location, or, in some cases, disappear altogether. To a species, *climate change* does not matter as much as *climate niche dislocation*. And if such dislocation occurs, the creature must "follow" its climate niche to the new location or adapt (first, behaviorally and then, in the longer-term, genetically) to the new climate where it is.

Species Versus Climate – The Arms Race of Climate Migration

In one of the first major syntheses of its kind, biologists Camille Parmesan and Gary Yohe undertook a meta-analysis of published studies, evaluating the recent movements and behaviors of over 1,700 species to determine if recent biological trends in range shifts and biological timing events matched climate change predictions. In these studies, Parmesan and Yohe documented poleward range shifts of 6.1 km per decade and the advancement of spring events by 2.3 days per decade.[23] Species in marine environments show the same trends. They are moving toward latitudes in more poleward directions or to greater depths. In 2013, ecologist Elvira Poloczanska and her colleagues conducted a similar large-scale review of published marine ecological observations. Reviewing 279 observed shifts documented in 36 published studies, the authors found that 83% of observed changes in studies of movements in individual species were in the direction expected under climate change.[24]

A species' movements to track its climate niche may not only be latitudinal, they may also be vertical (elevational). In birds, for example, recent research has indicated that direct effects of abiotic factors like temperature and precipitation probably do not constrain elevational ranges or drive upslope range shifts in most bird species. Biotic interactions, in contrast, appear to be much more important.[25] Bird distributions are associated with particular habitats and food resources,[26] but are these resources sensitive to climate change, and which ones are most important?

Worldwide, the overall direction of elevational shifts in birds indicates that it is warming temperatures, and the way such temperatures change the distribution of critical resources for birds, that are driving upslope movements.[27] Shifting elevation upward may work for species that have historically inhabited lower-elevation environments, but species already living at high elevations have nowhere to go. Among the environments that may be most vulnerable to climate change are the Central American cloud forests. Such forests are distinctive not only in the amount of rainfall they receive, but in the historically "misty" environment that predominates in them. Terrestrial amphibians, which depend on direct respiration through their skin as much as through their lungs, must keep their bodies moist for such respiration to occur. Cloud forests provide an environment of enhanced "skin respiration," and amphibian richness in cloud forests has historically been large (see Figure 5.5).

Alan Pounds and his colleagues, who have long studied cloud forest amphibians, reported changes in amphibian populations over decades of fieldwork. Based on direct measurements of conditions in the cloud forests, they noted, "Dry days have increased in frequency since the 1970s and have increasingly coalesced into dry periods. Whereas mist free periods in the 1970s rarely exceeded two days, they have recently lasted up to three weeks."[28] The outcomes of such changes were predictable.

> Massive declines of frogs and toads were apparent by 1990. A multispecies population crash in 1987 led to the disappearance of the endemic golden toad (*Bufo periglenes*)

Figure 5.5 The Monte Verde Cloud Forest of Costa Rica. <https://theblogofdimi.com/photos-monteverde-cloud-forest-costa-rica-2022/>.

[see Figure 5.6] and many other species ... Twenty of the 50 were missing throughout the surveys of a 30-km^2 area during 1990–1994, and there is still little sign of recovery 15 years after this crash.[28]

Figure 5.6 The Monte Verde Golden Toad (*Bufo periglenes*), now extinct in the wild, is a victim of the loss of its cloud forest habitat due to climate warming. <https://www.rainforest-trust.org/our-impact/rainforest-news/thirty-years-after-the-last-golden-toad-sighting-what-have-we-learned/>.

A similar "up and down" pattern is occurring in marine species. The global ocean is warming faster than adjacent land and faster than the global average.[29] This warming has a particular effect on the distributional responses of bottom-dwelling (demersal) fishes, with two-thirds of all North Sea fishes shifting mean latitude or depth,[30] a marine equivalent of the altitudinal migrations being observed in land animals. In the span of 25 years, European scientists have documented a 1.6℃ increase in the winter bottom temperature of the North Sea, with a corresponding deepening of all but one of 19 demersal fish assemblages by ~3.6 m per decade.[30]

How Climate Change Alters Biological Timing

Phenology is the study of "biological timing," such as the onset of flowering in plants or the initiation of migratory movements in birds. Changes in species phenology provide perhaps the clearest evidence of the reality of climate change effects on biological processes. In the Northern Hemisphere, the timing of the onset of the annual plant-growing season as detected by satellite observations advanced by 5.2 days per decade between the years 1982 and 1999.[31] In the Arctic, the advanced onset of flowering is greater, and phenology-related events in animals have become more frequent. Research scientist Toke Høye and his colleagues noted that, "Flowering dates in six plant species, median emergence dates of twelve taxa of arthropods, and clutch initiation dates in three species of birds have advanced, in some cases by over 30 days during the last decade."[32]

Changes in the phenology of individual species matter because they can disrupt the coordination of activities between that species and other species in any number of interactions, such as predator and prey, parasite and host, or pollinator and flower. David

Inouye and his co-workers at the Rocky Mountain Biological Laboratory in Colorado, USA found that, from 1976 to 1999, yellow-bellied marmots (*Marmota flaviventris*) (large ground squirrels that live mainly in montane environments of the western United States) had advanced their emergence date by 38 days as a result of warmer local spring air temperatures. Such advancement put them out of synchrony with their major food source – the surrounding alpine plant species. American robins (*Turdus migratorius*) advanced their arrival dates over about the same period by 14 days at the same study site. This pattern of early return has increased the difference between their arrival and the average first appearance of bare ground (which robins require for foraging) by an additional 18 days.[33]

Climate change also affects the flowering dates of many plants, advancing their first flowerings earlier in the spring than in the past.[34] Some pollinators, such as bees and butterflies, are also now appearing earlier in the year. Ecologists Jessica Forrest and James Thomson undertook a study within the Rocky Mountains of Colorado to assess whether or not plants and insects in these subalpine habitats were advancing their phenology in response to springtime warming similarly or asynchronously. Over three years, Forrest and Thomson collected data on air temperature, insect emergence (specifically, the cavity-nesting bees (*Hymenoptera*)), and flowering phenology. They found that one plant (*Lathyrus lanszwertii*) and one bee (*Hoplitis fulgida*) showed broad overlap in phenology at most sites in 2008 but only slight overlap at any site in 2009, such that, at all sites, *Lathyrus* had finished or nearly finished flowering by the time of peak *H. fulgida* emergence. This was due, in part, to generally earlier flowering by this plant species at warmer, low-elevation sites in 2009 compared to 2008, without a corresponding advance in the phenology of the bees.[35]

Every species will show unique responses to the climate change it experiences. But now, considering the full suite of latitudinal migrations, elevational migrations, and changes in phenology, we are seeing an unmistakably "globally coherent signal" of worldwide patterns – a "diagnostic fingerprint" of the impact of global climate change on biodiversity. There is one more unique set of climate change responses that is producing this kind of global signal and deserves an examination of its own – the response of coral reefs to warming oceans.

Marine Environments – The Coral Reef

The anthropogenic effects of climate change have led to the worldwide phenomenon of "bleaching" in corals, resulting in the death or degradation of many of the world's tropical and subtropical coral reefs (Chapter 3). Bleaching occurs when regional surface water temperatures exceed a fixed threshold, at which point corals expel their symbiotic zooxanthellae (single-celled algae that live in symbiosis with the corals, providing them with food through their own photosynthetic processes that they share with the corals). Such temperatures result in coral bleaching over large areas. In the Earth's past history, coral reef "gaps" (disappearance of corals from the fossil record) have been associated with all five of Earth's major extinction events, and in all five, the loss of coral reefs was correlated with the buildup of CO_2 in oceans – the foundational cause of increased water temperatures. Today 75% of coral reefs are threatened, and 95% are in danger of being lost by 2050. The amount of coral in the world's largest single coral ecosystem, the Great Barrier Reef, has declined by 50% since 1983, and in some areas 90% of its corals have experienced bleaching.[17] Given that coral reefs provide habitat for 25% of all ocean species, losses of coral reefs are consequential.

One mechanism of these losses is the increased temperature of the water itself, but other losses are just as deadly and systemic. As oceans become more acidic because of the greater diffusion of CO_2 into ocean surface waters, the addition of positive hydrogen ions allows their increased binding with pre-existing negatively charged carbonate (CO_3^{2-}) ions to form carbonic acid, HCO_3.[8] Coral reefs, cold-water reefs, and oyster reefs form where $CaCO_3$ production exceeds its dissolution (net availability of $CaCO_3$ in sea water). When the dissolution of $CaCO_3$ rises (as it does in the presence of increased CO_2), reef-building capacities of these organisms decline because increased levels of CO_2 enhance the breakdown of $CaCO_3$. In physiology, increased CO_2 causes changes in reproduction, affecting the fecundity and quality of gamete production, photosynthesis (increased rates), calcification (declines), energetic demand (which is increased), and growth.[36] Carbonate ions are what both coral and shellfish use to build their exoskeletal structures. With fewer carbonate ions free for the taking in the ocean waters, corals and shellfish both suffer, along with other species. One study of tiny pteropods (very small marine snails) in marine waters along the US West Coast found that 53% already had severely dissolved shells.[8] It was once thought that these changes affected only species that needed calcium carbonate in skeletal-building processes, but it is now apparent that such changes affect biogeochemistry, physiology, and behavior across a wide group of organisms.

Common Problems – Common Responses

Clarifying the Conservation Problems of Climate Change

If there is one thing that climate change does for conservation it is that it more precisely frames the problems to be solved by clarifying the *effects* of climate change (even if the *solutions* to the problem are anything but clear). The first revelation climate change brings to conservation is – this changes everything! Even the field's most foundational structures, methods, and principles are now often not only ineffective, but sometimes irrelevant. As Canadian ecologist Daniel Scott explained in speaking about the threat climate change posed to the national parks of his own country, "Climate change represents a fundamentally different threat to Canada's national park system. Never before has there been an ecological stressor that raised questions about the adequacy of the system to protect representative samples of Canadian ecosystems."[37]

Parks, protected lands, and wilderness areas (Chapter 9) can protect their non-human residents from many human disturbances, but are defenseless against a changing climate. The protected area (PA), no matter how well managed, is embedded in its climate and cannot escape it. *Thus, climate change presents the most difficult suite of challenges in conservation today.* It is an understandable cause for discouragement. Bioclimatologists Lee Hannah and Laura Hansen assess the situation this way:

> Climate change affects selection of [conservation] targets in two fundamental ways. First, it will alter ecoregion boundaries … [resulting] in different biome boundaries in the future, which may change the size or configuration of an ecoregion. … The second way in which climate change can affect target selection is by exacerbating existing threats … Therefore, climate change is likely to increase the number of conservation targets, through both changing the planning area and by increasing the number of species at threat.[38]

In some cases, climate is the predominant or singular force of change, while in other cases it acts as a compounding factor by exacerbating existing vulnerabilities. Conservation in the face of climate change must often begin with very simple responses.

Field Experiments to Assess the Effects of Climate Change

A first step for studying and modeling the effects of climate change on the Earth's biodiversity is as obvious as it is underemployed – the use of field experiments to determine correlations, physiological tolerances, and effects on biological timing associated with climate change that could be used to estimate the climatic tolerances of a given species. The second step follows logically – couple field experiments with climate models that can project future temperature changes within the species' range. If one can define the climate niche of a species and predict how the climate will change over time within its historic range, one should be able to project what will happen to the distribution of a species as climate changes and whether a species would have to migrate to remain within its climate tolerances.[39]

This kind of understanding has led to advances in predictive scientific modeling for responding to climate change. Scientists first integrated these kinds of climate tolerance data and concepts into Species Distribution Models (SDMs). SDMs are a particular form of modeling a species' distribution through its ecological niche. As climate has become a more important and changing factor affecting species' distributions, species' climate tolerances have been increasingly integrated with other aspects of their ecological niche to produce *bioclimate envelope models*.[40] Bioclimate envelope models function by discriminating between the climates of locations inside and outside the ranges of a given species based on particular climate variables, such as minimum and maximum temperature. This strategy of modeling, known as a "correlative approach," works to quantify the relationship between observed species' distributions and environmental variables (derived primarily from climatic data). Conceptually the model is created through a series of spatial overlay maps, each with different attributes. For example, when precipitation and temperature data are combined across space, these geographic domains can begin to shape distinct climatic ecoregions across the landscape and can be coupled with other input data layers to identify spatial areas that can provide a species' climate niche.[40]

The foundational assumption of these models is that species distribution is determined wholly or partly by climate. Two facts support this assumption: 1) distributional limits of species match particular combinations of climate variables; and 2) these limits shift through time in synchrony with changes in climate.[41] For example, a study by Walther and colleagues showed that the northern range limits of the European holly (*Ilex aquifolium*) were consistently associated with the 0°C degree winter isotherm, and the holly's range was shifting northward over time as it tracked increasing temperatures at higher latitudes.[42]

Establishing clear associations between climate and species distribution can be challenging because data are often incomplete or applied at inappropriate resolution. For example, at a continental scale, climate considerations will be dominant, such that only those areas where the species' climate tolerances match actual climate conditions can be considered potential occupiable range. However, that consideration alone may not produce accurate predictions of future species occurrence. At regional and landscape scales, topography and vegetation (habitat) may become a more critical detector of presence or absence. At even finer scales, one might examine biotic interactions and microclimate conditions, such as slope aspects (south or north), soil conditions (dry or hydric), or geologic factors (limestone, serpentine, or granitic rock bases) that could limit the occurrence of the species. Accurate modeling of landscape and topography is essential because, as conservation scientists Richard Pearson and Terence Dawson explain,

> the ability to migrate is a function not only of individual species' characteristics, but also the structure of the landscape over which dispersal is occurring, including the presence of natural barriers (such as mountain ranges) or the artificial fragmentation of habitats (through, for example, the growth of urban areas or deforestation).[40]

What Are We Going to Do? Conservation Responses to Climate Change

Designing the Climate-Sensitive PA Using the Bioclimate Envelope

What would happen to the distribution of species X if climate change happens according to scenario Y? Conservation outcomes have increasingly been evaluated using bioclimate envelope models, and such evaluations have included assessment of the effectiveness of PAs and networks. Climate-driven range shifts can cause some species to move out of reserves entirely, reducing species representation in PAs.[40] The same model-based assessments have also shown that additions of new PAs can be useful in counteracting this effect, with increases as small as 5–10% in national PAs networks restoring losses of species due to climate change.[43] Using knowledge of the climatic tolerances of individual species, bioclimate envelope models try to predict where certain species, especially plants, could occur in the future under projected changes in temperature and moisture or other climate variables. The model outcomes can often be honestly referred to as "best intelligent guesses," but they are nevertheless *informed guesses* that can predict areas where a species of conservation concern *would occur* in the future, and so plan a future network of PAs that should conform to a future climate that may match the species climate niche.

Modeling a species bioclimate envelope and predicting where it will occur in the future is a constructive first step toward conservation in a warming world. However, climate change not only has a spatial dimension, but a temporal one. What is the *speed* of climate change across landscape distances? Conservation scientists have addressed this problem using global circulation models (GCMs) that estimate the rate of mean annual temperature increase over time, combined with a topographic relief map to show change in elevation. Such models then make predictions of *climate velocity*, the speed of climate change over a landscape with defined features.

Using these models, scientists can relate the concept of climate velocity to the size of existing PAs. For example, ecologist Scott Loarie and his colleagues found that only 8% of an array of global PAs had "residence times" of more than 100 years, meaning that the majority of the PAs would experience entirely novel climates within this century. And size matters. Small PAs had small residence times for the type of habitats and ecosystems they were designed to protect. Larger PAs may be better able to retain some of the original protected biome within their borders. Loarie et al. suggest that these results have implications: 1) resident species will have to be able to migrate with equivalent speed (that is, the same speed as that of the climate change in km/y) to keep up with anticipated changes; and 2) existing PAs may no longer be sufficient in their current extent to protect assemblages that rely on contemporary climate conditions.[44]

In a world of changing climate, robust PA design and management will depend on four components: 1) a knowledge of the magnitude, rate, intensity, timing, and intensity of climate change and its impact on conservation targets; 2) identification of current gaps in protected-area design and management; 3) a working theory that can incorporate biophysical dynamics that are important to species persistence; and 4) innovative PA design and management approaches that translate this theory into effective practice for conservation. Effective habitat networks under the stress of climate change will require PAs that: 1) meet the habitat requirements of a wide array of species throughout their lifecycles; 2) accommodate different movement types, including short-range connectivity relevant to the persistence of megapopulations within the networks and long-range connectivity relevant to climate-driven range shifts across the networks; and 3) are robust to habitat loss and changing connectivity. The effects of even 1°C global warming (reached in 2018) make these implications and recommendations relevant because such warming has already induced species' range shifts, changes in the timing of biological phenomena, and the decoupling of interactions among co-evolved species.[45]

Is there an effective conservation strategy for climate change? Theoretically, yes. One approach developed by Dawson et al. begins with a "vulnerability assessment" of a species to climate change (see Figure 5.7), which begins with making a list of the "pressures" that a species is likely to face from expected climate change in the context of its own current range, population size, and movement (dispersal) ability.[46] Such analysis also requires an assessment of both the species' sensitivity to climate change (how much do a couple of degrees matter to this species?) and its adaptive capacity (has this species shown an ability to change its behavior in the face of changing climate?). These assessments help conservationists determine how vulnerable the species may be to the expected change in climate. Given that vulnerability, conservationists can then give their attention to two elements. One is preparedness. What is the best way to monitor and prepare for the expected climate change, and what are the most important things to monitor in the state of the species? The second is intervention – what are we going to *do*?

The general principles derived from current models have few concrete examples that have actually advanced conservation. Such climate modeling using species-specific bioclimate envelopes is a fascinating and rapidly developing area of conservation research, but, unless applied, bioclimate envelopes do not actually *do* anything in conservation. No species is saved simply because a model determined its personal bioclimate envelope. If species are to be saved, more direct action is needed. The fact is that, at this moment, no one really knows what to do or how to design reserves for climate change that will actually work, but some applications show promise.

The Marine Protected Area (MPA)

In the world's oceans, MPAs have been effective in improving the resilience of coastal ecosystems to climate change. In specific cases, they have been associated with the recovery of kelp forests, increasing stability of temperate reef fish communities and improving the resistance of marine invertebrates to mortality caused by climate-driven hypoxia. On the Great Barrier Reef, multiple studies have indicated that MPAs enhance the capacity of coral reefs to withstand warming, floods, and storms and achieve faster recovery from them. In the Bahamas, corals recovered from mass bleaching and hurricane impacts faster inside than outside MPAs.[14]

The performance of MPAs varies with multiple management-related factors. Well-enforced, older, larger MPAs achieve the best conservation performance, but these are only a small fraction of the MPAs established globally.[47] The performance of MPAs in enhancing resilience to climate change may vary with species tolerance to climate stress, but MPAs can have an effect. One of these was established by an Indigenous resident community in the Solomon Islands in 2023. One member of this community, Sabrina Pania, who was closely involved in this effort, tells their story here.

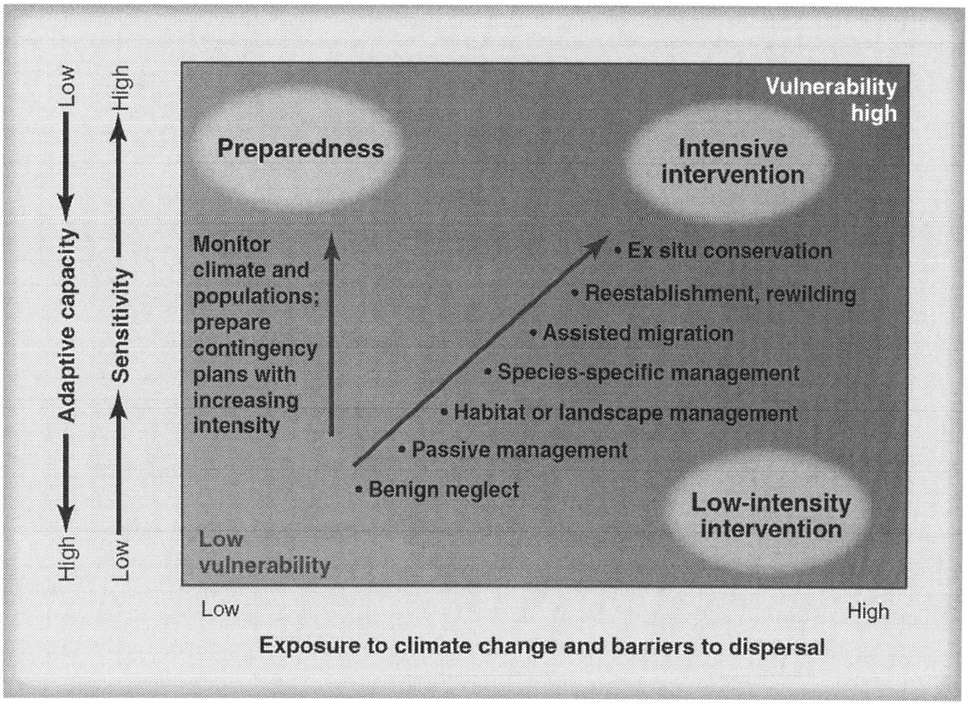

Figure 5.7 The vulnerability of a species or ecosystem is based on its exposure to climate change, its sensitivity, and its inherent capacity to adapt to change. The x-axis represents the degree of exposure to climate change faced by species and communities, largely determined by the species' or population's geographical location, the rate and magnitude of climate change anticipated for that region, and the size, cohesiveness, and connectivity of the species' habitat within and beyond that region (extrinsic factors). The other two measures from the vulnerability framework, adaptive capacity and sensitivity, are plotted together on the y-axis, primarily determined by biological characteristics of species that influence their mobility, specificity, and sensitivity (intrinsic factors). The relative position of species and ecosystems along the axes can inform decisions on appropriate research, monitoring, and management strategies. After Dawson et al. 2011. American Association for the Advancement of Science. Science 332:53–58. © 2011 AAAS.

Story Box 5.1 Climate Change in the Solomon Islands

Sabrina Pania

Wildlife Conservation Society – Solomon Islands

"The sea often comes and the whole Island will sink under water," Ivere, an elder and fisherman of Marapa Island, whispered. His eyes search the horizon as if he can see the Island when there was still vegetation on it. Now it's just a sandbank at the mercy of relentless waves. These tides not only claim our islands and our way of life, they change our perception of how we are going to go on living (see Figure 5.8).

I am a native Solomon Islander. After four years of studying Marine Science in the nearby Fiji Islands I had begun to understand climate change and its impacts, not only globally, but locally – in my own country. Climate warming is a

Figure 5.8 One of the Solomon Islands in the South Pacific, islands of extraordinary beauty and unique biodiversity. Local Indigenous communities of islanders live on individual islands and make their living from the sea. Now many of the smaller islands are shrinking or even in danger of disappearing due to rising sea levels caused by thermal expansion of ocean waters in the face of global warming. Photo by Douglas Junior Pikacha, courtesy of Sabrina Pania.

phenomenon that affects all of the Island states of Oceania in the Pacific. The Solomon Islands are among these, an archipelago of more than nine hundred islands with extensive coral reefs and lush tropical rainforests. But now, if you look more closely, you will see dying corals and felled trees. Today the untouched parts of the Solomon Islands can be found only in a few reserved areas.

In June 2023, I joined an ecological restoration project in Marau Sound off the northeast coast of the island of Guadalcanal. Marau Sound is a small body of water, but thriving with coral reefs, mangrove forests, and vibrant seagrasses in its turquoise waters. The smaller islands in its offshore seas are clustered close to one another and to the mainland. People living here and along the coast of the Marau Sound draw their livelihood from both sea and land. Like many Solomon Islanders, they depend on the coral reefs for food, income, identity, and even the land on which they live, which has been built over time by the reefs themselves.

Investigating the effects that climate change was creating locally, I spent time in the villages, listening to what my fellow Islanders had to say. Day after day I heard to their narratives of rising sea levels, extreme storms, and unstable weather patterns, all recounted as if these things were "supposed" to happen. I had a detailed knowledge of climate change, but what difference did that make? How could I help my fellow Islanders in ways that would really matter?

A few days later, my team visited Marapa Island, one of our study sites. As we made our way to one of the villages, I noticed two smaller islands further east. One was already heavily eroded, and both might be underwater within the next five years. As we sat near the beach at Marapa doing interviews, I saw that the island's adjacent sandbank was not far offshore. I asked Ivere about the sandbank, and the disappearing islands to the east. He replied simply, "It is the way of the ocean." The people of the Solomon Islands adjust to many of the effects of global warming by simply accepting the changes they bring. In this case they considered how the sandbank's reef would provide a rich fishing ground for their needs. But not all changes help them. Land-based resources like crops and fresh water have become scarce due to the increasing levels of salinity in the island's soil and freshwater that accompanies rising sea levels. In response, the Islanders sell their catch of ocean fish for store-bought goods like rice, and must develop new means to keep fresh rainwater longer for drinking.

My research team's collaboration with local islanders enabled us to observe and consider the importance of community effort in using and conserving their natural resources. More importantly, with our understanding of island culture, we have been able to gain the trust of people in local communities that can improve participation efforts in implementing practical solutions like coral planting, farming of saline-tolerant root crops, and constructing water retention dams. And these simple actions can be accommodated into an already established way of life.

One of the most important initiatives undertaken by the Marapa Island community is their locally managed marine areas. The community has agreed to designate some of the reefs as "no take zones" or a "marine protected area" (MPA). Surrounding reefs remain open to fishing and villagers from nearby areas can still fish on reefs bordering the MPA. Many villagers told us how their catches on the bordering reefs became more bountiful in only a short time after the MPAs were established. And their traditional subsistence fishing methods of spear fishing and hand-throwing individual fishing lines further limited the number of fish taken. These practices – shared together as a community – enable them to continue to abstain from fishing in the MPAs. We are pursuing our collaboration with the Islanders by monitoring fish species and their abundance to determine the size of fish populations, and by helping local residents with the assessment of coral regeneration in the MPA. Here, among these island communities I love, I see now, more clearly than ever, how we can combine traditional knowledge and practices with scientific investigation to make "conservation" an expression of the way we live, a way of life that will always be a part of our communities on these islands and the sea around us.

If You Cannot Control What You Are Changing, Stop Changing It

Climate change is not new. The current *speed* of climate change is new. For some, current (speedy) climate change is heralded as unmistakable evidence that human beings have now overwhelmed the "great forces of nature" and should assume their rightful role as the engineers and designers of the Earth System. Some of these same voices have begun to advocate "geoengineering" to stop the unfavorable effects of climate change, which have included proposals for "screens" around the Earth to reduce radiation input, adding sulfate aerosols to the atmosphere (remember Mount Tambora?) to induce cooler temperatures, or practicing "ocean fertilization" by adding chemical fertilizers to marine environments to support the growth of plankton that will convert more CO_2 into oxygen through photosynthesis and in the process reduce both ocean acidification and global warming.[48] Here, a recent word from the US National Research Council may be both instructive and sobering. The Academy deemed "geoengineering" an inappropriate term for combating climate change because, as they put it, "it is only possible to engineer systems that are understood."[49] Geoengineering should not be our first response to climate change because it treats the symptoms, not the cause.

Changing our climate, which we have already done and continue to do, is worlds away from *controlling* our climate, which we have no idea how to accomplish. We do know what adds to the problem of climate change, and we do know how to change these – fossil fuel emissions and global deforestation lead the list. If we were to change these two things, we would in the process be forced to change our intentions toward nature as well as our interactions with it – rejecting the idea that we should acquire energy without limit so that we can consume it without concern. Both premises are false. An intentional rejection of adopting a destructive (and thoroughly mistaken) pursuit of endless growth in energy supply would lead us to the search for both forms and uses of energy that we really *can* control, and that we could, in the power of that control, use to nurture Earth's life processes rather than destroy and degrade them.

We cannot, by any rational assessment, continue to consume fossil fuel energy without concern. The problems are at the door, and they are not leaving. At the human level, the loss of homes and livelihoods in island and coastal communities, the decline in agricultural production in traditionally fruitful farming areas, and the deaths of increasing numbers of people from levels of heat they are unable to endure strike hard and directly at the human condition. But we cannot *control* climate if we refuse to *limit* the causes of climate change. As the essayist Wendell Berry put it, "because we are always setting out to control something that we refuse to limit, we have made control a permanent and a helpless enterprise. If we will not limit causes, there can be no controlling of effects."[50] Understanding must begin with *limiting ourselves*, not with *controlling something else*. As Brendan Mackey expressed it precisely, "It is foolish and dangerous to confuse force with control. The Anthropocene, while an empirical fact, does not mean that humans 'run the show.' Rather, it means only that we can be powerfully disruptive."[51]

The highest and best interaction that humans can achieve with nature begins with our decision to *recognize our limits* for restraint and *nurture nature's potential for its good*, rather than pursuing our potential to *limit nature for our good*. The highest and best use of the human hand toward nature is not to take from it, but to form a place in it, one that will provide "room" for creatures other than ourselves. In this, we would learn the highest and best meaning of hospitality, and experience the greatest satisfaction of companionship, for nature and its creatures are the best of companions, and we alienate them from us at our peril.

Making a home for other people and other creatures begins with a commitment to fostering a good "environment" for them, and the foundation of that good environment is its climate. Many things that humans have done to climate will not be easily or soon undone, and much that has been done is evidence of greed, degradation, and neglect. Restoration, not control, should be our first response to climate change. Not that we can restore what we have, in ignorance, changed for millennia, but that we begin to nurture and repair what can be repaired now. We can amend those losses that, though caused by our actions, are not yet beyond our influence.

References

1. MacCracken, M. C. 2019. What is climate change? Pages 12–22 *in* T. E. Lovejoy and L. J. Hannah, editors. *Biodiversity and climate change: Transforming the biosphere.* Yale University Press, New Haven, Connecticut, USA, 28.
2. Karl, T. R., and K. E. Trenberth. 2005. What is climate change? Pages 15–30 *in* T. E. Lovejoy and L. J. Hannah, editors. *Climate change and biodiversity.* Yale University Press, New Haven, Connecticut, USA.
3. Mathez, E. A. 2009. *Climate change: The science of global warming and our energy future.* Columbia University Press, New York, USA.
4. Nordblad, N. 2021. On the difference between Anthropocene and climate change temporalities. *Critical Inquiry* 47:328–348.
5. Loeb, N. G., and B. A. Wielicki. 2015. Satellites and satellite remote sensing: Earth's radiation budget. Pages 67–76 *in* G. R. North, J. Pyle, and F. Zhang. *Encyclopedia of atmospheric sciences.* 2nd edition. Academic Press, Oxford, UK.
6. Fleming, J. R. 2020. John Tyndall, Svante Arrhenius, and early research on carbon dioxide and climate. Historical perspectives on climate change. Online edition, Oxford Academic, New York, USA. <https://doi.org/10.1093/oso/9780195078701.003.0011>. Accessed 17 November 2023.
7. Krauss, L. M. 2021. *The physics of climate change.* Post Hill Press, New York, USA.
8. Williams, J. W., and K. D. Burke. 2019. Past abrupt changes in climate and terrestrial ecosystems. Pages 128–141 *in* T. E. Lovejoy and L. Hannah, editors. *Biodiversity and climate change: Transforming the biosphere.* Yale University Press, New Haven, Connecticut, USA.
9. Lewis, S. L., and M. A. Maslin. 2015. A transparent framework for defining the Anthropocene Epoch. *The Anthropocene Review* 2:128–146. DOI: 10.1177/2053019615588792.
10. Schurer, A. P., G. C. Hegerl, J. Luterbacher, et al. 2019. Disentangling the causes of the 1816 European year without a summer. *Environmental Research Letters* 14:094019.
11. Oppenheimer, C. 2011. The last great subsistence crisis in the Western world. Pages 295–319 *in* C. Oppenheimer. *Eruptions that shook the world.* Cambridge University Press, Cambridge, UK. <https://doi.org/10.1017/CBO9780511978012.014>.
12. Mathez, E. A. 2009. *Climate change: The science of global warming and our energy future.* Columbia University Press, New York, USA, 175.
13. Keeling, C. D. 1960. The concentration and isotopic abundances of carbon dioxide in the atmosphere. *Tellus* 2:200–203.
14. Lindsey, R., and L. Dahlman. 2024. Climate change: Global temperature. National Oceanic and Atmospheric Administration. <https://www.climate.gov/news-features/understanding-climate/climate-change-global-temperature>. Accessed 25 March 2024.
15. Steffen, W., J. Rockström, K. Richardson, et al. 2015. Trajectories of the Earth System in the Anthropocene. *Proceedings of the National Academy of Science* 115:8252–8259. <https://doi.org/10.1073/pnas.1810141115>.
16. Stocker, T. F., D. Qin, G.-K. Plattner, et al. 2013. *IPCC, 2013: Climate change 2013: The physical science basis. Contribution of Working Group I to the Fifth Assessment Report of the Intergovernmental Panel on Climate Change.* Cambridge University Press, New York, USA. DOI: 10.1017/CBO9781107415324.
17. Hoegh-Guldberg, O. 2019. Coral reefs: Megadiversity meets unprecedented climate change. Pages 55–65 *in* T. E. Lovejoy and L. J. Hannah, editors. *Biodiversity and climate change: Transforming the biosphere.* Yale University Press, New Haven, Connecticut, USA, 56–57.
18. He, Q., and B. R. Silliman. 2019. Climate change, human impacts, and coastal ecosystems in the Anthropocene. *Current Biology* 29:R1021–R1035.
19. IPCC. 2014. *IPCC, 2014: Climate change 2014: Mitigation of climate change, Volume 3.* Cambridge University Press, Cambridge, UK.
20. MacCracken, M. C. 2019. What is climate change? Pages 12–22 *in* T. E. Lovejoy and L. J. Hannah, editors. *Biodiversity and climate change: Transforming the biosphere.* Yale University Press, New Haven, Connecticut, USA, 26.

21. Garcia, R. A., M. Cabeza, C. Rahbek, et al. 2014. Multiple dimensions of climate change and their implications for biodiversity. *Science* 344, 1247579.
22. Hutchinson, G. E. 1957. Concluding remarks. *Cold Springs Harbor Symposium on Quantitative Biology* 22:415–457.
23. Parmesan, C., and G. Yohe. 2003. A globally coherent fingerprint of climate change impacts across natural systems. *Nature* 421:37–42.
24. Poloczanska, E., C. J. Brown, W. J. Sydeman, et al. 2013. Global imprint of climate change on marine life. *Nature Climate Change* 3:919–925. <https://www.pnas.org/doi/full/10.1073/pnas.1810141115>.
25. Freeman, B. G., M. Strimas-Mackey, and E. T. Miller. 2022. Interspecific competition limits bird species' ranges in tropical mountains. *Science* 377:416–420. <https://doi.org/10.1126/science.abl7242>. PMID:35862538.
26. Louthan, A. M., D. F. Doak, and A. L. Angert. 2015. Where and when do species interactions set range limits? *Trends in Ecology and Evolution* 30:780–792. <https://doi.org/10.1016/j.tree.2015.09.011>. PMID:26525430.
27. Neate-Clegg, M. H. C., and M. W. Tingley. 2023. Building a mechanistic understanding of climate driven elevational shifts in birds. *PLOS Climate 2*, e0000174. <https://doi.org/10.1371/journal.pclm.0000174>.
28. Pounds, J. A., M. P. L. Fogden, and K. I. Masters. 2005. Responses of natural communities to climate change in a highland tropical forest. Pages 70–74 *in* T. E. Lovejoy and L. J. Hannah, editors. *Climate change and biodiversity*. Yale University Press, New Haven, Connecticut, USA, 70–71.
29. MacKenzie, B. R., and D. Schiedek. 2007. Daily ocean monitoring since the 1860s shows record warming of northern European seas. *Global Change Biology* 13:1335–1347.
30. Dulvy, N. K., S. I. Rogers, S. Jennings, et al. 2008. Climate change and deepening of the North Sea fish assemblage. *Journal of Applied Ecology* 45:1029–1039. <https://doi.org/10.1111/j.1365-2664.2008.01488.x>.
31. Post, E., and M. Avery. 2019. Phenological dynamics in pollinator–plant associations related to climate change. Pages 42–54 *in* T. E. Lovejoy and L. J. Hannah, editors. *Biodiversity and climate change: Transforming the biosphere*. Yale University Press, New Haven, Connecticut, USA.
32. Høye, T. T., E. Post, H. Meltofte, et al. 2007. Rapid advancement of spring in the High Arctic. *Current Biology* 17:12, R450.
33. Inouye, D. W., D. Barr, K. B. Armitage, et al. 2000. Climate change is affecting altitudinal migrants and hibernating species. *Proceedings of the US National Academy of Sciences* 97:1630–1633.
34. Calinger, K. M., S. Queenborough, and P. S. Curtis. 2103. Herbarium specimens reveal the footprint of climate change on flowering trends across north-central North America. *Ecology Letters* 16:1037–1044.
35. Forrest, J. R. K., and J. D. Thomson. 2011. An examination of synchrony between insect emergence and flowering in Rocky Mountain meadows. *Ecological Monographs* 81 :469–491. <https://doi.org/10.1890/10-1885.1>.
36. Kleypas, J. A. 2019. Impacts of ocean acidification on marine biodiversity. Pages 185–195 *in* T. E. Lovejoy and L. J. Hannah, editors. *Biodiversity and climate change: Transforming the biosphere*. Yale University Press, New Haven, Connecticut, USA.
37. Scott, D. 2005. Integrating climate change into Canada's national park system. Pages 342–345 *in* T. E. Lovejoy and L. J. Hannah, editors. *Climate change and biodiversity*. Yale University Press, New Haven, Connecticut, USA, 344.
38. Hannah, L., and L. Hansen. 2005. Designing landscapes and seascapes for change. Pages 329–341 *in* T. E. Lovejoy and L. J. Hannah, editors. *Climate change and biodiversity*. Yale University Press, New Haven, Connecticut, USA, 331.

39. Sanford E. 2002. Community responses to climate change: Links between temperature and keystone predation in a rocky intertidal system. Pages 165–200 *in* S. H. Schneider and T. L. Root, editors. *Wildlife responses to climate change: North American studies.* Island Press, Washington, DC, USA.
40. Pearson, R. G., and T. P. Dawson. 2003. Predicting the impacts of climate change on the distribution of species: Are bioclimate envelope models useful? *Global Ecology and Biogeography* 12:361–371, 365.
41. Araújo, M. B., and A. T. Peterson. 2012. Uses and misuses of bioclimate envelope modeling. *Ecology* 93:1527–1539. <https://doi.org/10.1890/11-1930.1>.
42. Walther, G.-R., S. Berger, and M. T. Sykes. 2005. An ecological footprint of climate change. *Proceedings of the Biological Sciences* 272:1427–1432. <https://doi.org/10.1098/rspb.2005.3119>.
43. Hannah, L., G. Midgley, S. Andelman, et al. 2007. Protected area needs in a changing climate. *Frontiers in Ecology and the Environment* 5(3):131–138. <https://doi.org/10.1890/1540-9295(2007)5[131:PANIAC]2.0.CO;2>.
44. Loarie, S. R., P. B. Duffy, H. Hamilton, et al. 2009. The velocity of climate change. *Nature* 462:1052–1055.
45. Marquet, P. A., J. Lessman, and M. R. Shaw. 2019. Protected-area management and climate change. Pages 283–293 *in* T. E. Lovejoy and L. J. Hannah, editors. *Biodiversity and climate change: Transforming the biosphere.* Yale University Press, New Haven, Connecticut, USA.
46. Dawson, T. P., S. T. Jackson, J. I. House, et al. 2011. Beyond predictions: Biodiversity conservation in a changing climate. *Science* 332:53–58.
47. Edgar, G. J., R. D. Stuart-Smith, T. J. Willis, et al. 2014. Global conservation outcomes depend on marine protected areas with five key features. *Nature* 506:216–220.
48. Kulkarni, S. 2022. Reversing climate change with geoengineering. Harvard Graduate School of Arts and Sciences. <https://sgkulkarni.com/Geoengineering/>. Accessed 23 July 2024.
49. National Research Council. 2015. *Climate intervention: Reflecting sunlight to cool Earth.* National Academies Press, Washington, DC, USA. <https://doi.org/10.17226/18988>.
50. Berry, W. 1984. Two economies. *Review & Expositor* 81:209–223, 219. <https://doi.org/10.1177/003463738408100204>.
51. Mackey, B. 2014. The future of conservation: An Australian perspective. Pages 126–136 *in* G. Wuerthner, E. Crist, T. Butler, et al., editors. *Keeping the wild: Against the domestication of the Earth.* Island Press, Washington, DC, USA, 129.

6 Conservation and the Novel Ecosystem

Appearance of the Novel Ecosystem

The ecologist Daniel Botkin remarked, "One can either preserve a 'natural' condition, or one can preserve natural processes, but not both."[1] This is a truth of authentic insight. In the Anthropocene, what happens if "natural" processes are increasingly influenced, even dominated, by human activity, and natural "conditions" are then altered to states that have never been seen before? This is the breeding ground of the "novel ecosystem," which has made its home in the Anthropocene.

The ecologists Richard Hobbs, Eric Higgs, and Carol Hall, pioneers in the study of novel ecosystems, offer this working definition of what a novel ecosystem is – "a system of abiotic, biotic and social components (and their interactions) that, by virtue of human influence, differ from those that prevailed historically, having a tendency to self-organize and manifest novel qualities without intensive human management"[2] and one that "consists of new combinations of species that have not previously coexisted, and/or new configurations of environmental factors such as changed climate or altered soil properties."[3] Its species occur in combinations and relative abundances that have not occurred previously within a given biome, and such novelty creates the potential for changes in ecosystem functioning.

The first use of the term "novel ecosystem" in peer-reviewed scientific literature occurred in the landmark paper, "Time lags and novel ecosystems in response to climate change in Arctic Alaska," by Stuart Chapin and Anthony Starfield, published in 1997 in the journal *Climate Change*.[4] Here Chapin and Starfield described an ecological simulation model they created to predict the future progress of an already ongoing phenomenon – the invasion of the Arctic tundra by trees (especially aspen, *Populus* spp.) due to climate warming. This invasion was not simply a change in species composition, but a change in the ecosystem from treeless tundra to a new and formerly non-existent system the authors described as "grassland-steppe." Chapin and Starfield grasped, if imperfectly, the significance of their findings in the words of their summary:

> Perhaps the most unexpected result of our simulations is the conclusion that climatic warming, in the absence of an increase in precipitation, will lead to the widespread distribution of grassland-steppe or other novel drought-resistant vegetation type that is currently rare or absent in the landscape.[5]

The Chapin and Starfield study proved to be only the first of many that documented ongoing change in ecosystem types and properties, not from one traditional ecosystem

DOI: 10.4324/9781003404415-6

category to another, but from a historic "natural" ecosystem to one with unique characteristics not previously encountered. The phenomenon became so frequent that it demanded the development of new terms and definitions. But such findings were not unpredicted. Prophetically, Michael Soulé, the principal "founder" of the discipline of conservation biology, foretold in 1991 that

> In 2100, entire biotas will have been assembled from (1) remnant and reintroduced natives, (2) partly or completely engineered species, and (3) introduced exotic species. The term natural will disappear from our working vocabulary. The term is already meaningless in most parts of the world because anthropogenic [activities] have been changing the physical and biological environment for centuries, if not millennia.[6]

Defining the Novel Ecosystem

The Role of Human Agency

Novel ecosystems may be novel, but they are no longer rare. Today's world is full of them, and their number is growing. In taking up the reality of novel ecosystems, and how conservation will respond to them, we are entrenched in the reality of climate change. Timescale matters. When it comes to climate change, the Earth has a long history of it. There has never been a time when (if you make the time scale long enough) the Earth's climate was *not* changing. In the same way, if we enlarge our time scale, every ecosystem is a novel ecosystem because few or no ecosystems today have the same species, ecological interactions, or distributions they did 12,000 years ago. Like climate change, the formation of new species assemblages and interactions is not new. The *speed* at which such new assemblages and interactions are forming *is* new, and so is the agent of such formation – the human disturbance of the Earth and its planetary systems.

Novel ecosystems represent an ecosystem-level adaptation to climate change, an ecosystem-level adaptive response to other kinds of anthropogenic disturbances, or both. As ecologist Joseph Mascaro and his colleagues note, "Novel ecosystems are not prescribed (designed) by humans but are the *response* of the biosphere to human influence"[7] (emphasis authors). All novel ecosystems arising today are the outcomes of human actions in which: 1) humans introduce individuals and populations of species into areas beyond their native range; 2) native species in a system are driven to extinction by humans, changing system interactions which alter available niches for other species; 3) humans erect barriers to species dispersal that prevent the re-establishment of former ("traditional") associations of species in areas within the species' range; and 4) humans alter climate and other environmental factors that have determined species distributions, provoking shifts in species ranges. Current estimates assess that between 28% and 36% of the Earth's ice-free surface is now covered by these novel systems.[8, 9]

Novel ecosystems are also distinguished by their origin, which is invariably rooted in human agency. Human action can create novel ecosystems in four ways (see Figure 6.1). Perhaps the most common is unintentional alteration (humans engaged in activities of which they did not understand the environmental effects) or intentional degradation (degrading the ecological features of a landscape to create human benefit, such as converting the complexity of species in a native prairie to the monoculture of a corn field). Humans may also generate a novel ecosystem by manipulating the system to maximize human

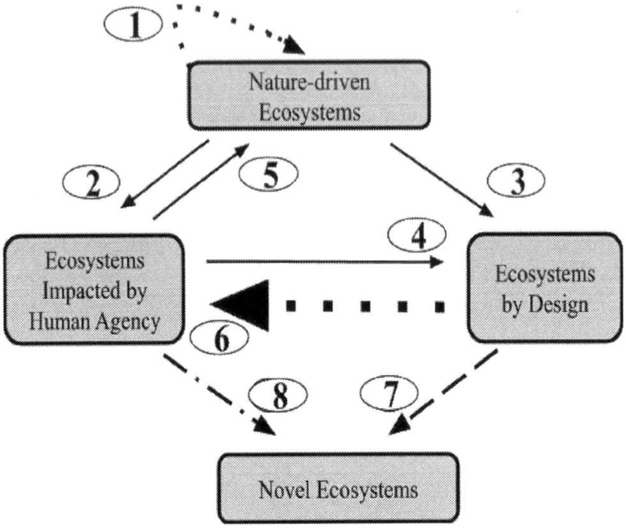

Figure 6.1 Human agency provokes the creation and maintenance of novel ecosystems. Arrow 1 designates natural ecosystem processes, but humans alter ecosystems through unintentional or intentional habitat alteration (Arrows 2 and 5), alteration of ecosystem services for human benefit (Arrow 3), environmental remediation or restoration of ecosystem services (Arrow 4), or full ecosystem restoration (Arrow 5). Arrows 6, 7, and 8 show pathways from designed (restored) or impacted ecosystems to novel ecosystems. After Morse et al. 2014. Used with permission from N. Morse.

gains from ecosystem services, such as damming a river to generate more water for irrigation in a local area than the flowing river would have provided. With further manipulation, humans may create a designed system that remediates a degraded system to provide ecosystem services that had been lost, such as planting trees on a vacated oil-drilling pad in a forest to restore wildlife habitat and mitigate climate warming. Finally, the altered system may transition to a natural state, either because the system still maintains the capacity to do so (no "thresholds" were crossed) or because humans actively restore the site through additional effort, such as planting grasses and wildflowers in an abandoned corn field to re-create a prairie ecosystem.[10]

There are two major pathways, working separately or together, through which human agency can stimulate the development of a novel ecosystem (see Figure 6.2). One can happen when human actions alter environmental or ecosystem conditions. These circumstances are particularly noticeable in urban and industrial settings, but they can also occur in agricultural or pastoral landscapes, in heavily used "natural" landscapes (such as intensively logged forests), or in aquatic ecosystems subject to systemic and ongoing human alteration (withdrawing water from a river to irrigate crops). A second, and not mutually exclusive, pathway develops when human actions drive changes in an ecosystem that require rapid adaptation by members of the biotic community.[10] These changes may alter plant and animal distribution, disrupt functions and behaviors in plants and animals, and redistribute the species composition of both plant and animal communities. All such changes can create a novel ecosystem and any combination of these can produce one, but all belong uniquely to the Anthropocene – an age of humans altering ecosystems.

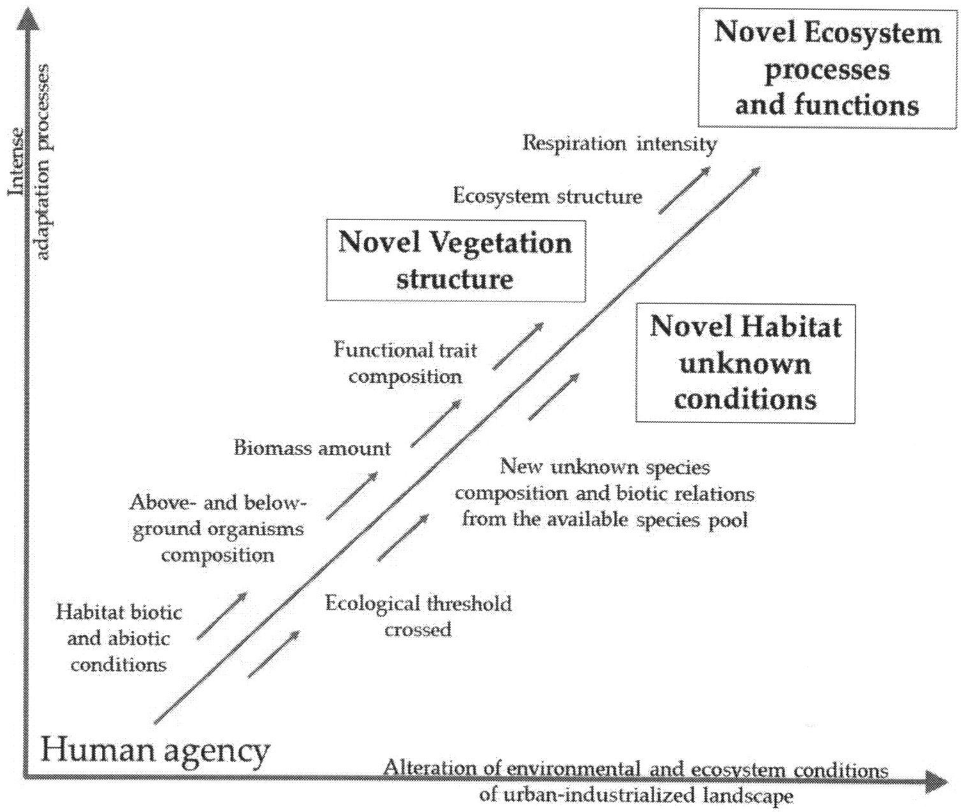

Figure 6.2 Pathways through which human agency (x-axis) creates novel ecosystems through landscape modification, or by engaging in actions (logging a forest) or producing systemic effects (climate change) which require rapid responses and intense adaptation processes (y-axis) of species that change the behaviors, functions, distributions and composition of the biotic community.

Ecological Thresholds, Species Composition, and Persistence

A novel ecosystem is further distinguished by the ecological "thresholds" it has crossed. That is, its changes in ecological processes, species interactions, and species composition have reached a point of "no return" and cannot be easily "restored" to historical conditions they might in some ways resemble or have replaced on a site. The species in the novel ecosystem are now operating under biophysical conditions and selection pressures different from those that existed prior to the human disturbance that initiated the development of the new system. A novel ecosystem is also one that has become self-sustaining. Although initiated by human action, it can now persist without it. It has become ecologically "self-sufficient," having developed populations, communities, and landscapes that continue without human intervention.

There are real novel ecosystems in the Anthropocene, and the conservation community does itself no favors by ignoring them. But how do these systems perform compared to historically similar systems, what makes them different, and what is it that they can be managed "for" (that is, what good are they – what species use them)? But first, what principles underlie the genesis of such systems?

The origin of novel ecosystems is rooted in some of the most foundational ecological laws of community assembly and composition. Far from "breaking" such laws, novel ecosystems follow them closely and predictably. In the 1940s, the ecologist Henry Gleason began to develop the "open" community concept in ecology. Gleason's views flew in the face of the established notion of ecological communities as stable and predictable "associations" or "formations" of species that were mutually dependent upon one another and functioned almost as a single unit or "ecological organism." But with intellectual perseverance and increasing empirical support, Gleason ultimately demonstrated, through his own studies and others, that associations of plants in communities were opportunistic relations of species with independent niches and environmental tolerances.[11] As Gleason expressed it, "It may be said that every species of plant is a law unto itself, the distribution of which in space depends upon its individual peculiarities of migration and environmental requirements."[12] Applying Gleason's principles to novel ecosystems, Mascaro et al. note that "Because the flora and fauna move independently of one another in response to their environment, communities are constantly in flux at some temporal resolution. Without this property of the biosphere, novel ecosystems would not be possible."[13]

It is also necessary to understand that, in novel ecosystems, as ecologists James Harris and his colleagues put it "the biotic and abiotic characteristics of an ecosystem are tethered [tied together]."[14] Although the associations are independent of one another, once formed, the interactions they produce connect the biological characteristics of the organisms to the biotic and abiotic processes of the ecosystem. These interactions make the new community a truly novel *ecosystem*, not simply a novel *association* of species. "Introduced species," note Mascaro et al., "are not static additions to the roster of ecological communities; they bring particular physiologies and life histories to bear on their environment."[7] Humans, like past massive disturbances to the biosphere, "cause changes in ecosystems that are directional and permanent"[15] and affect these associations, producing adaptive responses in plants and their communities that may reflect new associations that can persist in these altered conditions.

If a novel ecosystem is truly "novel," that novelty is a measurable attribute of ecosystem elements that are "different" from their predecessors. But how can we objectively detect "novelty?" There are three essential measurable features: 1) differences in ecosystem composition, structure, or function; 2) thresholds in these attributes that are currently irreversible; and 3) persistence or self-organization of the novel ecosystem over time.[2] Today the Earth's ecosystems are subjected to change more rapidly than ever before. As ecosystems are pushed into new arrays of species and new ensembles of relationships between them, the question for conservationists is not "Are novel ecosystems really there?" but rather "What do we do with them?" In the rapid, ongoing changes of the Anthropocene, it is little help (and shallow thinking) to simply proclaim that today "everything is novel." Just how novel is it? At what point should a conservation manager intervene in the development of a novel ecosystem, and to what end?

Restoration, Design, and Novelty

The Decade of Ecosystem Restoration

In 2020, the United Nations (UN) declared that 2021–2030 would be officially designated as the "Decade on Ecosystem Restoration." Through the United Nations Environmental Programme (UNEP), the UN identified ten "flagship initiatives" to illustrate the breadth and promise of restoration work that was already underway. These efforts aimed to restore more than 60 million hectares – an area about the size of Ukraine. But these initiatives, though impressive, were only a part of the global "fleet" of restoration efforts taking place all over the world. In recognition and support of this effort, the UN Development Programme's Learning for Nature curriculum created an "Open Online Course" in Ecological Restoration that enrolled nearly 17,000 students from around the globe.[15] In that offering, I worked as one of the course's "Discussion Facilitators," interacting with hundreds of "students" from every (ice-free) continent. I put "students" in quotes to indicate that, although that was their status in the course, I found myself conversing with researchers, practicing restoration ecologists, conservation activists, and others with extensive experience in ecological restoration. But I also discovered that ecological restoration very quickly encounters the need to answer difficult questions. Namely, where is the line between "traditional restoration" of systems disturbed by human activity intended to be returned to some historic "standard" or "baseline," and the "development" of new management plans for disturbed sites that are clearly transforming into *entirely new systems* that could not be easily (or advisably) returned to past ecological conditions?

The Forests of Hawaii and the Forests of Kelp

Today novel ecosystems are becoming a part of humanity's perception of "nature" as well as a subject of legitimate concern. But do novel ecosystems really "work?" Can these emerging associations of species and their interactions carry out the processes and functions that "real" ecosystems perform, or is calling something a "novel ecosystem" just a label that sounds better than "ruined and invasive species-infested landscape?" There are thousands of interactions, processes, and "services" provided by natural ecosystems. It will be some time before ecologists have thoroughly investigated all of these in novel systems, but some have been studied.

Mascaro et al. put the question of the ecological "functionality" of novel ecosystems to the test in their examination of ecosystem function and species diversity in novel forests on Hawai'i Island, part of the US state of Hawaii, a group of volcanic islands in the Pacific Ocean. Hawaii has higher regional plant richness than it did historically. It has experienced more than 1,000 plant species introductions and only about 71 known plant extinctions, resulting in a net increase in species richness of 100%.[16] As such, Hawaii is an ideal incubator for the breeding of novel ecosystems, many of which manifest as novel associations of non-native forest trees. Examining these systems, Mascaro et al. found that novel forests had more tree species richness and a higher diversity of dominant tree species, possessed greater levels of above-ground biomass, and had higher rates of productivity, nutrient turnover, and below-ground carbon storage compared to native forests. Mascaro et al. noted that

> Our results ... suggest basic ecosystem processes will continue even after dramatic losses of native species diversity if simple functional roles are provided by introduced species ... ecosystem function in novel forests would meet or exceed levels found in native forests in terms of aboveground biomass and productivity ..., nutrient turnover and below ground carbon storage.[17]

The Hawaiian forests are not a singular case. Functioning novel ecosystems can be found in marine environments as well as terrestrial ones. Novel associations of warm-temperate species of kelp in marine environments have led to the establishment of warm-temperate species associations that have higher densities of invertebrates than cold-temperate associations, and which, in terms of ecosystem function, have almost twice the rate of net primary productivity and rates of detritus composition 6.5 times faster than cold-temperate kelps. Such an increase in detritus composition rates decreases the residence time of detritus in the system, providing a larger pool of organic matter entering food webs of detritivores (detritus-feeding invertebrates) (see Figure 6.3).[18] Such findings in these two studies and others not only suggest that novel ecosystems exist, but some may be able to perform their ecological functions at least as well as the traditional systems they replace.

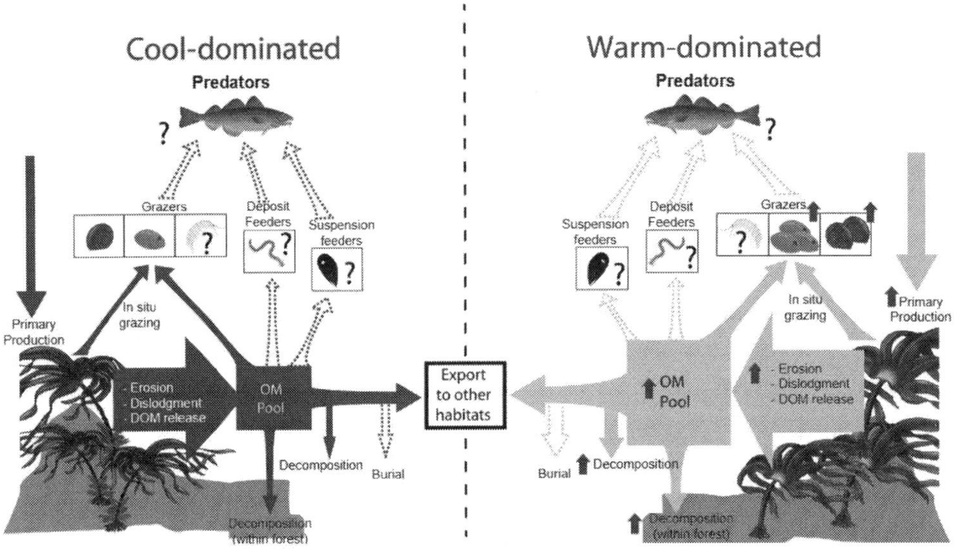

Figure 6.3 Expected fluxes of kelp organic matter as some kelp forests transition from cool- to warm-temperate foundation species. Solid arrows indicate changes in the processes measured; question marks and dashed, empty arrows denote mechanisms and pathways that were not evaluated. Warm-temperate kelps produced larger amounts of detritus, thus creating a large pool of organic matter (OM) either (a) consumed by suspension and/or deposit feeders, (b) decomposed by microbes, or (c) exported to other habitats. The kelp organic matter assimilated by primary consumers may flow to higher trophic levels via predation.

The Novel Forests of Puerto Rico

What is today the US Commonwealth of Puerto Rico was first the home of the Taíno, whose ancestors had settled on the island sometime between the first and fourth centuries. Ignoring the tenure of its Indigenous Peoples, Christopher Columbus "claimed" the island for Spain in 1493. Spanish colonial settlements began in 1508 and widespread colonization had commenced by the 1600s, and with it increasing deforestation, agricultural development, and land degradation.[19] Most of the island's 890,000 hectares remained covered by native dry forest ecosystems into the eighteenth century, but persistent forest removal, especially through the conversion of forests to pastureland, had reduced the total forested area to 81,000 hectares by the early 1930s, only 9% of the original forest cover.[20] Further conversions of forests to sugar cane, coffee, and tobacco plantations increased rates of deforestation, exacerbated by forest fragmentation caused by increasing road densities. By the 1990s, the Puerto Rican landscape held over 20,000 forest fragments, most with patch sizes of 1 hectare or less.[21]

From this point, agricultural land began to be abandoned in favor of other occupations and less land-intensive forms of economic development. Abandoned land began to reforest, and the coverage of forests in Puerto Rico had increased to 57% by 2013.[19] Such reforestation proceeded with little human intervention or design, and the "natural" processes driving reforestation did not produce forests of native trees. Today 75% of forests in Puerto Rico are "novel forests."[22, 23, 24] Compared to remnants of historic forests, the canopies of most forests in Puerto Rico are now dominated by introduced species,[25] although forest understories remain comprised of mostly native plants.[26] Introduced and naturalized tree species presently comprise 23% of forest flora. These non-native species grow in association with native trees and are becoming permanent and persistent components of the "new" Puerto Rican forests.[27] Structurally, the novel forests are similar to native forests, with structural attributes such as basal area, canopy height, tree density, and wood volume similar to those of the native forests they have replaced.[28, 29] Their net primary productivity, biomass accumulation, and nutrient fluxes are greater than those of the native forests,[28] and their species richness (trees per hectare) is also higher.[29] Vertebrate animals that use the novel forests are mostly native species,[30] but invertebrate communities consist of both native and introduced species.[31] The native trees, which cannot germinate and grow in habitats altered by human actions (such as degraded pasture lands), *can* germinate and grow in the new (novel ecosystem) forest of mostly introduced tree species.[19]

Ecologist Ariel Lugo, who has studied these Puerto Rican forests and other novel tropical forests extensively, has predicted that the future development of novel tropical forests like those in Puerto Rico will show four patterns: 1) novel forest types will emerge after disturbances within complex and diverse landscapes; 2) such forests will self-assemble and self-organize with novel combinations and proportions of plant and animal species; 3) novel forests will function similarly to current native forests in their delivery of vital ecological services to people; and 4) historically native species communities are unlikely to replace novel communities in the future because environmental conditions continue to change and the historical conditions that once favored traditional native species are not likely to persist.[19]

Like the novel Hawaiian forest ecosystems, these novel Puerto Rican forests "work" at an ecological level as well as the native forest systems. They clearly meet three of the most important criteria Hobbs and others have offered to distinguish novel ecosystems: 1) they arise as a result of human actions upon and disturbances of natural systems; 2) they are self-organizing, without human design or effort; and 3) they are persistent and self-sustaining.[2] The Puerto Rican forests do not merely tell us that novel ecosystems *are* persistent and self-sustaining, but *why* they are persistent and self-sustaining. Namely, the pre-existing native species groups are unlikely to replace the novel ones *because the naturalized introduced species regenerate under the prevailing (altered) environmental conditions, while the native species do not.* Thus, the novel forest system will remain self-sustaining precisely because humans have created conditions that favor its persistence. Although these novel forest ecosystems contain different plant species, they conserve and maintain native species of both plants and animals. In other words, they have conservation value for historic species, although they conserve them in a novel ecological context.

The Indicators and the Limits

The ecosystem performance of the novel forests of Hawaii and Puerto Rico, as well as kelp forests in marine systems, is not exceptional. Approximately 70% of studies of novel ecosystems have reported biological richness and diversity equivalent to or exceeding reference baselines.[32] Although they may replace "natural" systems, novel systems may themselves be the "natural" response of flora and fauna to human-induced changes. Novel ecosystems are the products of changes in nature by humans and remain because ongoing human disturbance causes such changes to persist. Novel ecosystems, then, are not just "new" ecosystems to be studied as interesting individual cases, but *systems that demonstrate that human activity and behavior must be incorporated into ecological theory and explanation*. Humans are not merely part of the ecological process in some "abstract" way, but *specific actors determining the development of ecosystems in new ways* through the conditions they create and persistently renew.

If novel ecosystems can perform ecological functions as well as or better than native systems, they should also be as amenable to principles and practices of ecosystem management as those native systems, although their management might require different approaches. Traditionally, restoration efforts have begun with careful assessment of past conditions to establish a reference ("baseline") that would provide the management target for the restoration endeavor and serve as the measure of its success. Restoration ecologists would then define a "reference model" informed by a real "reference system" (an existing ecosystem believed to represent the historic and "natural" system) and then use the presence or absence of abiotic and biotic attributes of that system (species diversity, nutrient recycling, rates of decomposition) to determine the "intactness" of that reference model (how closely did the reference system conform to the "true" natural ecosystem?). These attributes would then frame the goals of the restoration and determine which goals should be given the greatest emphasis.[15]

This is a good plan, and it begins with practices and protocols developed, endorsed, and used by restoration ecologists worldwide. But in a world now dominated by human action and effect, *traditional approaches to restoration have limits in their effectiveness*. The first limit is that, as valid, authentic, and well-researched as the historical baseline may be, the reference ecosystem may have developed under environmental conditions very different from present ones now operating in the same area (or even at the same site). Thus, *the reference baseline may be inappropriate to the new environment, or even unattainable in it*. The second limit is that the restored system may face the pressures of human actions – whether presence, use, or residence – that the reference site did not. Even if the restored system can approximate the reference system (it can never fully replicate it), it may not be sustainable under such pressures. A third limitation is that not only may environmental conditions be different for the new system than for the reference system, but even those current conditions may be in a state of rapid change. This means that restoring the system to one that can sustain itself must consider not only differences *between past and present conditions*, but differences *between present and future conditions*. Ecological restoration has become a moving target.

Such limitations do not mean that restoration should not be done, but is such an effort an exercise in ecological *replication* or ecological *design*? Ecologists Michael Perring, Rachel Standish, and Richard Hobbs declare that

> Novel ecosystems are not innovative planning. They simply are: they do not arise through deliberate planning, innovative or otherwise ... Today we face conditions in

> which design may be an appropriate, even a necessary, element of ecological restoration. And just as design has always been a component of historical restoration efforts, it can and should increasingly inform efforts associated with novel ecosystems.[33]

Thus, Perring et al. continue, "This basic (value-laden) tenet of restoration – i.e., that it is desirable to return to the historical reference – is increasingly questioned in an era of multiple, simultaneous, and rapid environmental changes."[33]

Novel Ecosystems and Designed Ecosystems

Many restoration ecologists assert that novel ecosystems and designed ecosystems are two different things, distinguished in three ways. First, designed ecosystems require intensive and repeated intervention to create and sustain. Novel ecosystems, in contrast, arise through initial, sometimes inadvertent, human disturbance, but then develop over time to form new, stable conditions in response to new mixes of species and environmental conditions. Second, designed ecosystems and novel ecosystems have different intentions. Designed ecosystems are just that –designed to serve human interests, including the provision of specific services valued by people. Novel ecosystems arise from human actions but retain value as relatively unmanaged or wild ecosystems. Third, designed and novel ecosystems have different trajectories of development.[34] Restoration ecologist Eric Higgs has said,

> Novel ecosystems can be distinguished from designed ecosystems in several ways: in their origin, intention, and developmental pathways ... Restoration straddles the self-assembled and designed ecosystem divide. As natural ecosystems shift, and interest increases in designing ecosystems to reflect specific functions, restoration must necessarily adapt to ensure it reflects classical restoration (restore to an historical trajectory), expanded forms of restoration ... and new contributions in restoring functional attributes of novel ecosystems and the design specifications for designed ecosystems.[35]

Degradation of historically continuous ecosystems is the starting point for novel ecosystems, which arise through natural processes and proceed to become self-assembled ecosystems. In contrast, designed ecosystems are conceived from their inception with positive value (usually for humans) and normally intended to be part of a larger landscape.[36]

Some shortcomings associated with the limits of restoration will not only have to be faced and accepted in their outcomes, but intentionally planned into and managed with them. In this truth, many who embrace conditions of the Anthropocene will find support for their affirmation that "nature is us" – that human intention and design will determine what nature will be. And herein lies the *greatest danger in modern restoration ecology.* If a designed ecosystem can approach the variety of species and functions characteristic of the historic "baseline" system, it can also be intentionally designed to reflect a novel ecosystem. However, the *power* to design the system provides no insight as to the *purpose* of that design. A purpose-driven design will be based on what people *value* in nature. This means that the values and judgments of people involved in the purpose-determining process become even more critical than in traditional restoration efforts. There is not only more choice about how to achieve the goal, but also the much more important – and potentially dangerous – choice of what goal to set.[37] But what kinds of novel ecosystems have humans designed (so far), what were their purposes, and what values were advanced in the effort?

Novelty by Intention – An Ecosystem Designed for Climate Change

Near the city of Boulder (population 104,000 in 2019) in the US state of Colorado, gravel had been mined from a 250-acre (100-hectare) land area originally covered by tallgrass prairie, a natural but now rare ecosystem throughout the United States. After gravel mining ceased in the early 1990s, native grasses were seeded onto the site. Following higher than normal precipitation in the seeding year, the site experienced three consecutive years of below normal precipitation, an increasingly common manifestation of climate change in the western and intermountain United States.

When gravel mining was initiated, soil removed from the site was stockpiled with the intention of replacing it after mining had been completed. However, soil stockpiled in this way, especially for a long time, changes in chemistry and structure. In this case, the "returned" soil was lower in organic matter, moisture, and organic nitrogen than had been characteristic of the original soil – less like soil typical of a tallgrass prairie and more like soil associated with arid steppes.[38]

Seeds sown on the restored site were all species native to the original site, but, recognizing that the soil had changed, managers intentionally planted some additional native species with wider environmental tolerances and greater resistance to drought than species uniquely associated with tallgrass prairie. The plant community that ultimately emerged on the old gravel pit included more drought-resistant species, especially one warm-season grass, salt sacaton (*Sporobulus airoides*).[39] Also known as alkali sacaton, the first name (alkali) reflects its ability to tolerate a wide range of acidic conditions in soil, including high pH levels (alkaline soils). Its alias, salt sacaton, attests to its capacity to thrive in dry, saline (salt-laden) soils. Alkali (salt) sacaton and other warm-season grasses typical of more arid mixed-grass and shortgrass steppes, rather than tallgrass prairie, eventually contributed over 90% of the restored gravel mine's plant coverage, and these resisted invasions by other species, both native and non-native.[39] As restoration ecologists Timothy Seastedt, Richard Hobbs, and Katherina Suding, who planned and executed this restoration, noted in retrospect,

> This successful revegetation effort resulted from the selection of a seed source containing representatives of short-, mixed-, and tallgrass species ... The diverse seed mix used by the revegetation team interacted with climate conditions and unusual soil characteristics to select the community type and produced an impressive, albeit novel, grassland community.[40]

Without such design, the gravel pit's restoration (or, perhaps, reclamation) would have failed.

What's the (Species') Use of Novel Ecosystems?

Many conservation efforts today feature animal species. Do animal species use novel ecosystems? Can they even thrive in them? Sara Martínez-Hesterkamp and her colleagues examined this question – do animals use novel ecosystems? – in diurnal raptors.[41] Like many predators, raptors are territorial and their territory *size* is often a good index of habitat quality. The higher the prey density, the smaller the area the raptor needs to sustain itself and its offspring. In Martínez-Hesterkamp et al.'s study, northern goshawk (*Accipiter gentilis*), Eurasian sparrowhawk (*A. nisus*), and common buzzard (*Buteo buteo*) were residing in a novel forest ecosystem in north-western Spain dominated by exotic tree species like Tasmanian blue gum (*Eucalyptus globulus*) and other non-native tree species that had become established, even dominant, in this area. Much of the land occupied by such forests had formerly been cultivated as Eucalyptus plantations by private landowners who harvested the trees at intervals (rotations) of varying lengths (years).[41]

Using extensive data collected on the territory sizes of these raptor species, Martínez-Hesterkamp and her colleagues compared them to the size of territories of the same species in other areas. More specifically, they focused not on the absolute size of the territory, but on the average distance of one raptor to another raptor of the same species (for example, goshawk–goshawk distance) as well as the average distance between nests of the same species.[41] In these forests, nest-to-nest distances for Eurasian sparrowhawks and common buzzards were similar to those documented in other studies in which these species were nesting in forests of native trees, but, for the northern goshawk, the largest of the three raptors, nest-to-nest distances in the Eucalyptus forests were among the shortest (smallest) ever recorded. All three species showed nesting location preference for more mature, more developed, and structurally more complex stands of Eucalyptus, rather than younger stands of Eucalyptus or remnant stands of native trees.[41] More complex forest structure has been preferred by the same species in other studies, probably, as those studies have demonstrated, because the more complex structure conceals raptor nests more effectively from predators and disturbance by humans.[42]

The northern goshawk is the world's most widely distributed accipiter (short, broad-winged, long-legged hawks adapted for fast flight and high maneuverability in forests) and is found throughout Europe and in northern Asia throughout the Himalayas and Japan, and similarly widespread in North America. Considered "sensitive to change" in habitat quality, goshawk abundance is often used as an indicator of habitat change as well as habitat quality.[43] The US Forest Service names the goshawk as one of only three animals designated as Management Indicator Species (MIS) for specific habitat types or conditions, with the goshawk designated an MIS of forests in mature condition.[44] Studies like those of Martínez-Hesterkamp et al. suggest that novel ecosystems can support wildlife populations with both average- and high-quality habitat, even for habitat-sensitive species like northern goshawk.

Creating and Managing Novel Ecosystems for Conservation

Sand Mining, Least Terns, and Piping Plovers

In the United States, two bird species of conservation concern, the Interior least tern (*Sternula antillarum athalassos*) (a subspecies of the least tern) and the piping plover (*Charadrius melodus*), have made use of novel ecosystems created by human activity. The least tern is widely distributed throughout North and South America, and most of its coastal populations are secure. However, one subspecies, the Interior least tern, was resident on the US Endangered Species List until 2021, when, after years of protection, it was removed.[45] The piping plover remains a federally protected species, and has (sadly) managed to gain the status of being both Threatened and Endangered because it is distributed in three regional populations – Northern Great Plains (Threatened), Great Lake (Endangered), and Atlantic Coast (Threatened).[46] The plover is a small sand-colored shorebird that raises its young on beaches of the US Great Lakes and US Atlantic coast, and on sand bars along rivers in the north-central United States, like the Platte River in

Figure 6.4 The piping plover (*Charadrius melodus*), a shorebird of sandy beaches on both coastal and inland (riverine) waterways in North America, whose regional populations have been threatened or endangered by excessive human use and alteration of its beach habitat.

the state of Nebraska (see Figure 6.4). It could be said to "nest" in these habitats, if we use the term loosely.

To attract a female, the male plover makes multiple "scrapes" (shallow depressions) in the sand. If the female finds one of the scrapes suitable, she will accept both the male and the scrape, then line the depression with small shells and stones to camouflage it (most expertly) so that it is almost impossible to discern on a beach or sand bar. Once hatched, the precocious young can walk, run, and feed themselves within hours, but nevertheless receive extended care, attention, and protection from both parents until they can fly.[47]

I have wonderful memories of watching piping plovers, young and adults, on Lake Michigan beaches in the Sleeping Bear Dunes National Lakeshore. Here the plovers received special protection – a tall mesh wire fence around the nest site and surrounding area, perhaps 20 by 20 meters square. It was an effective deterrent to dogs, the most common predator of plover eggs on these beaches, just in case any beach visitors missed the sign as they entered the beach that included the admonition "No Dogs Allowed." The tiny birds are marvelously quick – never appearing to "run," but walking with a gait speed that would win an Olympic gold medal. No wonder some have described plovers as resembling small windup toys, but ones which never "wind down." Fences and signs have helped plovers. They have increased in number, but remain under federal protection (and constant monitoring).

In the Northern Great Plains population, plovers and terns have long nested on sand bars along central Nebraska's Platte, Loup, and Elkhorn Rivers. The development of the bars is facilitated by the fact that these rivers are shallow with variable seasonal flows. Because such conditions create repeated locations of sand and gravel deposition within the river's channel, especially along the Platte, sand and gravel mining have long been human industries developed close to these sources. Mining companies haul the sand and gravel from the rivers to sites where both are processed and prepared for shipment. The "mines" which do this work are usually only 1 to 4 km (0.6 to 2.5 miles) from their source of sand and gravel in the river channel.[48]

In processing the sand and gravel, the mining companies create novel ecosystems. Traditionally mines stored excess or unsuitable ("waste") sand in "plumes" (manmade sand bars) along the edge of one or more lakes created as the mine excavated more sand from the ground to supplement the sand and gravel taken from the river. Such excavations often struck the water table, and the pit then filled with groundwater, along with precipitation and associated surface runoff, forming "mine pit lakes." Together, these activities formed relatively large expanses of sandy shorelines close to the rivers from which the sand and gravel came. The plovers and terns, discovering such inviting beaches, use them as nesting habitat.

Although traditional mining practices have created habitat for both plovers and terns, modern mining practices do not. Today most sand and gravel mines store waste sand only in very limited quantities along the edge of the pit lake. They deposit the rest in large piles further away, with no access or connection to water (no sandy shoreline, just a big pile of sand). Older mines also extended the suitability of habitat for piping plovers because, when mining operations were phased out, the mine areas were often adapted in preparation for housing development. The pit lake and its surrounding beach were retained as "scenery" and the houses built around it. Such transitions often provided as much or more nesting for plovers and terns as the original mine site (see Figure 6.5),[48] but that is no longer the case. Modern mines do not make these transitions.

As traditional mining practices have declined, so has the habitat they once created. The benefits from sand mining were an unintended, but cost-free, consequence. As Joel Jorgensen of the Nebraska Game and Parks Commission and his colleagues, who have long studied these systems, explain,

> The local and regional benefits that off-river sites in LPRS [Lower Platte River Systems] provided to plovers and terns over several decades have been a by-product of industrial and commercial processes. Habitat creation and maintenance have occurred with no planning and essentially no cost to government agencies and conservation organization.[48]

Jorgensen et al. note that piping plovers and least terns "are expected to nest in our study sites for the foreseeable future, but overall numbers are expected to be smaller than what has been observed in previous decades."[48]

Sometimes it is better to be lucky than good, or even unplanned than planned, but neither is a good strategy for success. Instead, suppose that a novel ecosystem, unplanned (or poorly planned), without conservation intention could be incorporated into a larger conservation effort around it. Could that novel ecosystem add value to the conservation endeavor? Two conservation scientists, Thabang Sibiya and Izak Smit, faced this problem when a novel ecosystem became incorporated into South Africa's Garden Route National Park (GRNP).

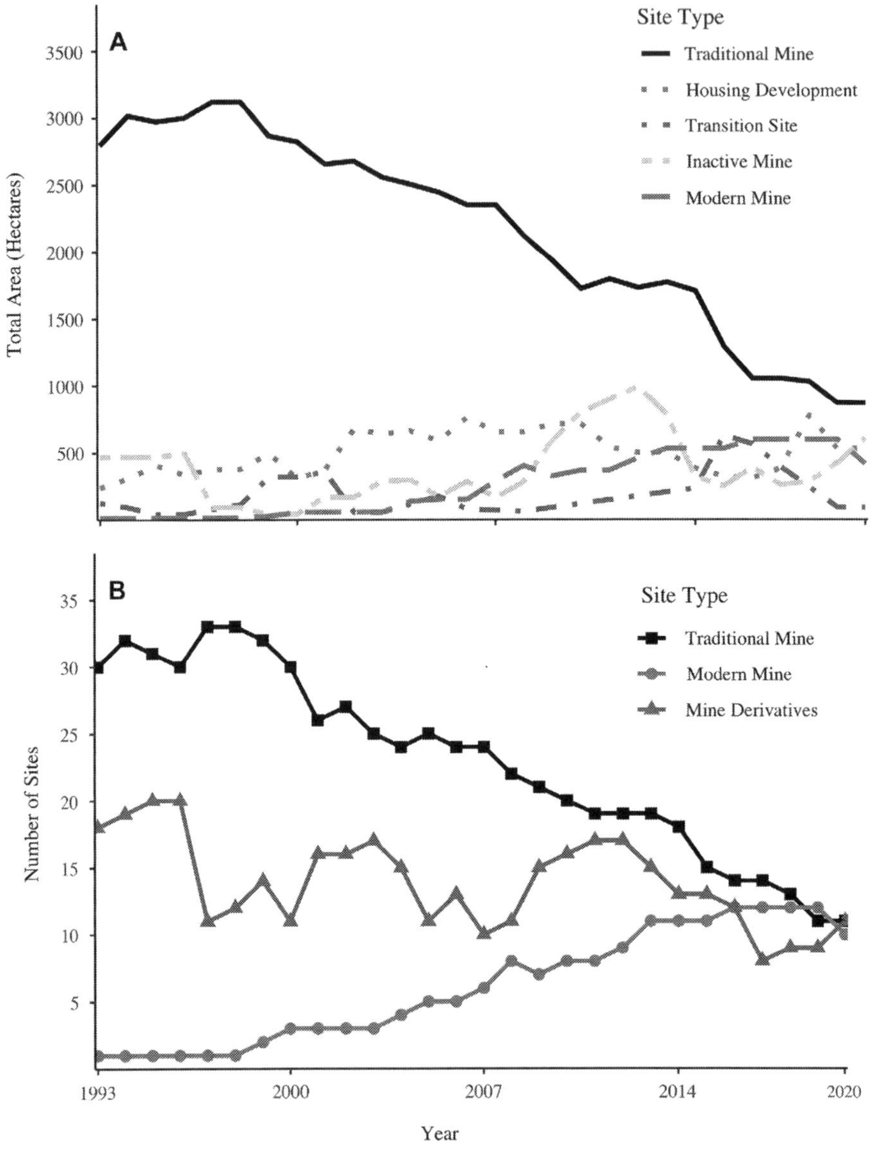

Figure 6.5 (A) Total area of off-river nesting sites for piping plovers and least terns by site type in the lower Platte River system from 1993 to 2020. (B) Total number of traditional mines (squares), modern mines (circles), and traditional mine derivatives (triangles; inactive mines, housing developments, and transition sites) by year from 1993 to 2020. After Jorgensen et al. 2021. Reprinted with permission from *Ecosphere*.

Story Box 6.1 Confronting the Anthropocene: Rehabilitating South Africa's Transformed Fynbos Biome

Thabang E. Sibiya and Izak P. J. Smit

South Africa National Parks

Situated in the Western and Eastern Cape provinces of South Africa, the Garden Route National Park (GRNP) is a coastal reserve renowned for its indigenous forests, dramatic coastline and unique fynbos plant communities. The "fynbos" biome is a natural shrubland and heathland vegetation community that contains one of the greatest concentrations of endemic (and endangered) plant species on Earth. However, in recent years that park has incorporated 18,000 hectares of clear-felled pine plantations within its boundaries. Now, in contrast to the historic plant biodiversity of the fynbos, GRNP scientists were directed to manage what many had described as a "transformed wasteland of little conservation value."

The incorporated areas had initially been densely cultivated with monoculture pine plantations. Such pine stands were originally created to meet South Africa's timber demands and alleviate strain on limited and over-exploited indigenous forests. Over the years, afforestation (planting trees on formerly non-forested land) became a means of addressing post world war unemployment and poverty, resulting in the expansion of plantations into marginal areas of slower growth and smaller yields. Such expansion in turn rendered the pine plantations financially unsustainable, necessitating substantial government subsidies as management costs exceeded returns from timber sales. Furthermore, the introduced pines had a higher water demand compared to native species, exacerbating water scarcity across the entire ecosystem. Under these conditions, the environmental impact and economic inefficiencies of the plantations could no longer be ignored. Priorities and management strategies changed, and, in 2000, the government began to de-emphasize commercial forestry in the region, transferring the pine lands to the GRNP.

In these newly added areas, the past wholesale removal of the pines created an Armageddon-like landscape (see Figure 6.6). But surprisingly, it was not long before a breath-taking kaleidoscope of colors rose from the soils of these degraded sites. In the span of only a few years, the mountainside landscapes were transformed in shades of pink, red, and white characteristic of the native fynbos vegetation. The sight stirred us to the depths of our souls as we witnessed this testament to the resilience and beauty of nature's rebirth and renewal. But even as the pines were replaced by the fynbos, traces of the plantations and commercial harvesting still lingered, especially in the sprawling network of logging roads that had been constructed across the plantations to facilitate the removal of trees as they were cut. And amongst the vibrant bushes of native fynbos species, small pine stumps defiantly poked out, silently demonstrating their own resilience amidst the restoration of the native flora.

Figure 6.6 The "Armageddon-like" landscape that faced conservation scientists Thabang Sibiya and Izak Smit in the Garden Route National Park of the Republic of South Africa after logged pine plantations were added to the national park subsequent to the wholesale removal of their trees.

Our perception of the fragile nature of these restoration efforts increased when a massive fire engulfed the mountains within the park in October 2018, burning 100,000 hectares of fynbos. Although the fire encouraged the sprouting and re-seeding of native – and fire-dependent – fynbos species, it also triggered the emergence of invasive acacias and conifers, which now lined the road networks and were scattered amongst the fynbos shrubs. Once again, the recovery of the fynbos was threatened by the shadows of the past.

We saw that the past three centuries of afforestation could not be reversed in only a few decades. We turned to the Working for Water (WfW) program, an effort that addresses pressing environmental issues while intertwining them with socio-economic considerations. Rather than striving to restore the sites to their original pristine state, we worked with WfW to establish essential biotic and abiotic conditions that could foster a sustainable trajectory of recovery – the formation of a new plant community. But first came the work of clearing vast areas in the GRNP from invasive plant species, an effort that involved active collaboration with economically disadvantaged residents from surrounding communities. Such collaboration was aided by the financial opportunities for residents involved in and employed by WfW in this effort, but there was also an increased appreciation for the restoration program among local people, as well as an empowerment of individual residents with skills in restoration that extended far beyond the immediate job at hand. Today, thanks in great part to the support of the local community, we find ourselves caring for a novel ecosystem that was not present when the GRNP was created (see Figure 6.7). Now that "new" system flourishes here, providing both natural resources (like water) and cultural ones (like ecotourism) for local community members, and engaging rather than excluding people in the care of this landscape.

Figure 6.7 The resilient native and endemic species of South Africa's fynbos vegetation reappeared in the former pine plantations that had been added to Garden Route National Park, but, in the wake of fires, stressed water conditions, changes in soil and vegetation inherited from the pine plantations, and socioeconomic considerations of local residents, Thabang Sibiya, Izak Smit, and others intentionally developed a "novel ecosystem" better adapted to changing environmental conditions and community needs.

Emerging Principles for Designing and Managing Novel Ecosystems

Examples like South Africa's GRNP show that novel ecosystems can intentionally be made part of larger conservation efforts. Around the world, protected areas (PAs) and wildlands are experiencing human-caused disturbances at unprecedented rates. Traditional restoration efforts have attempted to mitigate the disturbance by moving the *system* back toward its previous or historical condition. Could it be a better strategy to move the *trajectory* of a system's development – creating a new pathway that could result in a system better adapted to the new and still-changing conditions? The restoration of the Boulder gravel pit is one example of such an effort. Are there emerging precepts and recommendations that can provide a guide to designing and managing novel ecosystems?

Plant scientists Thomas Jones and Thomas Monaco offer four principles for success in restoring altered ecosystems in ways that would better adapt them to novel and changing conditions. The plant species assembled in the altered ecosystem should: 1) reflect general historical evolutionary patterns of the ecosystem; but 2) be particularly suited to the modified environment; 3) be able to adapt to contemporary selection pressures; and 4) contribute to the restoration of ecosystem structure and function.[49] A changing world, and the novel ecosystems it produces, forces us to recognize that future conservation will be about intervention, and intervention is not easy (Chapter 10). But how can we predict where novel ecosystems might arise or even where it might be well to plan for their design? US Forest Service conservation scientist Kevin McKelvey and his colleagues offer three recommendations. First, conduct landscape-level assessments that will identify potential vulnerabilities of key ecosystem components to various disturbances, and assess multiple alternative development trajectories for the ecosystem that could happen following disturbances, with assessments of how disturbance frequency and severity may influence the resiliency of these landscapes. Second, enhance the adaptive capacity of managed systems by keeping, as Aldo Leopold famously prescribed, as many "pieces" (species and ecological processes) as possible in place and maintaining as many ecosystem services as possible.[50] Ways to do this might include sustaining native species diversity, sustaining functional diversity (a wide variety of ecological roles performed by different species and functional groups), maintaining and enhancing above- and below-ground food webs, reducing soil loss, and maintaining water quality. Third, emphasize management actions and strategies that create landscape diversity ("mosaics") of plant composition and structure, especially those that retain disturbance-resilient and genetically adapted vegetation.[51]

The Dangers and Values of Novel Ecosystems

The assessment we make of the values, and dangers, of novel ecosystems will affect the way in which conservation is practiced. One concern is that the acceptance of novel ecosystems will replace (or displace) traditional conservation and restoration efforts. Standish et al. answer,

> The incorporation of novel ecosystems into established management frameworks … calls for a broader framework for traditional goals and practice rather than suggesting their replacement … the common and traditional practice of biodiversity conservation remains a possible and perhaps overlooked utility of novel ecosystems.[52]

Another concern is that accepting novel ecosystems into management frameworks will add more complexity to management strategies and decisions. The straight answer to this concern is – yes! And it's about time! A chronic failing of conservation management is that it often devolves into a "one best solution" approach and subsequently overlooks alternatives that might be better, or, at the very least, provide a safeguard against overdependence on a single strategy. Ideally, all conservation efforts should employ *adaptive management* – the framing of management decisions as hypotheses implemented and tested in an experimental framework. But many management strategies drift into a "passive adaptive management" approach[53] constructed to learn only from the implementation of a historically informed precedent ("we've always done it this way before") judged to be the "best" practice. Such passive adaptive approaches limit the use of other options and reduce the flexibility that can be gained from having multiple management goals. Intelligent management of both species and systems must invariably evaluate multiple strategies and foresee alternative outcomes that can arise from them. The increasing presence of novel ecosystems, especially ones that may provide value for endangered species, requires complex assessments – and they should! More complex assessments, including complex assessments of novel ecosystems, represent assessments of *real conditions*, not simplified assessments of imagined conditions that fit only one strategy. These are the kinds of assessments that have the greatest hope for success.

Yet another consideration addresses human perception. Will people care about novel ecosystems? We value natural ecosystems not only for their present state, but for their "pedigree" – their natural history and pattern of development that is the story of "what they have been through" and "how they came to be." For many, such knowledge "authenticates" an ecosystem, deepening the perception of what it now is because of our understanding of what it once was and what it experienced to get here. Novel ecosystems lack these histories, or, if they have a history at all, it is often a brief tale of human ignorance expressed in intervention and disturbance with unintended consequences. Again, to quote Standish et al.,

> Perception of natural heritage encompasses the flora and fauna of a region as well as the geological forms on land or sea as people understand themselves to have received it. What people value as part of their natural heritage is therefore set within a narrative structure.[54]

We value species and ecosystems around us through our relational connections to them. When we lack such connections, we do not value them. They are "just there," and all too often, and all too soon, they are "in the way" of what we do value. Forests are cut down, marshes are drained, and prairies are bulldozed to be replaced with housing, roads, and stores. But, as Standish et al. point out, "The fact that novel ecosystems are unique expressions of a place therefore means there is potential for them to be ascribed a heritage value."[54]

This truncated history of novel ecosystems can spawn another related problem – the problem of hubris (excessive pride or self-confidence). Will a novel ecosystem's "separation" from "deep ecological history" make it easier to justify and impose human ambitions upon the ecosystem? Will ecosystems then increasingly resemble what humans want them to be (efficient service providers) rather than what they would or could be "on their own" – a self-willed system being and becoming what it is *without* intentional human intervention? This problem is deeply embedded in human nature and has reared its ugly head throughout our history. In the nineteenth century, the (British) Naturalization Society in New Zealand attempted (and, for the most part, succeeded) in re-creating the ambiance of the English countryside in many parts of New Zealand by releasing common British songbirds, often to the detriment of native species.[55] A more recent example might be even more harmless in its appearance, but all the more sinister in its intent.

In 1972, the US city of Los Angeles made an innovative proposal for urban beautification – line the median strip of a major boulevard with plastic trees constructed of factory-made "leaves" and "branches" wired to plumbing pipes. The trees were "planted" in aggregate rock coated with epoxy, and, after their installation, an unknown person or persons added plastic birds.[56] Proponents of the plan contended that only plastic trees could survive the soil-deficient, smog-ridden, and increasingly warmer environment of downtown Los Angeles, whereas real trees could not (this may have been one of the first novel ecosystems designed to accommodate climate change). The urban research planner Martin Krieger, in an article in *Science*, asserted that "the demand for rare environments is a learned one" and "conscious public choice can manipulate this learning so the environments which people learn to use and want reflect environments that are likely to be available at low cost."[57] And, added Krieger,

> It is possible that by manipulating memory through the rewriting of history, environments will come to have new meaning ... In order to create substitutes, we must endow new objects with significance by means of advertising and social practice ...

> Much more can be done with plastic trees and the like to give most people the feeling that they are experiencing nature.[58]

Here is hubris at its highest. Not only is the system to be remade (with plastic trees instead of real ones), but history is to be rewritten to give this change "new meaning." The new (plastic) trees will be endowed with significance and connection to people, not because of what they are, but through "advertising and social practice." In this argument we are presented with advocacy for: 1) manipulating memory (influence without consent); 2) rewriting history (deception); 3) substitution and simulation of natural environments with man-made ones (devaluing nature); and 4) creating value and meaning through advertising and "social practice" (confusing "value" with "benefit" and "meaning" with "demand"). The scheme to install the plastic trees was ultimately abandoned. However, the plan itself shows the self-deception the human mind is capable of. Given means, motive, and opportunity, people will commit the crime of treating human needs as the ultimate frame of reference for what nature will be. And their resulting "values" for nature will be embedded in nothing but human self-interest.

If we design ecosystems primarily, or even solely, to deal with stressed landscapes, are we turning a blind eye to the stressors themselves, and to what has caused them, which is us? The Canadian conservationist Harvey Locke spoke to this problem with commendable bluntness:

> ...the death of the wild in favor of the garden with Homo sapiens triumphant is no vision for those who proclaim to love nature. It will also inevitably be disastrous for the human species. We do not know how to run the world. It is time for our species to become humble and wise and to stop being greedy and clever.[59]

Novel ecosystems demand careful evaluations of values, particularly in the conversation of restoration and conservation. "When a place is categorized as irredeemably degraded," note Laurie Yung et al., "it is dismissed. When something is categorized as a novel ecosystem, the next questions become 'what values can it provide?', and 'what roles can it play?'"[60] We must be more intentional in choosing our values, but is the value of nature and all it contains simply and only a matter of human choice? Are we choosing to conserve nature and its biodiversity as a means to some other end, or as an end *in itself* that should be valued *for itself* (Chapter 11)? Because novel ecosystems may sometimes be designed ecosystems, they will require an element of choice in what to design. That choice will reflect the values and virtues of the designer.

Novel Ecosystems in Conservation Planning

Novel ecosystems are part of nature today and their legitimacy in conservation is increasingly recognized. But in any time and place they occur, the novel ecosystem represents a loss – a place where a historic, reference, "natural" ecosystem once thrived. The first stage of grief from such loss is denial, a failure to see what is really there. Assertions that novel ecosystems do not exist, that anything that "appears" to be a novel ecosystem can be restored to "the real thing" with enough skill, effort, and resources, express this kind of obfuscation. Novel ecosystems are real. They are adaptive responses of plants and animals to changes that have come about in the world, the causes of which are overwhelmingly human. They are "novel" (without analog) because conditions have changed. They are self-generating because the species within them adapt to the conditions we create. And novel ecosystems are persistent – self-sustaining – if present human actions foster the genesis of plant communities unlike any historical precedents. If this is the case, conservation managers must work hard to retain or return to such reference points. But even with such effort, the reference system may not persist under current conditions.

Some novel ecosystems can perform ecosystem functions and provide ecosystem services more efficiently and in greater quantity than the systems they replaced. They are sometimes able to exceed the range and distribution of the former system precisely because they may be better adapted to colonize the disturbed conditions around them. Novel ecosystems can support biodiversity – a diversity that may be the product of *both* native species (plants and animals that can use the system) *and* new species which can thrive under the novel conditions but could not survive in the former system.

Human interactions with nature often cause natural systems to lose their intrinsic functions and capacities. That is why restoration ecology today is one of the most thriving disciplines in conservation science – complete with its own society, journals, certification programs, academic curricula, and even its own "decade." Novel ecosystems will increasingly be elements of ecological restoration, and restoration has drawn so many people to it precisely because it is a conservation practice embedded in hope – and its hope is embedded in action. But no matter how great the effort expended, restoration in the modern world will face a difficult question from the very inception of its endeavor – can this patient (system) be saved? What we value in and from nature will have a prevailing force in our answer. In some cases, the answer must be "no" – a frank admission that current conditions are too different from former ones to sustain the same community of plants and animals that once thrived in them.

Conservationists will have to make important decisions rooted in wise perception. Restoration ecologists James Miller and Brandon Bestelmeyer offer a logical algorithm (see Figure 6.8).[61] First, if an ecosystem is relatively free of adverse human impacts, it may indeed be one of the "last best places" and its conservation should be undertaken immediately. If, on the other hand, human influences have altered it, the question is – do we want to restore the system to its historic norms? Many factors may be embedded in determining what we "want," because human interactions with nature and the things we value in it will be part of the ecological shaping process. But if a traditional restoration is desired, is it feasible (reasonably possible)? If so, we begin a "traditional" restoration effort. If a full recovery is not feasible, a "partial" restoration may still have value, either by engaging in *revegetation* or *remediation* of the site with plants that, even if not part of the original system, can yet maintain some of its function. It may be possible to achieve *remediation*, maintaining the most essential ecological processes despite the loss of many of the original species of the system while minimizing harmful effects that might have occurred from altering the site. Such efforts, whether they can achieve full or partial restoration, have a strong record of success and a wealth of practical examples and scientific theory to support the efforts. But desire, example, and theory may not be enough.

If a restoration, in whole or in part, is not feasible, the manager faces a *new* vegetation community on an altered site. Is this new community one that can maintain itself – is it a novel ecosystem? If the answer is "yes," then this new system may be amendable to conservation management for many of the same goals as the former system, if perhaps not in the same way. If the answer is "no," the manager might attempt to "engineer" the new ecosystem toward specific goals for ecosystem services through more intensive human input and manipulation.[55] Prescribed fire (in altered prairie systems), prescribed

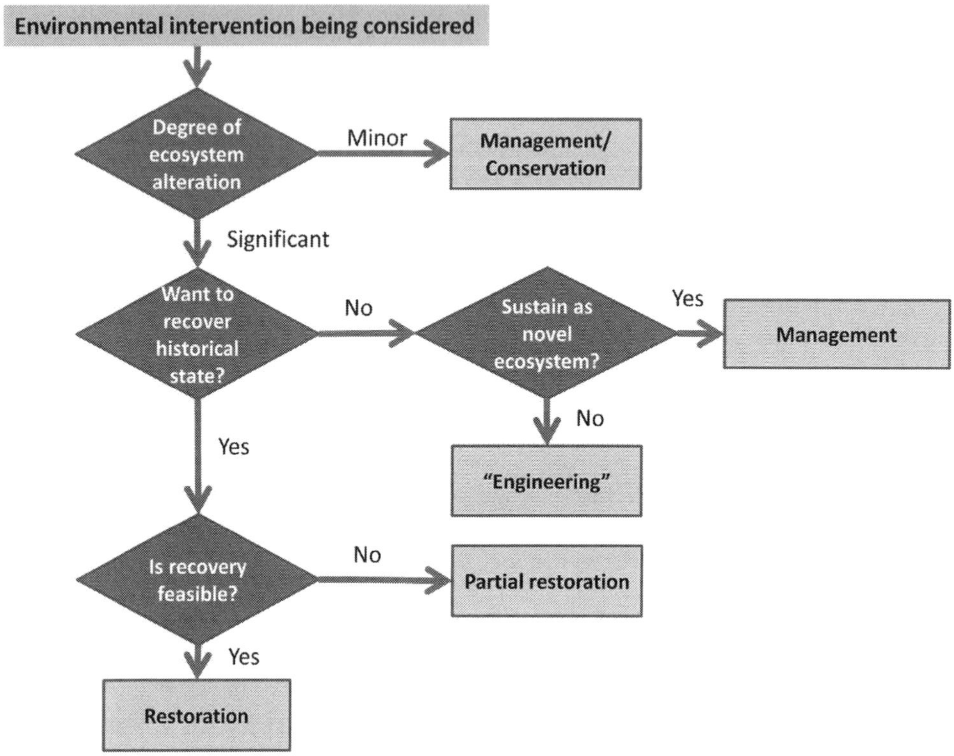

Figure 6.8 A decision tree integrating elements of the novel ecosystem framework with restoration options.

flooding (in altered wetland systems), or prescribed species removal (in systems altered by non-native species) are some of the alternatives, and, in terms of reaching their goals, they often "work," but their success is a full-time job. If we want a community of plants and animals to persist under the new conditions, it may have to be a new community, or at least have some new members, like the grassland restored on the Boulder, Colorado gravel pit. But restorations like this one are restorations *of* and *for* nature – not restorations of and for "ecosystem services." Humans do impose their own needs and values on nature – in farms and pastures and cities and suburbs – but there is need for reciprocity as well as imposition. Humans are beginning to consider the questions of how to live *in* nature, or even *with* nature. Are they finally ready to ask, "How can we find ways of living *for* nature?"

In the Anthropocene, the novel ecosystem is both an objective reality (something that is really "there") and a human social construction (something we chose to design that reflects the values we hold and the choices we make). The novel ecosystem, whether accidental or intentional, must be part of the context of conservation endeavor. But just as the *reality* of novel ecosystems should be taken seriously, the *design* of novel ecosystems should not be undertaken carelessly. Standish et al. note that, "While the creation of novel ecosystems may be a sensible approach to ecosystem management in some cases, ecologists, restoration practitioners and planners must be mindful of hubris or overestimating the ability to understand and manage complex ecosystems."[32] But just as care is warranted, risk is necessary. Go boldly where no one has gone before. Past ecological precedents may no longer be the best guide for future conservation planning. Novel ecosystems will be a growing component of ecological restoration in the years to come. They will be the ecosystems of the Anthropocene.

References

1. Botkin, D. B. 2001. The naturalness of biological invasions. *Western North American Naturalness* 61:261–266, 261.
2. Hobbs, R. J., E. S. Higgs, and C. M. Hall. 2013. Defining novel ecosystems. Pages 58–60 *in* R. J. Hobbs, E. S. Higgs, and C. M. Hall, editors. *Novel ecosystems: Intervening in the new ecological world order*. Wiley, Chichester, UK, 58.
3. Lunch 14. 2019. Novel ecosystems. Conducted by T. Wiens, editor. Illustrations by Aisha Sawatsky. Interviews with R. Hobbs, J. Zedler, and P. Del Tredici, with annotation by A. Felson. *Lunch* 14:55–67. <https://tarynwiens.com › novel-ecosystems-interviews>.
4. Chapin, F. S. III, and A. M. Starfield. 1977. Time lags and novel ecosystems in response to climate change in Arctic Alaska. *Climate Change* 35:449–461.
5. Chapin, F. S. III, and A. M. Starfield. 1977. Time lags and novel ecosystems in response to climate change in Arctic Alaska. *Climate Change* 35:457–458.
6. Soulé, M. 1991. Conservation biology in the twenty-first century: Summary and outlook. Pages 297–303 *in* D. Western and M. Pearl, editors. *Conservation in the twenty-first century*. Oxford University Press, New York, USA, 301.
7. Mascaro, J., J. A. Harris, L. Lach, et al. 2013. Origins of the novel ecosystem concept. Pages 46–56 *in* R. J. Hobbs, E. S. Higgs, and C. M. Hall, editors. *Novel ecosystems: Intervening in the new ecological world order*. Wiley-Blackwell, West Sussex, UK, 49.
8. Perring, M. P., and E. C. Ellis. 2013. The extent of novel ecosystems: Long in time and broad in space. Pages 66–80 *in* R. J. Hobbs, E. S. Higgs, and C. M. Hall, editors. *Novel ecosystems: Intervening in the new ecological world order*. Wiley-Blackwell, West Sussex, UK.
9. Ellis, E. C. 2015. Ecology in an anthropogenic biosphere. *Ecological Monographs* 85:287–331.
10. Morse, N. B., P. A. Pellissier, E. N. Cianciola, et al. 2014. Novel ecosystems in the Anthropocene: A revision of the novel ecosystem concept for pragmatic applications. *Ecology and Society* 19(2):12. <http://dx.doi.org/10.5751/ES-06192-190212>.
11. Gleason, H. 1926. The individualistic concepts of the plant association. *Bulletin of the Torrey Botanical Club* 53:7–26.
12. Gleason, H. 1926. The individualistic concepts of the plant association. *Bulletin of the Torrey Botanical Club* 53:26.
13. Mascaro, J., J. A. Harris, L. Lach, et al. 2013. Origins of the novel ecosystem concept. Pages 46–56 *in* R. J. Hobbs, E. S. Higgs, and C. M. Hall, editors. *Novel ecosystems: Intervening in the new ecological world order*. Wiley-Blackwell, West Sussex, UK, 46.
14. Harris, J. A., S. D. Murphy, C. R. Nelson, et al. 2013. Characterizing novel ecosystems: Challenges for measurement. Pages 192–211 *in* R. J. Hobbs, E. S. Higgs, and C. M. Hall, editors. *Novel ecosystems: Intervening in the new ecological world order*. Wiley-Blackwell, West Sussex, UK, 47.
15. United Nations Development Programme. 2022. Ecosystem Restoration 2022 [United Nations Development Programme Global Open Online Course]. <https://www.learningfornature.org/en/courses/ecosystem-restoration-2022/>. Accessed 22 November 2023.
16. Mascaro, J., R. F. Hughes, and S. A. Schnitzer. 2012. Novel forests maintain ecosystem processes after the decline of native tree species. *Ecological Monographs* 82:221–238.
17. Mascaro, J., R. F. Hughes, and S. A. Schnitzer. 2012. Novel forests maintain ecosystem processes after the decline of native tree species. *Ecological Monographs* 82:221.

18. Pessarrodona, A., A. Foggo, and D. A. Smale. 2019. Can ecosystem functioning be maintained despite climate-driven shifts in species composition? Insights from novel marine forests. *Journal of Ecology* 107:91–104. DOI: 10.1111/1365-2745.13053.
19. Lugo, A. E. 2013. Novel tropical forests: Nature's response to global change. *Tropical Conservation Science*. Special Issue 6(3):325–337. <www.tropicalconservationscience.org>.
20. Birdsey R. A., and P. L. Weaver. 1987 (February). Forest area trends in Puerto Rico. Research Note SO-331. Southern Forest Experiment Station. US Forest Service.
21. Lugo, A. E. 2002. Can we manage tropical landscapes? An answer from the Caribbean perspective. *Landscape Ecology* 17:601–615.
22. Brandeis, T. J., E. H. Helmer, and S. N. Oswalt. 2007. The status of Puerto Rico's forests, 2003. USDA Forest Service, Southern Research Station Resource Bulletin SRS-119, Asheville, NC.
23. Lugo, A. E., and E. Helmer. 2004. Emerging forests on abandoned land: Puerto Rico's new forests. *Forest Ecology and Management* 190:145–161.
24. Brandeis, T. J., E. Helmer, H. Marcano Vega, et al. 2009. Climate shapes the novel plant communities that form after deforestation in Puerto Rico and the U.S. Virgin Islands. *Forest Ecology and Management* 258:1704–1718.
25. Lugo, A. E., and T. J. Brandeis. 2005. A new mix of alien and native species coexist in Puerto Rico's landscapes. Pages 484–509 *in* D. F. R. P. Burslem, M. A. Pinard, and S. E. Hartley, editors. *Biotic interactions in the tropics: Their role in the maintenance of species diversity*. Cambridge University Press, Cambridge, UK.
26. Francis, J. K., and H. A. Liogier. 1991. Naturalized exotic tree species in Puerto Rico. General Technical Report SO-82, USDA, Forest Service, Southern Forest Experiment Station, New Orleans, USA.
27. Abelleira, O. J. 2011. Flooding and profuse flowering result in high litterfall in novel *Spathodea campanulata* forests in northern Puerto Rico. *Ecosphere* 2:105. <https://esajournals.onlinelibrary.wiley.com/doi/full/10.1890/ES11-00165.1>.
28. Lugo, A. E., O. J. Abelleira, A. Collado, et al. 2011. Allometry, biomass, and chemical content of novel African tulip tree (*Spathodea campanulata*) forests in Puerto Rico. *New Forests* 42:267–283.
29. Abelleira Martínez, O. J., M. A. Rodríguez, I. Rosario, et al. 2010. Structure and species composition of novel forests dominated by an introduced species in north central Puerto Rico. *New Forests* 39:1–18.
30. Abelleira-Martínez, O. J. 2008. Observations on the fauna that visit African tulip tree (*Spathodea campanulata* Beauv.) forests in Puerto Rico. *Acta Científica* 19:37–43.
31. Lugo, A. E., O. J. Abelleira, S. Borges, et al. 2006. Preliminary estimate of earthworm abundant and species richness in *Spathodea campanulata* Beauv. forests in northern Puerto Rico. *Caribbean Journal of Science* 42:325–330.
32. Ellis, E. C., E. C. Antill, and H. Kreft. 2012. All is not loss: Plant biodiversity in the Anthropocene. *PLOS One* 7(1):e30535. <https://journals.plos.org/plosone/article?id=10.1371/journal.pone.0030535>.
33. Perring, M. P., R. J. Standish, and R. J. Hobbs. 2013. Incorporating novelty and novel ecosystems into restoration planning and practice in the 21st century. *Ecological Processes* 2:18.
34. Higgs, E. 2017. Novel and designed ecosystems. *Restoration Ecology* 25:8–13. DOI: 10.1111/rec.12410.
35. Higgs, E. 2017. Novel and designed ecosystems. *Restoration Ecology* 25:8. DOI: 10.1111/rec.12410.
36. Hobbs, R. J., E. S. Higgs, C. M. Hall, et al. 2014. Managing the whole landscape: Historical, hybrid, and novel ecosystems. *Frontiers in Ecology and the Environment* 12:557–564.
37. Light, A., A. Thompson, and E. S. Higgs. 2013. Valuing novel ecosystems. Pages 258–265 *in* R. J. Hobbs, E. S. Higgs, and C. M. Hall, editors. *Novel ecosystems: Intervening in the new ecological world order*. Wiley, Chichester, UK.

38. Seastedt, T. R., R. J. Hobbs, and K. N. Suding. 2008. Management of novel ecosystems: Are novel approaches required? *Frontiers in Ecology and Environment* 6:547–553.
39. Cherwin, K. L., T. R. Seastedt, and K. N. Suding. 2009. Effects of nutrient manipulations and grass removal on cover, species composition, and invasibility of an adventive grassland in Colorado. *Restoration Ecology* 17:818–826. <https://doi.org/10.1111/j.1526-100X.2008.00418.x>.
40. Seastedt, T. R., R. J. Hobbs, and K. N. Suding. 2008. Management of novel ecosystems: Are novel approaches required? *Frontiers in Ecology and Environment* 6:551.
41. Martínez-Hesterkamp, S., S. Rebollo, L. Pérez-Camacho, et al. 2018. Assessing the ability of novel ecosystems to support animal wildlife through analysis of diurnal raptor territoriality. *PLOS One* 13(10):e0205799. DOI: 10.1371/journal.pone.0205799. PMID:30325967; PMCID:PMC6191124.
42. Rebollo, S., S. Martínez-Hesterkamp, G. García-Salgado, et al. 2017. Spatial relationships and mechanisms of coexistence between dominant and subordinate top predators. *Journal of Avian Biology* 48:1226–1237. <https://doi.org/10.1111/jav.01337>.
43. Pajerski, L. 2005. *Accipiter gentilis* (online), Animal Diversity Web. <https://animaldiversity.org/accounts/Accipiter_gentilis/>. Accessed 24 November 2023.
44. US Forest Service. 2009. Management Indicator Species (MIS). <https://doi.org/10.1111/jav.01337>. <https://www.fs.usda.gov/Internet/FSE_DOCUMENTS/stelprdb5319176.pdf>. Accessed 24 November 2023.
45. Cornell Lab. 2023. All about birds. Least tern. <https://www.allaboutbirds.org/guide/Least_Tern/overview#:~:text=By%202021%2C%20following%20years%20of,over%20deserted%20American%20Flamingo%20nests≥. Accessed 2 December 2023.
46. American Bird Conservancy. 2023. ABC's Bird Library. Piping plover, the "bird of the beaches." <https://abcbirds.org/bird/piping-plover/?gad_source=1&gclid=EAIaIQobChMInbnwr7bfggMVZlhHAR2qdA72EAAYAiAAEgIYpPD_BwE#breeding-feeding>. Accessed 25 November 2023.
47. Wildlife NYC. Piping plovers. 2023. <https://www.nyc.gov/site/wildlifenyc/animals/piping-plovers.page>. Accessed 2 December 2023.
48. Jorgensen, J. G., S. J. Brenner, L. R. Greenwalt, et al. 2021. Decline of novel ecosystems used by endangered species: The case of piping plovers, least terns, and aggregate mines. *Ecosphere* (4):e03474. DOI: 10.1002/ecs2.3474.
49. Jones, T. A., and T. A. Monaco. 2009. A role for assisted evolution in designing native plant materials. *Frontiers in Ecology and the Environment* 7:541–547.
50. Leopold, A. 1953. *Round river*. Oxford University Press, Oxford, UK, 146–147.
51. McKelvey, K. S., W. M. Block, T. B. Jain, et al. 2021. Adapting research, management, and governance to confront socioecological uncertainties in novel ecosystems. *Frontiers in Forests and Global Change* 4:644696. DOI: 10.3389/ffgc.2021.644696.
52. Standish, R. J., A. Thompson, E. S. Higgs, et al. 2013. Concerns about novel ecosystems. Pages 296–309 *in* R. J. Hobbs, E. S. Higgs, and C. M. Hall, editors. *Novel ecosystems: Intervening in the new ecological world order*. Wiley-Blackwell, West Sussex, UK, 298. <https://doi.org/10.1002/9781118354186.ch37>.
53. Walters, C. J, and C. S. Holling. 1990. Large-scale management experiments and learning by doing. *Ecology* 71:2060–2068.
54. Standish, R. J., A. Thompson, E. S. Higgs, et al. 2013. Concerns about novel ecosystems. Pages 296–309 *in* R. J. Hobbs, E. S. Higgs, and C. M. Hall, editors. *Novel ecosystems: Intervening in the new ecological world order*. Wiley-Blackwell, West Sussex, UK, 302. <https://doi.org/10.1002/9781118354186.ch37>.
55. Godfray, H. C. J., and M. J. Crawley. 1998. Introductions. Pages 39–65 *in* W. J. Sutherland, editor. *Conservation science and action*. Blackwell, Oxford, UK.
56. Tribe, L. H. 1974. Ways not to think about plastic trees. *The Yale Law Journal* 83:1315–1348.
57. Krieger, M. H. 1973. What's wrong with plastic trees? *Science* 179:446–455, 451.

58. Krieger, M. H. 1973. What's wrong with plastic trees? *Science* 179:446–455, 453.
59. Locke, H. 2014. Nature needs half: A necessary and hopeful new agenda for protected areas in North America and around the world. *The George Wright Forum* 31:359–371, 369.
60. Yung, L., S. Schwarze, W. Carr, et al. 2013. Engaging the public in novel ecosystems. Pages 233–256 *in* R. J. Hobbs, E. S. Higgs, and C. M. Hall, editors. *Novel ecosystems: Intervening in the new ecological world order*. Wiley, Chichester, UK, 248–249.
61. Miller, J. R., and B. T. Bestelmeyer. 2016. What's wrong with novel ecosystems, really? *Restoration Ecology* 24:577–582.

7 Conservation in the Urban Ecosystem

History and Growth of the Urban Landscape

The Growth of Urbanity amidst Shrinking Biodiversity

Over the next 30 years, the global human population is expected to increase by 2.5 billion. More than 50% of the world's people live in urban landscapes covering 10.2% of the planet's coastal land area,[1] and during the next 30 years that proportion will rise. Increases in human populations in cities will be accompanied by urban land conversions that will contribute to habitat and biodiversity loss.[2] Today many of the Earth's smaller nations, such as Kuwait, Monaco, and Singapore, are 100% urban. But larger countries are not far behind. Belgium, Qatar, and Uruguay have urbanization levels of 96% or higher.[3] Countries such as Canada, Spain, France, Finland, Mexico, the United Kingdom, and the United States all have urbanization levels greater than 80%, and so do those often perceived to have the greatest remaining "natural resources" and biodiversity such as Australia, Brazil, Columbia, New Zealand, and Venezuela.[3] Eight of the ten countries with the most rapidly increasing urban populations are in sub-Saharan Africa, led by Burundi whose urban population is growing at nearly 5.5% per year.[3] Already the total size of urban areas in many countries exceeds that protected for conservation. In Singapore, urbanization had encompassed more than half the country's total land area by 1990[4] which contributed to the disappearance of three-quarters of Singapore's native species.[5]

Many native species cannot thrive in urbanized areas where environmental changes are both rapid and dramatic.[6] Urban growth has profound implications for conservation, and we do not need to gaze into the future to understand the problem. Cities typically have relatively low scores in biodiversity rankings compared to non-urban environments. As a result, conservationists, policymakers, and the public often view urbanized areas as ecologically "irredeemable." Efforts to conserve and restore cities are avoided, not only because conservationists may perceive that little natural value remains, but also because such effort would come at a disproportionately high expense. Most urban development initiatives address problems more immediate to the city's human residents, many of whom are economically and socially marginalized and frequently excluded from discussions about conservation science.[7] As an African-American resident of inner-city Chicago told one of my students working on biodiversity conservation in that city, "That's a white man's issue." If urban residents have little interaction with, concern for, or awareness of nature, how can efforts in urban biodiversity conservation succeed, and who even wants them to? Under these conditions, species loss in urban areas becomes nothing more than the expected outcome of a self-fulfilling prophecy. Yet a recent study of cities in the

DOI: 10.4324/9781003404415-7

US state of California found that urban areas often hold substantial numbers of native biodiversity indicator species.[8] Nevertheless, under regimes of the consistent and drastic environmental change that urban areas typically experience, it may be impossible to protect or reestablish a viable ecosystem that looks and functions like the native system the urban area replaced. There are daunting obstacles to overcome, but there is urgency to save what is now being lost.

By 2030 urban land cover is projected to increase by 1.2 million km^2, nearly three times urban coverage at the beginning of 2000.[9] Such growth will not simply be an expansion of current urban landscapes but will include the development of new urban areas, replacing today's "natural" environments. This conversion of natural areas into urban, human-dominated zones is happening in many of the world's biodiversity hotspots.[9] Urban ecologist Rohan Simkin and his colleagues project that urban expansion will add to habitat loss for about one-third of the species which have been assessed thus far and up to 855 species will be directly imperiled.[10] Urban expansion is expected to encroach upon or destroy habitat for 139 amphibian species, 41 mammalian species, and 25 bird species on the Critically Endangered or the Endangered Lists of the IUCN.[9] In taxa that have been carefully studied, species richness in the urban core is less than 50% of that found in rural, more natural areas.[11]

Of even greater concern is that the geographic size of urban areas is increasing faster than urban populations. Historically, urban areas have grown slowly, often in a pattern of concentric spread. Urban growth and development in the twentieth and twenty-first centuries have accelerated dramatically. Within the last 30 years urban areas have grown twice as fast as urban populations.[11] And unless significant changes are made, cities in the Anthropocene will continue to be what they have always been – the most human-dominated and human-centered environments on Earth. In such environments non-human species are typically only considered if they provide resources that might be used to accumulate wealth or if they are pests that could threaten the city's commercial activity or the health and well-being of its human residents.

The Slaughter of the Egrets

In Carrolton, Texas, USA (population 134,000), a suburb of the much larger Dallas-Fort Worth metropolis of over 7.6 million, a nesting colony of snowy egrets (*Egretta thula*) (see Figure 7.1), which was also used by great blue herons (*Ardea herodias*) and other migratory birds, was razed to the ground while the young egrets were still in their nests being nurtured by parent birds. The decision to destroy the rookery was made in response to complaints by human residents about the noise, odor, and excrement produced by the egrets – and in their neighborhood! Bulldozers and other "removal equipment" arrived in the predawn hours on the morning of July 23, 1998 and leveled the rookery to the ground by removing all trees in a matter of hours. A conservative count put the number of birds killed at over 300. Volunteers rescued many more injured adults and abandoned (orphaned) juveniles, some of which recovered and were later returned to the wild.

The "massacre of the egrets" in Carrolton was, in our current culture, an exceptional case that sparked, first, outrage and, then, legal action against the city, which was ultimately fined $70,000 for killing the birds. Snowy egrets, great blue herons, and other species prominent in the rookery were not endangered species, but they were protected by the Migratory Bird Treaty Act, an international agreement which prohibits killing, capturing, selling, trading, and transporting protected migratory bird species without

Figure 7.1 A rookery of snowy egrets (*Egretta thula*) was destroyed by bulldozers in the US city of Carrolton, Texas while rearing their young in July 1998, killing at least 300 individuals. The decision to destroy the rookery was made in response to complaints by human residents about the noise, odor, and excrement produced by the egrets in their neighborhood. Jean Beaufort. CCO Public Domain. Available at PublicDomain-Pictures.net.

prior authorization of the federal government. It was not that there was no legal protection for the birds, or lack of outcry from the public against their slaughter. The tragedy of the Carrolton Massacre was that the lives of animals, even in the care and raising of their young, was nothing more than an inconvenience to the quality of human life. That inconvenience was removed by the quickest and most efficacious method possible to ensure that such "quality" could be maintained.

Urban residents may be accustomed to a high standard of life because, at many levels, cities are exemplars of human progress and have advanced in their technologies to deliver goods and services to their residents, construct transportation systems to permit the rapid movement of millions of people, and provide systems of government responsive to human need. Cities have excelled in creating infrastructures of business and commerce that make urban residence attractive and profitable, or, at least, economically livable, for millions of people. Concurrent with such development cities have become centers of culture, advanced education, artistic expression, historical preservation, and places that promote and protect some of the greatest achievements of human effort. Yet despite centuries of development, increasing efficiency, and cultural achievement, urban centers are only in their infancy in exploring ways for humans, plants, and animals to live together in sustainable, non-destructive, and beneficial interactions.

The destruction of the egret rookery at Carrolton was, by today's standards, remarkable for its insensitivity and cruelty to other creatures, but it was not the unauthorized action of a few disgruntled urban residents. The Carrolton city government approved, planned, and financed the eradication of the Carrolton rookery. It was authorized, legitimated, and endorsed by the city's governing structures in response to complaints of residents that elected officials believed might produce political consequences. This was in every way a civic action that was "business as usual" and its structure of human incentives led predictably and inexorably to the destruction of nature. Given the public outcry and scandal that followed the Carrolton Massacre, it is unlikely that this kind of authorized destruction of nature – and particularly of nesting birds – will ever happen again, at least not in the state of Texas. But these kinds of actions are authorized every day by urban governments all over the world. A deliberate act directed to destroy nature in an urban area is taking place as you read these words, and it may be happening near you.

Cities as Ecological Systems

Urban Linear Metabolism

The ecology of the natural world is often described as a "circular metabolism" in which nutrients are recycled but energy is not. The nutrients and other essential materials are limited in quantity (in economic terms, "stock limited"), but can be moved quickly in ecological processes. The energy driving natural cycles is flow abundant (for example, sunlight) but also stock limited (only small amounts arrive and can be used in ecological processes at a given place and time). Nutrients and other materials in their biogeochemical cycles pass through both abiotic and biotic compartments (Chapters 3, 4, and 5) in a "circular" pattern, and can be used again and again (the ultimate "recycling" process). In contrast, material flow in cities has a "linear metabolism" in that nutrients and essential materials are often not recycled.[12] Therefore, there must be constant inputs of them (stock abundant), and there is a constant output of unused or degraded nutrients and material as "waste" (waste abundant). The material elements are usually not processed by flow-limited solar energy, but by (for the time being) "flow-unlimited" inputs of energy from fossil fuels.

Urban areas depend (for both material and energy) on sources (systems) outside the city itself. For this reason, urban areas have sometimes been referred to as "parasitic" ecosystems, "feeding off" the material and energy of other ecological networks. Although such a description may be unflattering, it is essentially correct. In a parasitic relationship, the parasite (the urban area) lives on or inside another – the host organism (natural or agricultural ecosystems surrounding the city) – such that the host is harmed rather than benefitted. The entomologist E. O. Wilson characterized parasites as "predators that eat prey in units of less than one."[13] So cities "consume" natural ecosystems incrementally but relentlessly, such that the harm they impose on such systems may not be immediately apparent but is nonetheless persistently destructive.

Parasitic or not, ecological processes that transform energy and matter and mediate species' interactions occur in cities with the same inexorable certainty that they occur in wildlands and rural areas. Urban ecologists Steward Pickett and Mary Cadenasso note that even ecologists frequently fail to see that these processes are the same processes being manifested in the urban environment, if often in different ways.

> Ecologists are quite comfortable studying complex systems, comprising biotic and abiotic components, which are linked by flows of matter, energy, and information. The basic concept is as applicable to urban settlements as it is to areas in remote wilderness or uninhabited locations.[14]

Or, as the American conservationist Aldo Leopold put it more memorably, "the weeds in a city lot convey the same lesson as the redwoods."[15] Leopold is correct. Ecologists must learn to perceive the operation of ecological processes in both redwoods and weeds, especially what they have in common. In urban areas, the weeds outnumber the redwoods because weeds do better in human-modified environments, and the city is the most modified of all. The urban ecosystem, then, is simply a more intense and manifested version of a truth the Anthropocene is forcing upon all of us – *the study of ecosystems must include human sociological, political, and economic processes as critical elements that determine the development and maintenance of these same ecosystems.* The place to begin understanding what this means is the city.

What looms larger in cities than in non-urban environments are people – their higher densities and greater impacts. As such, cities also immediately create a problem for the traditionally educated ecologist who has been trained to exclude human effects from a "correct" or "pure" understanding of how the ecosystem is "supposed to work" (Chapter 3). Ecologists, and the science of ecology, are beginning to appreciate and embrace the reality of the urban ecosystem. In the United States, the National Science Foundation's (NSF) Long-Term Ecological Research (LTER) program now includes two major US metropolitan areas – Baltimore, Maryland and Phoenix, Arizona. Studies in Baltimore, for example, have led to new insights about how the urban environment changes (and, in some cases, does not change) the basic ecological processes that affect urban biodiversity. Pickett and Cadenasso, who have studied the Baltimore system for many years, note that, in the case of nitrogen cycling,

> Agriculture retains 85% of N applied to that land cover type, while suburban watersheds retain 71% of N inputs. The suburban retention rate is surprisingly high, and suggests that the combination of maintained lawns, volunteer vegetation, and engineered retention basins as biophysical infrastructure trap substantial amounts of N. This suggests that human structures have not obliterated all biological processes that determine ecosystem function in this suburban catchment.[14]

The Physical Structure of the City

Physically, urban systems have a predictable physical structure. Common features include human dwellings, factories, office buildings, warehouses, roads, pipelines, power lines, railroads, channelized waterways, reservoirs, sewage disposal facilities, dumps, gardens, parks, cemeteries, and airports. Crops, ornamentals, domestic pets, pests, and disease organisms are usually the dominant biotic elements. Socioeconomic features such as financial and material wealth, diverse lifestyles, high resource use, abundant waste production, and the prevalence of (or perhaps, obsession with) human-centered values are often the most powerful array of forces shaping the structure, function, and development of the urban ecosystem.[16]

The physical dimensions of cities are not environmentally neutral. They occupy air, space, land, rivers, and estuaries, and impose conditions and processes that change natural vegetation, atmospheric makeup, and water quality.[17] Compared to non-urban areas, the reduced coverage of vegetation, and, in its place, the expanding coverage of impermeable surfaces, such as asphalt or concrete, increase rates of runoff and decrease heat capacity (the ability to absorb energy without temperature change). As a result, urban areas discharge high levels of water into surrounding streams, rivers, and lakes – water often contaminated by the impervious surfaces it cannot penetrate but from which it unavoidably absorbs nutrients and chemicals, some of them toxic, which are part of the urban environment. Instead of allowing infiltration of rainfall and other forms of precipitation through the soil and the return of moisture to the atmosphere through transpiration from vegetation, urban surfaces create rapid runoff and discharge into existing water systems through drains and sewers. Because these impermeable urban surfaces are often made "dirty" from a variety of human activities, discharge moving over and through these polluted sources modifies the quality of the water systems, polluting channels, inlets, streams, lagoons, wetlands, and near-shore environments.[17] The same surfaces that alter the hydrology of cities also alter their climate, such that urban areas are often "heat islands" with average temperatures higher than surrounding landscapes, and so contribute disproportionately to regional climate change.

Cities modify climate in more ways than generating above-average heat levels. They create unique climates of their own by: 1) replacing grass, soil, and trees with asphalt, concrete, and glass; 2) replacing rounded soft shapes of trees and bushes with blocky angular buildings and towers; 3) releasing artificial heat from buildings, air conditioners, industry, and automobiles; 4) disposing of precipitation in drains, sewers, and gutters, preventing surface infiltration; and 5) emitting contaminants from a wide range of sources, which with resultant chemical reactions can create an unhealthy and dangerous urban atmosphere.[17] The urban atmosphere created in this way has two layers: 1) the *urban canopy layer* (UCL) consisting of the air from the ground to the rooftops, dominated by conditions specific to individual locations within the city; and 2) the *urban boundary layer* (UBL) or "mixed layer" above the UCL, with conditions more representative of the urban area as a whole. The UBL also influences the nearby rural atmosphere of surrounding areas.[17]

The mechanisms of heating in the urban atmosphere are not fundamentally different from those warming the global atmosphere (Chapter 5) but are distinct from rural environments. Longwave radiation emissions in rural environments are determined mainly by the temperatures of natural vegetated surfaces or soil. In the city, more surfaces, like pavements and buildings, release longwave energy. As with the global atmosphere, such longwave radiation is more fully absorbed by the urban atmosphere, especially in light of its higher levels of carbon dioxide that absorb more longwave radiation (Chapter 5). As such, longwave radiation released from the city is absorbed by the city's atmospheric contaminants and by its physical structures and materials. The contaminants include industrial emissions, auto exhaust and engine leakage, air conditioners, leakage from poorly insulated homes and other buildings, and heat from the metabolism of the city's human residents. This last element – heat generated from human metabolism – goes up because there are not only more people, but more energy use per person.

Apart from people, the physical structures and their high density possess a low heat capacity and thus generate higher temperatures per unit of radiation absorbed than vegetation, soil, or water. Thus, more longwave radiation (heat) is leaving the city's surface even as the urban atmosphere is changed so that it can absorb increasing amounts of the longwave radiation being emitted. The process of radiative forcing now becomes operative. The higher levels of longwave radiation in the urban atmosphere change the amount of energy the upper atmosphere can emit into space. More radiation (energy) is trapped below, such that the surface temperature of the system (the city) *must* rise to keep the system in equilibrium. The heat is a threat to the health and well-being of human and non-human creatures in the city. The presence of chemical pollutants in such a heated atmosphere also creates an additional threat – one that increases at higher temperatures.

Once in the atmosphere, some pollutants may interact with one another, and with natural atmospheric components, to create other pollutants. Under specific weather conditions, such as conditions that reduce or completely stop normal circulation of air over the city, such pollution may increase to levels that can be physically detrimental or even lethal. However, as circulation resumes over the city, the pollutants in the urban atmosphere may be carried downwind, affecting the air quality of rural areas, even some that may be thousands of miles away.[17]

This increased heat within the city also provokes a human response that creates a detrimental positive feedback loop – the increased use of energy-driven indoor cooling (air conditioning). Although air conditioning units cool indoor spaces, they emit heat into external surroundings. Salamanca and his colleagues demonstrated, for example, that, at night, heat emitted from air conditioning systems in some urban areas increased the mean air temperature around those systems by more than 1°C.[18] Further, they noted that the air conditioning systems had greater and more systemic effects on the urban atmosphere. Waste heat emitted from the air conditioning units

> modified the thermal stratification of the urban boundary layer, promoting vertical mixing during nighttime hours. Our results demonstrate that releasing waste heat [from air conditioning units] into the ambient environment exacerbates the nocturnal urban heat island and increases cooling demands.[19]

Cooling indoor areas makes people feel better in those spaces, but increases the overall temperature of the urban area, and its effects on global climate warming.

Biodiversity in the Urban Landscape

Because urban landscapes have become increasingly large, they are now an important component of regional and global biodiversity. In fact, many cities were historically established in riparian areas, ecological transition zones, or other naturally species-rich locations.[20] Today, however, environmental conditions in most cities pose threats to many species. The common patterns of changes are: 1) a reduction in overall species richness; 2) a decrease in species evenness (plant and animal communities dominated by a few abundant species rather than many equitably abundant species of lower densities); 3) an increase in abundance of larger species in a given taxon; and 4) an increase in the abundance of introduced species. Although the increased prevalence of non-native species in urban areas is widely accepted as a leading cause of biodiversity loss in cities, much evidence for this claim is based on the dominance of exotic species and the decline of native ones.[21] As urban ecologist Ingo Kowarik explains, "The decline of previously occurring native species is often due to habitat loss and transformation associated with urban land uses, which in turn provide beneficial conditions for introduced species."[22] Broad generalizations about cities and non-native species are inaccurate, but cities remain not only important points of entry for many non-native species but also points of secondary release of such species to other areas.[21]

There are exceptions to these patterns, and some globally declining taxa can attain high densities in urban habitats. Urban parks in the US city of San Francisco support higher mean abundances of bumblebees (*Bombus* spp.) than did two parks beyond the city boundary.[23] Populations of the common frog (*Rana temporaria*) in Great Britain have experienced declines in rural areas but increased in urban parks and gardens.[24] Other researchers in the UK found that the density of bumblebee nests in suburban gardens (~36 nests/ha) was comparable to that found in countryside habitats such as hedgerows (20–37 nests/ha).[25] If there are species unique to cities that deserve conservation protection, can urban areas make unique contributions to global biodiversity? Yes. In a study of Australian cities, Christopher Ives and his colleagues determined that cities contained more nationally threatened animal and plant species than non-urban areas on a unit–area basis, with 30% of threatened species found in cities.[26] Likewise, 22% of endangered US plant populations are found in the 40 largest urban areas.[26]

Over 20% of the world's human population now resides within areas designated as biodiversity hotspots, which includes many living in dense urban areas.[27, 28, 29] Some cities have none, but the natural environments they do possess may be the only ones that urban residents will ever know, and the only places in which they will encounter nature's biodiversity in some form. Sophie Parker, a senior scientist with The Nature Conservancy, puts it well:

> The conservation of biological diversity is closely linked with the fate of the world's cities. While the protection of sensitive and threatened species and habitats has often taken place in natural landscapes largely devoid of people, strategies for preserving the Earth's biodiversity that can be employed within cities are likely to become more common as urban areas continue to increase in size and number.[30]

Beginning in the eighteenth century in the UK, and subsequently throughout Western Europe, city dwellers were among the first to call for the conservation of nature, initially expressed in the "Back to the Land" movement which idealized and romanticized rural areas.[31] In more recent times, the same perspective has been shown in the tendency of urban residents to, on average, place higher value on species conservation than residents of rural communities.[32] Such views provoke legislators representing urban constituents to be more supportive of conservation legislation than those representing people from rural communities.[33] These potential advantages for biodiversity conservation, however, may be offset, in part, by the fact that urban residents tend to have less knowledge about nature and species than rural residents,[34] and may be less directly affected by conservation legislation and its enforcement. Such differences between rural and urban voters on conservation issues have existed for decades, but that division has been widening. In Europe, for example, voters supporting candidates of the Green Party, which is growing rapidly in political influence and parliamentary representation, are disproportionately concentrated in cities.[35]

Such differences present one of the most enigmatic and frustrating paradoxes of the Anthropocene – that urban residents, especially in Western nations, are supportive of ascribing high values to nature and taking political action to enforce these values in conservation law, while at the same time experiencing an increasingly diminished interaction with the natural world itself. Support for abstract ideals that one does not experience – or seek to create – in real life is intellectually discordant, and dangerously close to moral hypocrisy. This dilemma makes the critical question more urgent – what can and should be done to integrate the support of ascribing ethical value and political will to nature conservation while at the same time fostering the creation of actual experience and interaction with nature in an urban environment?

Incorporating People into Urban Conservation

Changing the "Nature" of Environmental Education

Urban environments can lead to species endangerment, but the greatest threat arising from them is not the *extinction of species* but the *extinction of experience* – the loss of human interaction with nature. Urban ecologist Clarke Adams noted, "Decades and generations of human isolation from the natural world through urbanization have produced a society that lacks a connection to the natural world and a relationship to that world."[16] The problem is now so systemic it has developed its own pathology, the "Nature Deficit Syndrome"(NDS) or "Nature-deficit Disorder" (NDD), a condition increasingly observed in children raised in urban environments. As one child explained his preference for indoor recreation, "I like to play indoors 'cause that's where all the electrical outlets are."[36] That is a funny remark describing a tragic condition symptomatic of a terrible ailment. In the Anthropocene, human interaction with nature is increasingly detrimental to nature and the absence of interaction with nature is increasingly detrimental to humans.

The cultivation of human experience with nature is critical to the human relationship with nature. Aldo Leopold expressed the goal of his wildlife ecology course this way: "The objective is to teach the student to see the land, to understand what he sees, and to enjoy what he understands."[37] But one of the most pervasive obstacles to genuine experience with nature is deeply embedded in the traditional curricula of education itself. From the earliest grades, nature is presented as an *external object to be studied*, not as an *interactive element of human experience*.

One effort to change this perception was developed at Texas A & M University (USA) through a program fittingly named NATURE (Needed Activities to Understand Resource Ecology). Taught in the university's Department of Wildlife and Fisheries Sciences, NATURE was not designed for students majoring in some field of natural resource management, but crafted for elementary and secondary school teachers. Enrolling over 2,000 teachers from 1987–2010, NATURE taught these educators how to present nature as a field of human interaction incorporated into courses in math, social studies, history, and science. Learning via interaction occurred through activity-based teaching strategies, field trips to various facilities and natural environments, gaming strategies, and computer simulations.[38]

At about the same time a state agency, the Texas Parks and Wildlife Department, had already established the Texas Urban Wildlife Program (TUWP) as part of the agency's infrastructure. In 1997, the TUWP created the Texas Master Naturalist Program, a practical curriculum that culminated in recognized certification of "Master Naturalists" – individuals skilled in the teaching and interpreting of natural history. TUWP became the vehicle for creating a core of committed, well-informed volunteers equipped to offer education, outreach, and service. In the words of the Parks and Wildlife Department, these people were "dedicated to the beneficial management of natural resources and natural areas within their communities."[39] By 1998, the program had grown to a statewide initiative. Private conservation organizations in Texas have also begun partnering with Texas A & M and Texas Parks and Wildlife to further develop this corps of Master Naturalists and expand their influence. The most dramatic effects were in urban areas, where biologists from the TUWP were able to multiply urban conservation efforts with the help of the growing number of Master Naturalists. By 2018, the Master Naturalist Program had not only increased to 48 chapters in Texas, but had been adopted by 28 other US states.[39]

The Delights and Dilemmas of Small Landscapes

In urban environments, most private landowners own little or no land on which they can make private decisions, yet such small but numerous landowners function with near complete autonomy, "managing" their lawns and yards with little regard for government authorities or even their neighbors. But there can be advantages to such autonomy. When I lived in suburban Chicago, I developed a "prairie perennial garden" of native Illinois prairie plants along the edge of my (very small) veranda at the back edge of my front lawn. My garden was perhaps 15 feet long and 4 feet wide, but, with intention, diligence, and modest financial investment, it came to hold stable and reproducing species of over 40 native prairie plants, including some, such as rattlesnake master (*Eryngium yuccifolium*) (one of my favorites), which had become difficult to find anywhere in northern Illinois. In my subdivision, I knew of only one other homeowner who also had a prairie garden (they had a Rattlesnake Master too, but mine was much better looking). Such an extravagance of autonomy as I and other homeowners exercised can lead to what might be called "the tyranny of small decisions." These decisions might be intended to benefit nature, yet still have little effect. I may have been able to cram 40 species of native plants in front of my porch, but my 60 square feet of "prairie" would not save any of them.

Nevertheless, decisions on small properties do matter. Gardens are an important element of the urban environment – small as individual units, but large as a collective "landscape" embedded within and among urban residents. Estimates of the proportional coverage of gardens in cities range from 16% in Sweden,[40] 22–27% in the UK,[41] and up to 36% in New Zealand.[42] Governments cannot (or usually do not) legislate gardening, but they can provide incentives for landowners to be gardeners by providing recognition, even subsidies, for creating gardens that contain specific plants or physical elements that benefit wildlife. Although the garden habitat of any individual urban resident is almost always small, structure can matter more than size. In an urban area, biodiversity often depends on the maintenance of "informal open space,"[43] which may be present in backyards, vacant lots, rail lines, and highway medians and boundaries (I once watched a coyote (*Canis latrans*) make a very skillful crossing of six lanes of 70 mph traffic on Interstate 355 in west suburban Chicago). These areas can be critical to conserving biodiversity but are often not addressed in urban management plans.

An urban resident's private garden, if sufficiently complex, can make a difference. The three-dimensional structure of garden vegetation is a strong predictor of both vertebrate and invertebrate diversity.[44] While governments typically can do little to affect the gardening decisions of individual property owners, NGOs are often much more influential. In the UK, the Royal Society for the Protection of Birds (RSPB) "Homes for Wildlife" scheme encouraged more than 25,000 people to undertake over 300,000 tailored management actions in their homes and gardens in its first year, and the US National Wildlife Federation Backyard Habitat Certification Program now includes more than 100,000 certified backyards.[44] The most important contribution of efforts like these is often not their contribution to increasing the number of individuals or species that are saved, but in changing the property owner's intentions toward and interactions with nature – not "nature" in some geographically distant location or abstract intellectual concept, but "nature" that is part of their everyday experience, their own efforts, and their own private property. One example of such changed interaction has been produced in Nairobi, Kenya, through the work of the Dandora Transformational League, as told by Silva Namalwa.

Story Box 7.1 Living Green in Urban Life

Silva Mgunda Namalwa

Maasai Mara University, Kenya

Urban green spaces are a sight to behold – a feast for the eyes, a peaceful retreat, and an integral part of a city's effort to connect people with nature. Unfortunately, the current trend of increasing urbanization and population growth is rapidly encroaching upon these precious spaces, leaving city dwellers bereft of their benefits. As someone born and raised in Dandora, Kenya, one of Nairobi's most densely populated slums, I saw firsthand how urban green spaces were systematically destroyed by developers, who prioritized economic advancement over environmental and social sustainability. I saw children playing on the roads or rooftops of abandoned buildings because their playing fields had been converted to other forms of land use. In my youth there was a profusion of biodiversity in these spaces – I encountered peacocks and crested cranes in their natural habitat. But as years passed, I observed a disturbing trend: the trees dwindled, high-rise buildings mushroomed, and once-vibrant green spaces shrank. These experiences motivated me to pursue a degree in environmental science and conduct research which confirmed what I had suspected – urbanization and population growth were responsible for these unsustainable changes.

In the Dandora community, Community-Based Organizations (CBOs) have emerged as influential advocates for the protection of urban green spaces, actively combating land-grabbing and highlighting their significance to policymakers. Among these, the Dandora Transformational League (DTL) recognized the value of these green spaces for the public. The DTL petitioned for changes in Kenya's land laws to prohibit the allocation of these public green spaces for other uses. CBOs like the DTL have worked to protect these spaces by sponsoring environmental management events and, in these events, engaging urban residents in conversations about the value of green spaces and mobilizing them as advocates for their protection.

The DTL, not the Kenyan government, has reminded Dandora residents of the value of surrounding themselves with beautiful and well-maintained environments. And people have been persuaded, working with DTL to transform degraded public spaces into beautiful community gardens. One example (see Figure 7.2) has been the Mustard Seed Courtyard (MSC), a serene and delightful haven within the Dandora slums where people can appreciate and connect with the beauty of nature which otherwise might be absent from their surroundings, and in doing so be part of the global effort to advance United Nations Sustainable Development Goal 11 – creating sustainable cities and communities. Its name comes from Jesus' parable of the mustard seed as being like the Kingdom of Heaven – one of the smallest of the garden seeds, but which grows into one of the largest of the garden plants and provides a place for birds to rest and nest in its branches. This first MSC is itself a model of a larger concept, envisioned by DTL's founder Charles Gachanga, to create more gardens like it throughout Dandora, each managed by different individuals in place-specific neighborhoods. Alongside these projects, DTL is coordinating the work of local residents in starting forests in neighborhood open space, as well as clean-up efforts, especially in both streets and green spaces throughout the neighborhood. These initiatives have not only enhanced the beauty of our surroundings in Dandora, but have also restored safe and engaging play areas for children and so reduced the risks of accidents to them on the busy roads.

Figure 7.2 The Mustard Seed Courtyard (MSC) in Dandora, one of the poorest sections of Nairobi, Kenya. Developed by local community organizations and now managed by Dandora residents, the MSC provides a place to care for and enjoy nature in an urban environment. Photo by Silva Namalwa.

By reclaiming and revitalizing these public spaces, we are not only improving our own quality of life but also setting a remarkable example for future generations to emulate. It is through our collective and ongoing efforts that we can continue to transform Dandora into a more vibrant and flourishing community, distinct from its previous state. The world around us is a beautiful place of vast, but not infinite, resources. The current demands pressed upon the Earth by population growth, urbanization, and economic development are not sustainable. In this human-dominated age of the Anthropocene, we must normalize the idea that even spaces we have degraded, like those in Dandora, can be reclaimed, that their reclamation can be part of our work together as a community. By doing so, we can encourage people to take ownership of their own environments and protect them for the sake of generations yet to come.

People as Contributors to Conservation – Citizen Science

People who are not part of the scientific community may have little regard for the findings, pronouncements, and recommendations of conservation scientists about the impending "extinction crisis" or "loss of biodiversity." But perception can be changed by participation. With training and guidance, people who are not scientists can make important contributions to scientific research and discovery. This is "citizen science," *the involvement of citizens from the non-scientific community in academic research.*[45] Citizen science has made important contributions in many fields, including conservation. The man most responsible for initiating the efforts to save the formerly endangered Kirtland's warbler, Dr. Lawrence Walkinshaw, was not a PhD. He was a DDS – a dentist! But his contributions to save the warbler, as well as many species of cranes, were greater than many PhDs could ever boast. Citizen science has made important contributions to conservation efforts around the world, but it is often in urban environments that it can have its greatest effects. Citizen science is not only one of the most important and foundational ways to connect people with nature, but also a means to show people how a greater understanding of nature leads not only to greater protection for nature but to an increasing enjoyment of nature.

Citizen science does more than collect data. It can be the primary conduit for increasing the number of urban residents directly involved in interaction with nature, as well as in positive connections among people across the divides of income, race, or heritage. It is a form of social interaction to advance conservation. In Michigan, for example, the Department of Natural Resources (DNR) oversees the Michigan Clean Water Corps (MiCorps) through which it trains volunteers to collect and identify bottom-dwelling (benthic) invertebrates from streams (see Figure 7.3). Based on the ratios and relative abundances of different orders of such insects, and their tolerance to pollution, volunteers compute a stream quality "score" that can be assigned to the sampled stream, a first

Figure 7.3 Volunteers monitor stream quality through the Michigan (USA) Clean Water Corps (MiCorps), a program of public participation in citizen science supervised by the Michigan Department of Natural Resources (DNR). Volunteers engage with one another, DNR scientists, and nature in collecting stream invertebrates to assess the condition of the stream. Public domain.

step in assessing its condition. Although MiCorps operates throughout the state, its "socialization" of conservation education, monitoring, and action is particularly effective in urban settings. Participants learn about nature from DNR scientists through the different kinds of "bugs" inhabiting their neighborhood stream – how to identify them and what their presence indicates. The participants interact with nature by actually entering the stream, nets in hand (and usually waders on legs), to collect the invertebrates. They form relationships with one another and with scientists in identifying the insects and "scoring" the stream. They perceive the state of the stream through their effort, understand its condition near their own home, and enjoy the understanding they experience.

Interaction with nature in an urban environment can be even more personal – and even more beneficial to non-human creatures. In spring in the city of Utrecht in the Netherlands many fish species migrate through the city via the River Vecht, a branch of the Rhine heading further inland to their spawning areas. They are blocked by the Weerdsluis Lock in central Utrecht, which is rarely opened in this season. To aid the fish in their migration and reduce their "waiting time" (when they are vulnerable to predators, especially fish-eating birds like cormorants and grebes), the Municipality of Utrecht, the Waterboard De Stichtse Rijnlanden, and the regional public water authority (Water Authority Amstel, Gooi, en Vecht) have collaborated to create the "fish doorbell." The doorbell is an underwater camera at the lock, with a live feed streamed to a special "homepage" of the doorbell that is internet accessible. If a person sees fish waiting at the lock, they "press" the digital fish doorbell to signal the lock operator to open the lock and let the fish pass. When a sufficient number of fish are assembled, the operator opens the lock and the fish swim upstream.[46] By using the doorbell to aid the movement of fish through the city, the residents of Utrecht have the experience of aiding fish migration. They accomplish what Aldo Leopold set out to achieve in his wildlife ecology course: to see the landscape of the stream, understand what they see, and enjoy together what they understand. The fish doorbell is one example of how urban conservation must create a social ecology for people to frame their efforts to conserve nature so that human beings become positive determinants shaping the ecology of urban environments.

Cities of Refuge – The Urban Areas that Saved Wildlife

The Urban Attraction

Wild animals are nothing if not adaptive and they adapt to the urban environment like any other. Such responses collectively have been described as *synurbanization* – the response of wildlife to the ecological changes associated with urban development.[16] Many features of urban areas, which we have reviewed previously, are the same features that draw animals to urban areas – homes, factories, office buildings, warehouses, roads, pipelines, power lines, railroads, channelized waterways, reservoirs, sewage ponds, garbage dumps, gardens, parks, cemeteries, and airports. These features and others provide unique structures, juxtapositions of resources and refuge, and, in some cases, specialized habitat that individual species may be uniquely adapted to exploit. Wildlife may also be drawn by crops, ornamental plantings, and domestic pets (coyotes are fond of small dogs, and cats are a favorite of great horned owls). Speaking particularly of buildings, Adams puts it precisely:

> Wild animals occupy urban buildings because they have been invited by the structural design, are opportunistic in seeking shelter, or need an area to rest (e.g., hibernate) or

> raise young … The assemblage of vertebrate animals that have been observed on or about structures and features (e.g., lawns and gardens) [in the United States] included 101 amphibians, 84 reptiles, 457 birds, and 98 mammals.[47]

Some attractions of wildlife to urban structures have become spectacles. The structure attracted the wildlife. Now the wildlife is attracting the people. Each evening, from March through October, in the city of Austin, Texas, about 20 minutes before sundown, over 1 million Mexican free-tailed bats (*Tadarida brasiliensis*) (see Figure 7.4) emerge en masse from beneath the Anne W. Richards Congress Avenue Bridge spanning the Guadalupe River. The origin of this enormous bat colony was an unintended consequence of a 1980 reconstruction that involved the addition of crevices on the underside of the bridge, structures which the bats began to exploit. The rapid increase in the number of bats caused concern. Some citizens petitioned to have them eradicated. However, through an intensive public education campaign led by Bat Conservation International, local residents were eventually persuaded that the bats were not only harmless but fascinating, not to mention useful. In one night's work, the "Bridge Bats" would consume 10,000 to 20,000 pounds of insects during their evening foraging along the Guadalupe, an action appreciated by residents and a welcome help to local farmers. The bats were spared and their population has grown to 1.5 million individuals, making it the largest urban bat colony in the world. Today over 100,000 people visit Austin each year to see the bats make their spectacular evening departure, and one summer night I was one of them, part of the crowd, but more importantly a witness to this amazing display of vibrant life "exploding" from beneath the bridge in a span of just three or four minutes. Small wonder the city constructed the Bat Observation Center adjacent to the Congress Bridge that gives visitors a dedicated area to view the nightly emergence. And the bats have now "spawned" their own cottage industry of businesses specializing in "bat tours" to enhance bat viewing. Austin now has a bat statue, an annual bat festival (with live music and over 50 arts and crafts vendors), and a bat costume contest.[48]

Austin has become a good place to be a Mexican free-tailed bat, but many urban landscapes remain areas of high mortality for a multitude of species. Population modeling

Figure 7.4 Mexican free-tailed bats (*Tadarida brasiliensis*) now sustain a population of over 1.5 million individuals living beneath the Anne W. Richards Congress Avenue Bridge over the Guadalupe River in the US city of Austin, Texas. Photo courtesy of Ann Froschauer, U.S. Fish and Wildlife Service. Public Domain.

studies by Jeffrey Hepinstall-Cymerman et al. indicate that future bird communities in urban areas, for example, will be less diverse and more vulnerable to future losses than at present.[49] And rates of bird mortality have been particularly devastating. Scott Loss et al. estimated that bird losses due to building collisions ranged from 365 to 988 million birds annually.[50] In North America, assessments of population losses in avifauna have climbed steadily over a recent 48-year period, approaching 3 billion birds by 2017.[51] Birds, however, have been the most studied, and in some cases the most welcomed of all wildlife in urban environments, and sometimes the most protected from extinction.

Urban Raptors

The first modern urban bird study published in the scientific literature was by Frank Pitelka, "High populations of breeding birds within an artificial habitat," in the journal *Condor* in 1942, in which Pitelka compared bird populations in a southern California resort town to the surrounding rural hills.[52] Since then studies of urban birds have become legion, and in no group has this been more evident than raptors (from the Latin *rapere*, "to seize or plunder"), a group of birds from multiple taxa that share the morphological features of hooked beaks, grasping talons, and predaceous lifestyles that often feature sudden, swooping attacks on their prey.

Raptors are among the most intelligent of birds, displaying complex behaviors in foraging, courtship, nest building, migration, and (especially) prey selection and preference. Not surprisingly, raptors have often proved the quickest to adapt to and exploit the unique characteristics of urban habitats. They have rapidly perceived that urban areas offer structural elements that provide them with safe nesting sites and favorable feeding areas or, as Bildstein and Therrien put it, "a safe bedroom and a well-stocked pantry."[53] With these benefits in hand (or talon), urban areas can provide birds in general, and raptors in particular, with an increase in reproductive success, a reduction in mortality, and even significant additions to overall regional and worldwide populations.

The story of urban adaptation is particularly intriguing in Cooper's hawks (*Accipiter cooperii*). Historically a "second-growth" forest hawk, Cooper's populations had begun declining by the mid-1900s, and, by the mid-1980s, Cooper's hawks were included on the imperiled or special concern lists of 15 US states. Populations began to increase after the 1972 ban on DDT and the inclusion of the species under the protection of the Migratory Bird Treaty Act, with trends from the North American Breeding Bird Survey (BBS) from 1966 to 2013 indicating a survey-wide increase of about 2% annually.[54] During this period, Cooper's hawks increased in traditional second-growth forest habitat, but also persisted in traditional habitat encroached by suburban development around cities. They even began to colonize core urban areas where traditional breeding habitat did not exist.[54] Over time Cooper's hawks became successful breeders in urban and suburban landscapes throughout North America, with some of their highest reported nesting densities and reproductive rates occurring in these same urban and suburban environments. "What a pleasant irony," noted Robert Rosenfield et al., "that a raptor traditionally regarded as a forest species and deemed to be headed for extinction in the last century … is now perhaps the most common backyard breeding raptor in (nonforested) cities throughout North America."[55]

I found this to be true in my own backyard – literally. When I lived as a suburbanite in the Greater Chicago Area, we had a "neighborhood Cooper's hawk," well known to all the local residents. "Cooper" could be seen most mornings flying low and (very) deliberately from block to block. He was especially fond of my yard. We not only had a backyard bird feeder, but a small outdoor deck, complete with an outer railing and interior benches. The hawk seemed to regard this arrangement as some kind of "backyard bird bistro" and often dined on the railing. His favorite repast was mourning dove (*Zenaida macroura*), and, in the early hours, I often found small piles of soft gray breast feathers, the remains of his breakfast, on the deck platform. It may be that no human gourmet ever had such opportunity to begin each day with squab (in human cuisine, a farm-raised young pigeon that is, according to connoisseur Ryan Grim "the tastiest bird you can legally eat").[56] In urban areas, the Cooper's hawk has shown itself to be a master of behavioral plasticity and habitat flexibility, with an admirably discriminating palate. Squab anyone?

The Peregrine Falcon – How the City Saved a Species

The most dramatic rescue of an endangered species not merely *in*, but arguably *because of* an urban environment is the peregrine falcon (*Falco peregrinus*). A charismatic species if ever there was one, the peregrine is not only the world's fastest bird, but its fastest animal, reaching speeds of over 300 km (186 miles) per hour in its diving attacks on prey. Yet even such breathtaking velocity cannot outrun the pursuit of extinction, especially from pesticides. Peregrine researcher Joel Pagel and his colleagues note that

> Peregrines declined drastically in the mid-20th century from the effects of dichlorodiphenyldichloroethylene (DDE), the principal metabolite of the pesticide dichlorodiphenyltrichloroethylene (DDT). High levels of DDE in the blood caused eggshell thinning by inhibiting function of the shell gland, blocking calcium deposition on the outer layer of eggshells. These broke during incubation, and the result was inevitable population decline.[57]

At this level of pesticide poisoning, peregrine reproduction had all but stopped by the 1960s. The Nature Conservancy reports that, by then, there were no peregrines left in the eastern United States, and by the mid-1970s western US populations had declined by up to 90%. An estimated 3,875 nesting pairs were found in North America prior to the 1940s, but, by 1975, only 324 pairs were left in the US. The use of DDT was banned in the United States by the US Environmental Protection Agency in 1972, and by 1974 The Peregrine Fund was formed, a private conservation effort dedicated solely to saving the peregrine falcon.[58] But now the falconers themselves became the foundation of the peregrine's recovery.

Before training a peregrine in *hunting*, a falconer must first learn the techniques of *hacking* – a process in which nestlings and fledglings are kept and fed for several weeks at "hack boards" where food is left for them as they transition to independent living. It was this technique that was used to introduce captive-bred peregrine falcons into cities, where they would find an abundance of both prey (pigeons!) and "cliff-nesting habitat" (the sheer walls and narrow ledges of man-made skyscrapers!). For peregrines, these things constituted the "safe bedroom and well-stocked pantry" that urban wildlife look for. Such efforts were largely successful. One of their outcomes was the release of over 7,000 peregrines in the United States and Canada from the late 1970s through the early 1990s. By then, 34% of introduced peregrine falcons in the eastern United States were nesting in cities, as were 58% of Midwestern US peregrine populations. In 1999, the peregrine falcon was removed from the US Endangered Species List, and with good reason. Peregrines were displaying a level of behavioral plasticity and habitat adaptability in urban environments that was nothing short of incredible. They constructed nests and reared young on nuclear energy contaminant buildings, bell towers of churches, flowerpots on high-rise apartments, the mast of a mothballed aircraft carrier, and *operating* shipbuilding cranes and cement counterweights on drawbridges (which move vertically whenever a large boat passes under the bridge).[57] Peregrines have now been found on human-made structures in 43 of 50 US states and all but 3 provinces and territories in Canada. Worldwide, they are now present in cities in 26 countries on 6 continents (see Figure 7.5).[59]

Figure 7.5 A peregrine falcon (*Falco peregrinus*) lands near the top of a skyscraper. Peregrines were rescued from the US Endangered Species List through the banning of the pesticide DDT and by their conservation in urban areas, including inducing them to nest on skyscrapers, which are functionally analogous to their preferred nesting habitats of cliffs. Ohio Department of Natural Resources. Public Domain.

Such success has not only changed the status of the Peregrine's populations, but has also altered the interactions of humans with it. Now urban building owners want to learn how to *attract* peregrines to their buildings, and scientists have provided such owners with attachable nest trays and loads of pea gravel to make their skyscraper ledges (on which the nest trays are placed) more attractive as nest sites. Urban residents have become members of "peregrine patrols" that monitor the status of resident falcons. Others participate in "Fledge Watch Squads" that monitor peregrine nesting and rearing success. There has been public observation and participation in banding peregrines and in the use and observation of visual identification bands and digital photography.[57] Today the IUCN lists the peregrine in its category of "Least Concern."

Could the City Be the Well-Used Landscape of the Anthropocene?

The Social and Regulatory Environment Needed for Effective Urban Conservation

Even more important than the ecological effects of urban conditions are the effects on human behavior and human interaction with nature in those conditions. The primary means of advancing biodiversity conservation – the "pillars" upon which conservation in an urban environment must stand – are: 1) governing structures that incorporate conservation into standard policies and regulatory mechanisms; 2) a commitment to "knowing" the state of biodiversity in the city through standardized methodologies; 3) cultivating a curriculum of public education that emphasizes learning how to interact with nature and enjoy the interaction; and 4) building a social culture of conservation that defines and rewards private and neighborhood decisions to enhance biodiversity, creates community participation in conservation, and emphasizes actual changes in human behavior as the most meaningful expression of environmental care.

In an urban environment, human decisions are the primary force governing the nurture and development of the ecosystem. Municipal governing structures frame the priorities and boundaries of those decisions. The city's governing structure must define biodiversity conservation as a specific and intentional objective, complete with a budget allocated to reach it. Only by identifying conservation as a *goal* for the urban environment will subsequent regulations, policies, and financial allocations make sense. Foundationally, city zoning ordinances must not merely allocate "green space," but also include areas where biodiversity is the first intention. Human use can then be integrated into such "conservation zones," but in ways that further, rather than detract from, the primary objective of species protection.

Structurally, the *development* of physical structures and spaces within the city must be required to consider their effects on biodiversity conservation because one of the greatest losses of biodiversity in urban areas arises from such development – both residential and commercial – that replaces natural habitat with human commercial structures, residences, and the imperviously surfaced roads that connect them. In this dimension of urban life, only a change in the *behavior of the developers* can make any difference in these outcomes. Some municipalities have begun programs that make developers "pay first" for any environmental damage they might do, and receive their payment back only if they do no harm.

Many local governments worldwide have begun to adopt programs that implement both the principles of paying first, especially in promoting positive behavior to protect their "urban forests." One mechanism is *tree bonds*, which require the developer to pay for trees they might remove or alter in advance of their construction activities. Another example is tax rebates allocated to developers for maintenance provided to trees in a homeowner's backyard or private greenspace. But an even more positive and systemic method of building biodiversity conservation into development is through the use of municipal funds and bonds to support infrastructure that sustains ecosystem processes, fosters and protects wildlife habitat, and identifies, maintains, and adds to existing urban structures that can be attractive to wildlife. Just as the wall of a skyscraper is more than an upright slab of cement to a peregrine falcon, regulatory mechanisms in development must provoke developers to think creatively, seeing structures they design in development the way different species of wildlife would see them. Bonds or tax subsidies that promote permeable to semipermeable surfaces in the construction of new parking lots, sidewalks, and neighborhood streets can stabilize runoff from precipitation in ways that preserve groundwater reservoirs beneath cities as well as the health of wetlands and waterways within them. Nest boxes for cavity-nesting birds and bats in natural or semi-natural habitats within city parks and preserves foster population growth in these species. Lighting systems – and lighting restrictions – that reduce bird disorientation in migrations and lower bird–building collisions make cities regional and global contributors to increasing wildlife populations. These kinds of examples multiply and, in fact, suggest themselves, when there is a commitment to use physical structure to enhance nature, and gratefully "give back" what nature has given the city.

The State of Urban Biodiversity – Seeing What Is There

For biodiversity conservation to succeed in an urban environment, cities must develop ways of assessing the state of biodiversity conservation in the city and communicating that state to its human residents. Daniel Cooper and his colleagues, who evaluated biodiversity in 112 urban areas in the United States, have noted that the first step is developing a city's "biodiversity score" which evaluates the presence and abundance of native plants and animals. Cooper et al. developed such a metric for the cities they evaluated, and found that

> The fact that biodiversity management score [of a city] ... is not related to the scale variables of city area, income, or population size ... suggests that great potential exists for a wide range of cities to enact meaningful policies that could enhance biodiversity, and for large cities to address gaps in "neighborhood biodiversity"...[60]

Cities can be important centers of native biodiversity, and such biodiversity can be measured. As the old saying goes in business, "what gets measured gets managed." Some cities, recognizing their lack of a good metric to evaluate their urban biodiversity, have invented their own. For example, because of a lack of biodiversity indicators for its own urban environments, Singapore developed the Cities Biodiversity Index in 2008. This self-assessment tool is composed of a qualitative background profile of the city and a quantitative scoring of 23 indicators which measure changes in native biodiversity, provision of ecosystem services, and the presence of governance mechanisms for biodiversity. Worldwide, cities such as London, Durban, Montréal, Los Angeles, and Hyderabad have applied the Singapore Index. From this assessment, these and other cities have initiated processes that have increased their capacities for biodiversity conservation, with the biodiversity indicators serving as guidelines, and the quantitative scoring helping set their priorities for budget allocation and action.[61]

Educating for Interaction

If all of a person's life experiences have occurred within an urban environment, they cannot have the experience of the conservation of nature unless it takes place in that environment. As ecologists James Miller and Richard Hobbs note,

> From the perspective of someone who lives in a city or suburb, conservation is too often something that happens somewhere else—in a national park, wilderness area, or rainforest—and is experienced second-hand (if at all) on television or in a magazine.[62]

If, as Aldo Leopold observed, "the weeds in a city lot convey the same lesson as the redwoods,"[15] then urban educators must ask, "What are these weeds teaching us?"

I remember as a boy in Columbus, Ohio, how amazed and delighted my friends were when, on one summer's day for lack of anything better to do, we wandered into a vacant lot behind our neighborhood shopping center and began exploring. Having lived in the country before my sojourn to the city, I had the advantage of experience in knowing where to look and what to look for. Under every rock, loose plank, and rotting piece of cardboard we found the most incredible living treasures – centipedes and sowbugs (that would roll into balls in our hands), earthworms and millipedes, toads, beetles, ant colonies, and colorful caterpillars crawling up the scraggly plants making a living on the "weeds" thriving in the lot. What a wonderful time we had, and nobody wanted to go home! I think all of us were late for dinner. My tardiness earned a verbal rebuke, but no physical punishment. I hope my friends received equally light sentences. Urban conservationists must perceive weeds and vacant lots with the same insight – and passion –formerly devoted to redwoods and wildlands. They must make urban environments places where nature flourishes and where people understand and enjoy the flourishing it offers.

One of the best educators I ever knew was Dr Harold Snyder, whose vision for interaction and education with nature led to the formation of the Au Sable Institute of Environmental Studies, where I served as Executive Director for nine years. Before Harold was founding environmental institutes he was teaching high school biology in a Michigan city – Grand Rapids. Ever in love with nature in every form, Harold defied conventional thinking in his educational approach. Instead of first developing lectures and then creating labs to support them, Harold first planned the "labs" (which were invariably excursions into nearby forests, prairies, and wetlands) and then built his lectures around his students' interactions with nature. Small wonder so many of his former pupils described these "labs" as transforming experiences. Many ultimately devoted their lives and careers to nature study and conservation science.

The first focus of "environmental education" for students in the city should be, "How can I interact with nature?" And the second focus should be, "How can I enjoy this interaction?" Cities provide exceptional resources for seeing nature through zoos, museums, and botanical conservatories, but the educational aim of all experiences must be directed to "meeting and greeting" the natural world. Learning how to press and preserve plants, create an insect collection, identify birds by sight and song, or use a botanical key to identify shrubs and trees you never knew before are the first pathways to changes in human behavior toward nature, and the means through which such nature can be immensely enjoyed.

The Culture of Urban Conservation

Sociologist Nicki Lisa Cole defined "culture" as "the values and beliefs, language, communication, and practices that are shared in common by a group of people ... our knowledge, common sense, assumptions, and expectations."[63] It is the way we always do things and the way we learn what things we should always do. Urban culture is human culture. Given that people – and interactions between people – form a critical element of urban life, how can conservation become a part of such interaction – an integral element of urban human culture?

The most important quality necessary for involvement in any activity is *commitment*, the force that stabilizes individual behavior under conditions in which a person might otherwise be tempted to alter that behavior.[64] Conservation social scientists Stanley Asah and Dale Blahna assessed this problem by examining what *functions* are served in an individual's life by changing their attitudes and behaviors, and how those changes occur in a social context. The key motivator that created commitment in conservation volunteers was not "to save nature," but, as Asah and Blahna put it,

> The motivation to help the environment was a marginally significant predictor of affective commitment and did not significantly predict normative and overall commitments ... [rather] The more volunteers wanted to socialize with others, the more they were emotionally attached to, identified with, and got involved with volunteering.[65]

In other words, volunteers were more motivated to engage in urban conservation activities if they could socialize with others. *Motivation* is the driver of behavioral change fostered through *social interaction*, especially in the city. The best mechanism to foster conservation, then, is to bring people together in social activities and events that *encourage positive conservation behaviors*, even if the motivating influences are something other than saving species. Asah and Blahna conclude,

> Conservation initiatives are increasingly dependent on volunteers to accomplish goals. Traditionally, conservation practitioners used environmental problems to incite people's involvement in conservation initiatives. Practitioners could be more effective and efficient if they were to use personal and social functions, especially community motivations, as a means to secure volunteers' commitment to conservation issues. These personal, social, and community motivations could be planned to occur during, or related to, voluntary conservation efforts.[66]

In an urban environment, these four elements – a regulatory environment that assigns value to biodiversity conservation, a city that monitors and evaluates the state of its urban biodiversity, education *about nature* based on *interaction with nature*, and a culture of urban sociology that integrates the enjoyment of such interaction (between people and nature) with the conservation of nature – must form the basis of conservation in the urban ecosystem. In the Anthropocene, conservationists must practice and perfect their craft in the urban landscape, dedicating themselves to making urban environments places where nature flourishes and people understand and enjoy its flourishing. If they succeed, the city will become a refuge for biodiversity, not the cause of its demise.

References

1. McGranahan, G., P. J. Marcotullio, X. Bai, et al. 2005. Urban systems. Pages 795–825 *in* R. DeFries and S. Pagiola, editors. *Millennium Ecosystem Assessment. Ecosystems and human well-being: Current state and trends*. Island Press, Washington, DC, USA.
2. Simkin, R. D., K. C. Seto, R. I. McDonald, et al. 2022. Biodiversity impacts and conservation implications of urban land expansion projected to 2050. *Proceedings of the National Academy of Sciences* 19(12):e2117297119.
3. World Population Review. 2024. Most urbanized countries. Central Intelligence Agency. <https://worldpopulationreview.com/country-rankings/most-urbanized-countries#Top%2010%20Countries%20with%20The%20Highest%20Urbanization%20Rates>. Accessed 1 April 2024.
4. Corlett, R. T. 1992. The ecological transformation of Singapore, 1819–1990. *Journal of Biogeography* 19:411–420.
5. Brook, B. W., N. S. Sodhi, and P. K. L. Ng. 2003. Catastrophic extinctions follow deforestation in Singapore. *Nature* 424:420–423.
6. Pouyat, R. V., D. E. Pataki, K. T. Belt, et al. 2007. Effects of urban land-use change on biogeochemical cycles. Pages 45–58 *in* J. G. Canadell, D. E. Pataki, and L. F. Pitelka, editors. *Terrestrial ecosystems in a changing world*. Springer, Berlin, Germany.
7. Evans, M. C. 2021. Re-conceptualizing the role(s) of science in biodiversity conservation. *Environmental Conservation* 48:151–160. DOI: 10.1017/S0376892921000114.
8. Cooper, D. S., E. M. Wood, N. D. Katz, et al. 2021. Large cities fall behind in "neighborhood biodiversity." *Frontiers in Conservation Science* 2:(734931). DOI: 10.3389/fcosc.2021.734931.
9. Seto, K. C., B. Güneralp, and L. R. Hutyraca. 2012. Global forecasts of urban expansion to 2030 and direct impacts on biodiversity and carbon pools. *Proceedings of the National Academy of Sciences* 109(40):16083–16088. <https://doi.org/10.1073/pnas.1211658109>.

10. Simkin, R. D., K. C. Seto, R. I. McDonald, et al. 2022. Biodiversity impacts and conservation implications of urban land expansion projected to 2050. *Proceedings of the National Academy of Sciences* 19(12):e2117297119.
11. McKinney, M. K. 2002. Urbanization, biodiversity, and conservation. *BioScience* 52:883–890.
12. Girardet, H. 2023. The metabolism of cities. *Ecologist Special Series: Megamorphosis*. <https://theecologist.org/sites/default/files/2023-03/Megamorphosis-Cities-FINAL.pdf>. Accessed 3 April 2024.
13. Wilson, E. O. 2014. *The meaning of human existence*. W. W. Norton & Company, New York, USA, 112. ISBN 978-0-87140-480-0.
14. Pickett, S. T. A., and M. L. Cadenasso. 2006. Advancing urban ecological studies: Frameworks, concepts, and results from the Baltimore Ecosystem Study. *Austral Ecology* 31:114–125, 115.
15. Leopold, A. 1949. *A sand county almanac*. Oxford University Press, New York, USA, 174.
16. Adams, C. E. 2016. *Urban wildlife management*. 3rd edition. CRC Press, Boca Raton, Florida, USA, xxxix.
17. Bridgman, H., R. Warner, and J. Dodson. 1995. *Urban biophysical environments*. Oxford University Press, Melbourne, Australia.
18. Salamanca, F., M. Georgescu, A. Mahalov, et al. 2014. Anthropogenic heating of the urban environment due to air conditioning. *Journal of Geophysical Research: Atmospheres* 119:5949–5965. <https://doi.org/10.1002/2013JD021225>.
19. Salamanca, F., M. Georgescu, A. Mahalov, et al. 2014. Anthropogenic heating of the urban environment due to air conditioning. *Journal of Geophysical Research: Atmospheres* 119:5949. <https://doi.org/10.1002/2013JD021225>.
20. Kuhn, I., R. Brandl, and S. Klotz. 2004. The flora of German cities is naturally species rich. *Evolutionary Ecology Research* 6:749–764.
21. Kowarik, I. 2011. Novel urban ecosystems, biodiversity, and conservation. *Environmental Pollution* 159:1974–1983.
22. Kowarik, I. 2011. Novel urban ecosystems, biodiversity, and conservation. *Environmental Pollution* 159:1980.
23. McFrederick, Q. S., and G. LeBuhn. 2006. Are urban parks refuges for bumble bees *Bombus* spp. (*Hymenoptera: Apidae*)? *Biological Conservation* 129:372–382.
24. Carrier, J.-A., and J. T. C. Beebee. 2003. Recent, substantial, and unexplained declines of the common toad *Bufo bufo* in lowland England. *Biological Conservation* 111:395–399.
25. Osborne, J. L, A. P. Martin, C. R. Shortall, et al. 2008. Quantifying and comparing bumblebee nest densities in gardens and countryside habitats. *Journal of Applied Ecology* 45:784–792.
26. Ives, C. D., P. E. Lentini, C. G. Threlfall, et al. 2015. Cities are hotspots for threatened species. *Global Ecology and Biogeography* 25:117–126.
27. Cincotta, R. P., J. Wisnewski, and R. Engelman. 2000. Human population in the biodiversity hotspots. *Nature* 404:990–992.
28. Balmford, A., J. L. Moore, T. Brooks, et al. 2001. Conservation conflicts across Africa. *Science* 291:2616–2619.
29. Parker, S. S. 2015. Incorporating critical elements of city distinctiveness into urban biodiversity conservation. *Biodiversity and Conservation* 24:683–700.
30. Parker, S. S. 2015. Incorporating critical elements of city distinctiveness into urban biodiversity conservation. *Biodiversity and Conservation* 24:683.
31. Guha, R. 2000. *Environmentalism: A global history*. Longman, New York, USA.
32. Kellert, S. R. 1996. *The value of life: Biological diversity and human society*. Island Press, Washington, DC, USA.
33. Mehmood, S. R., and D. W. Zhang. 2001. A roll call analysis of the Endangered Species Act amendments. *American Journal of Agricultural Economics* 83:501–512.
34. McKinney, M. K. 2002. Urbanization, biodiversity and conservation. *BioScience* 52:883–890.
35. Haffert, L., and R. Mitteregger. 2023. Cohorts and neighbors: Urban–rural conflict along the age gradient. *Electoral Studies* 86:102705. <https://doi.org/10.1016/j.electstud.2023.102705>.

36. Louv, R. 2005. *Last child in the woods: Saving our children from Nature-Deficit Disorder*. Allen and Unwin, New South Wales, Australia. Quoted in C. E. Adams. 2016. Urban wildlife management. 3rd edition. CRC Press, Boca Raton, Florida, USA.
37. Leopold, A. 2013. *Aldo Leopold: A sand county almanac and other writings on conservation and ecology*. Library of America #238. Library of America, New York, USA, 626.
38. Texas A & M Agrilife Extension and Texas Parks and Wildlife. 2023. Texas Master Naturalist Program. Texas A & M University, College Station, Texas, USA. <https://txmn.tamu.edu/>. Accessed 1 April 2024.
39. Davis, J. M. 2018. Urban raptor case studies: Lessons from Texas. Pages 246–257 *in* C. W. Boal and C. Dykstra, editors. *Urban raptors – ecology and conservation of birds of prey in cities*. Island Press, Washington, DC, USA.
40. Colding, J., J. Lundberg, and C. Folke. 2006. Incorporating green-area user groups in urban ecosystem management. *Ambio* 35:237–244.
41. Loram, A., K. Thompson, P. H. Warren, et al. 2007. Urban domestic gardens (X): The extent & structure of the resource in five major cities. *Landscape Ecology* 22:601–615. <https://doi.org/10.3170/2008-8-18373>.
42. Mathieu, R., C. Freeman, and J. Aryal. 2007. Mapping private gardens in urban areas using object-oriented techniques and very high-resolution satellite imagery. *Landscape and Urban Planning* 81:179–192. <https://doi.org/10.1016/j.landurbplan.2006.11.009>.
43. Soanes, K., L. Taylor, C. E. Ramalho, et al. 2023. Conserving urban biodiversity: Current practice, barriers, and enablers. *Conservation Letters* e12946. <https://doi.org/10.1111/conl.12946>.
44. Goddard, M. A., A. J. Dougill, and T. G. Benton. 2009. Scaling up from gardens: Biodiversity conservation in urban environments. *Trends in Ecology and Evolution* 1175:1–9. DOI: 10.1016/j.tree.2009.07.016, 1.
45. Vargo, T. L., O. D. Boyle, C. A. Lepczyk, et al. 2012. The use of citizen volunteers in urban bird research. Pages 113–124 *in* C. A. Lepczyk and P. S. Warren, editors. *Urban bird ecology and conservation*. University of California Press, Berkeley, USA.
46. The fish doorbell. 2024. Gemeente Utrecht, Die Strichtse Rijnlanden, and OAK consultants. <https://visdeurbel.nl/en/the-fish-doorbell/>. Accessed 30 June 2024.
47. Adams, C. E. 2016. *Urban wildlife management*. 3rd edition. CRC Press, Boca Raton, Florida, USA, 204.
48. Van Dyke, F., and R. L. Lamb. 2020. *Conservation biology: Foundations, concepts, applications*. Springer, Cham, Switzerland.
49. Hepinstall-Cymerman, J., J. M. Marzluff, and M. Alberti. 2012. Predicting avian community responses to increasing urbanization. Pages 223–248 *in* C. A. Lepczyk and P. S. Warren, editors. *Urban bird ecology and conservation*. University of California Press, Berkeley, USA.
50. Loss, S. R., T. Will, and P. P. Marra. 2014. Refining estimates of bird collision and electrocution mortality at power lines in the United States. *PLOS ONE* 9(7):e101565. <https://doi.org/10.1371/journal.pone.0101565>.
51. Rosenberg, K. V., A. M. Dokter, P. J. Blancher, et al. 2019. Decline of the North American avifauna. *Science* 366:120–124.
52. Pitelka, F. A. 1942. High populations of breeding birds within an artificial habitat. *Condor* 44:172–174.
53. Bildstein, K., and J.-F. Therrien. 2018. Urban birds of prey: A lengthy history of human–raptor cohabitation. Pages 3–17 *in* C. W. Boal and C. Dyksra, editors. *Urban raptors – ecology and conservation of birds of prey in cities*. Island Press, Washington, DC, USA.
54. Rosenfield, R. N., R. W. Mannan, and B. A. Millsap. 2018. Cooper's hawks: The bold backyard hunters. Page 93–109 *in* C. W. Boal and C. Dyksra, editors. *Urban raptors – ecology and conservation of birds of prey in cities*. Island Press, Washington, DC, USA.
55. Rosenfield, R. N., R. W. Mannan, and B. A. Millsap. 2018. Cooper's hawks: The bold backyard hunters. Page 93–109 *in* C. W. Boal and C. Dyksra, editors. *Urban raptors – ecology and conservation of birds of prey in cities*. Island Press, Washington, DC, USA, 93.

56. Grim, R. 2022. The tastiest bird you can legally eat. Food and Wine. <https://www.foodandwine.com/cooking-techniques/the-tastiest-bird-you-can-legally-eat>. Accessed 2 April 2024.
57. Pagel, J. E., C. M. Anderson, D. A. Bell, et al. 2018. Peregrine falcons: The neighbors upstairs. Pages 180–195 *in* C. W. Boal and C. Dyksra, editors. *Urban raptors – ecology and conservation of birds of prey in cities*. Island Press, Washington, DC, USA, 180.
58. The Nature Conservancy. 2018. The peregrine falcon: Nature's finest flying machine. The Nature Conservancy. <https://www.nature.org/en-us/about-us/where-we-work/united-states/indiana/stories-in-indiana/peregrine-falcons/>. Accessed 2 April 2024.
59. White, C. M., S. A. Sonsthagen, G. K. Sage, et al. 2013. Genetic relationships among some subspecies of the peregrine falcon (*Falco peregrinus* L.), inferred from mitochondrial DNA control-region sequences. *Auk* 130:78–87. <https://doi.org/10.1525/auk.2012.11173>.
60. Cooper, D. S., E. M. Wood, N. D. Katz, et al. 2021. Large cities fall behind in "neighborhood biodiversity." *Frontiers in Conservation Science* 2(734931):6. DOI: 10.3389/fcosc.2021.734931.
61. Urban20. 2020. Resilience in the Anthropocene: Mainstreaming nature-based solutions to build resilient cities. Nature-Based Urban Solutions, Riyadh, Saudi Arabia.
62. Miller, J. R., and R. J. Hobbs. 2002. Conservation where people live and work. *Conservation Biology* 16:330–337, 334.
63. Cole, N. L. 2019. So what is culture, exactly? ThoughtCo. <thoughtco.com/culture-definition-4135409>. Accessed 2 April 2024.
64. Asah, S. T., and D. J. Blahna. 2013. Practical implications of understanding the influence of motivations on commitment to voluntary urban conservation stewardship. *Conservation Biology* 27:866–875.
65. Asah, S. T., and D. J. Blahna. 2013. Practical implications of understanding the influence of motivations on commitment to voluntary urban conservation stewardship. *Conservation Biology* 27:871–872.
66. Asah, S. T., and D. J. Blahna. 2013. Practical implications of understanding the influence of motivations on commitment to voluntary urban conservation stewardship. *Conservation Biology* 27:873.

8 Indigenous Peoples and Conservation

Who Are Indigenous Peoples?

They are 6% of the world's population, 476 million individuals in 90 countries.[1,2] They are members of more than 5,000 identifiable ethnic groups, speaking nearly 4,000 of the world's 7,000 languages.[2] They comprise 19% of the world's "extreme poor"[2] (living on the equivalent of 2023 US$2.15 per day). They reside on 22% of the planet's land surface which contains 80% of the Earth's terrestrial biodiversity.[1,2] They are *Indigenous Peoples*. Their designation as "indigenous" is rooted in the Latin word *indigena*, "sprung from the land, native."

It is 1491. This is the last year that Indigenous Peoples would be thriving in their own societies in North and South America without foreign influence. These were continents that, based on the best available evidence, were densely populated by a plethora of human cultures, living in landscapes that their own hands had shaped.[3] Their agriculture featured terraced farms, irrigated fields, fertilized soils, and domesticated crops. They were experienced in the use of prescribed fire, planted desirable trees and shrubs, and selectively removed others. Most were intentional planners and managers of landscapes rich in the resources they desired to obtain with ample and intentionally created habitats for the animals they preferred to hunt. They had been interacting with surrounding biota and environments in ways that shaped evolutionary dynamics, ecosystems, and landscapes for thousands of years.[4] They planned and created cities. The Aztec city-state of Tenochtitlan, the largest metropolis of the pre-Columbian Americas, was larger than any European city of its time – the Venice of America. Built on an island in Lake Tetzcoco in the Valley of Mexico and connected to the mainland from the north, south, and west, its causeways were spanned by bridges of sufficient height that allowed canoes and other water traffic to pass freely beneath them. The bridges themselves could be pulled away at need to protect the city from attack. Interlaced with human-constructed canals, all sections of the city could be visited either on foot or by boat.[5]

The best scientific estimates place the Indigenous American population at between 43 and 65 million. The population of Europe at the same time was perhaps 70 million.[3] Their numbers would be decimated by the military conquests of European invaders, and even more by diseases spread from these invaders that ravaged Indigenous Peoples and aided the conquests of their civilizations, without which they might not have succumbed to such incursions. In Central and South America, most Indigenous Peoples had been subjugated by the end of the seventeenth century. In North America, such conflicts took longer. The "Indian Wars" of the nineteenth century in the United States were not completed until 1886, when the last military campaign against Native Americans ended with the

DOI: 10.4324/9781003404415-8

surrender of Chiricahua Apache Geronimo and his band of 24 warriors, women, and children to 5,000 US troops in Skelton Canyon, Arizona.[6]

The late fifteenth and early sixteenth centuries began a period of European colonization by conquest of Indigenous Peoples in the Americas, Africa, Asia, Australia, and the Pacific Islands, one that continued, in some areas, until the mid-twentieth century. Indigenous Peoples in all these places understood themselves as "society in nature," an integrated whole, what we might call a "nature culture" without separation between humans and the natural world. Colonists saw "nature" in these "new" lands as something to be inventoried and then "organized" to the end of maximum production of specific commodities for market-based profit, leading to a potentially limitless accumulation of wealth (Chapter 2). Indigenous Peoples saw the same land as a relationship between themselves and nature, the objective of which was mutual benefit. They often managed nature to this end, but also engaged in rites, rituals, and practices intended to increase the capacity of the land and its ecological health. They did not possess a concept of land "ownership," but rather of land "use" and "habitation," such that community, identity, and land were one thing, inextricably intertwined. Among the Anishinaabe people of North America's Great Lakes region, for example, the standard greeting to a newcomer was not "How are you?" but "Where are you from?"[7]

The fact that Indigenous Peoples were not maximizing their use and extraction of all the land's "resources" was viewed by colonists as evidence of ignorance. The observed failure of the Indigenous Peoples to exert the effort needed to increase the taking of resources they did use displayed a lack of industry – laziness. The (supposed) lack of cultivation was evidence of absence of ownership, since colonial societies treated cultivation as a prerequisite to claim the land as private property – the grounds for possession and legal title. The apparent lack of institutional governance was evidence that Indigenous Peoples had no "society." Indigenous residents therefore were not "people" – they were simply part of nature. Such thinking was explicit in British colonizers of Australia who declared that land *terra nullius*, a "territory without a master," a legal term used to describe a "space" that can be inhabited but does not belong to a state, and thus has no owner. The British made this declaration even though the best estimates put the number of Indigenous Peoples (Australian Aboriginals) somewhere between 750,000 and 1 million at the time of the British invasion of the continent.[8] But although formally stated in Australia, the concept of *terra nullius* was systemic in colonization throughout the world, and provided the foundation for claiming new land for colonists and removing Indigenous Peoples from it.[9]

The concept of *terra nullius* was supported by the *Inter Caetera* or Doctrine of Discovery, a principle of public international law which declared that when a nation "discovered" land, it directly acquired rights to that land and whatever it found in it. Formalized as early as 1493 in a Papal Bull (pronouncement) of the Catholic Church, the Doctrine of Discovery issued by Pope Alexander VI on May 4, 1493 played a central role in the justification of the Spanish conquest of the New World, supporting Spain's strategy to ensure its exclusive right to the New World lands Columbus had reached the previous year.[10] The Bull stated that any land not inhabited by Christians was available to be "discovered," claimed, and exploited by Christian rulers and declared that "the Catholic faith and the Christian religion be exalted and be everywhere increased and spread, that the health of souls be cared for and that barbarous nations be overthrown and brought to the faith itself."[11] This Doctrine of Discovery would become the basis of all European claims in the Americas as well as the foundation for the United States' western expansion. In the 1823 US Supreme Court case of *Johnson v. McIntosh*, Chief Justice John Marshall's stated opinion in the court's unanimous decision was that "the principle of discovery gave European nations an absolute right to New World lands."[11] The Court decreed that American Indians had only a right of occupancy, which could be abolished.

As armed conflicts with Indigenous Peoples in North America declined near the end of the nineteenth century, the government's plan to prevent further uprisings was to forcibly relocate Native American Indians to defined land holdings (reservations). After such relocations, the government sought to assimilate Indians into a "civilized" way of life by turning them to farming and other occupations compatible with surrounding white culture around the reservations, and accomplish the same in their children through forced education in "Indian schools."[12] Restriction to a relatively small area of land, without freedom of movement, effectively destroyed Indigenous subsistence economies, and education in "formal" schooling was intended to destroy Indigenous culture. Created and funded in the United States by the Civilization Fund Act (or, as it was also called, the Indian Civilization Act) of 1819, the federal government called upon religious institutions to "eliminate Native military resistance and suppress Native traditional practices" of Native American children through such boarding schools, the majority of which were run by the Catholic Church and operated from 1819 until as recently as 1977.[13] By 1926, 83% of all Native American children in the United States and Alaska (then not a state in the US) were enrolled in an Indian boarding school. The Canadian equivalent was the "Indian residential school," established in 1831. A similar pattern was followed in Australia, often executed through the forced (physical) removal of Aboriginal children from their mothers in Aboriginal communities. In all of these cases, the schools determined that the Indigenous culture of the pupils would not be taught, their ceremonies and rituals would not be practiced, and their stories, songs, and poems would not be told, sung, or recited. In the United States and Canada, the stated mission was to "kill the Indian in the child."[13] But often it was the child who died, as crowded conditions promoted the spread of infectious diseases and treatment of children included severe forms of corporal punishment as well as sexual abuse. In 2016, the graves of 215 Indigenous children were found at a former residential school near Kamloops, British Columbia. In Saskatchewan, 751 graves were discovered at another institution.[14] US records are sketchy, but the most recent investigation found that at least 973 Native American children died in boarding schools in the United States.[15] Generations of children subjected to these conditions were emotionally and physically scarred, and, in some tribes, language and culture were almost completely eradicated, to the point that few remaining tribal members could speak the language or teach the traditional rituals or customs of their culture.[16]

In Canada, in 1986, the United Church of Canada apologized to the First Nations People of Canada. "We tried to make you be like us," the church's apology confessed, "and in so doing we helped to destroy the vision that made you what you were."[17] In 2008, Canada's Office of the Prime Minister issued an apology to former students of Canada's residential school system. In 2021, Canada held its first day of National Truth and Reconciliation to address the scandal of the residential schools. In Australia, on February 13, 2008, Prime Minister Kevin Rudd apologized to the Aboriginal People on behalf of the Australian Government and specifically to the "Lost Generations" of Aboriginal children forcibly taken from their families between 1910 and 1970 as a result of "the laws and policies of successive Parliaments and governments" which "inflicted profound grief, suffering and loss on these our fellow Australians."[18] The apology, presented as a motion to the House of Representatives, was unanimously adopted by all members present.[18] In 2009, US President Barack Obama signed an apology for the "violence, maltreatment and neglect inflicted on Native Peoples by citizens of the United States," but his words were buried in a multibillion-dollar defense spending bill, and without public announcement.[19]

Such experiences are but three examples of the abuses and attempted assimilations and genocides perpetrated against Indigenous Peoples throughout the world. Those things that Indigenous Peoples have suffered at the hands of the government, the Christian church, and other institutions in colonized countries are part of their identity today. It is one of the most important problems to be addressed and understood, not only for the sake of justice, but in order to fully apprehend that "conservation," as enacted by colonial governments, was not an innocent bystander to such abuse. It was often the rationale for it.

The Tip of the Spear – The Genesis of Colonial Conservation

Protected Areas (PAs) and Indigenous Peoples – The US Experience

The first conception of a US national park did not come from a political leader, but from a portrait painter, George Catlin. From an early age Catlin was fascinated by stories about Native Americans. His mother, Polly Parton Catlin, told him tales based on firsthand experience, for she and her mother had been captured by Iroquois Indians in Pennsylvania when their settlement along the Susquehanna River was raided. In 1805, at the age of 9, George met an Oneida Indian in woods he was exploring in New York. Catlin recalls being terrified, but the Indian lifted his hand in a gesture of friendship. Catlin never forgot the Indian's kindness.[20]

Fascinated with the lives and cultures of Native Americans, Catlin moved west to Saint Louis, Missouri in 1830, then the gateway to the US frontier. From there he would make repeated excursions to visit and live among Indigenous Peoples in their villages, often spending all of the non-winter months among them. By the end of 1836, Catlin had visited at least 48 tribes, completed more than 300 portraits of Native Americans, and painted over 100 scenes of their landscapes, villages, and rituals. His portraits were the first to depict Native Americans with dignity, as fellow human beings rather than as "savages" (see Figure 8.1).[20] Through his experiences, Catlin developed an admiration for Native Americans and empathy for their misfortunes. He called his portraits "a collection of Nature's dignitaries" and wrote of their moral quality with respect, with negative comparison to his own race. "No Indian," wrote Catlin, "ever betrayed me, struck me with a blow, or stole from me a shilling's worth of my property."[20] "Native Americans," Catlin said, "had been invaded, their morals corrupted, their lands wrested from them, their customs changed, and therefore lost to the world."[20]

Catlin's convictions about Native Americans were foundational to his uniquely original vision of an American national park, one that would not only preserve the magnificent ecosystems and landscapes of the American West, but the Indigenous Peoples and their cultures within them. He advocated the establishment of a reserve across a vast area of the Rocky Mountains from Canada to Mexico and envisioned making most of the Great Plains a sanctuary where Native Americans would enjoy their traditional ways of life, and where vast herds of buffalo, elk, and antelope could continue with their seasonal migrations.[21] It would be, wrote Catlin, "A nation's park containing man and beast, in all the wild and freshness of their nature's beauty."[22]

Figure 8.1 The White Cloud, Head Chief of the Iowas, 1844/1845. This painting by George Catlin, American portrait artist, is believed to be one of over 400 paintings Catlin composed of US Native American individuals, their way of life, and their sacred rituals. Creative Commons Zero (CC0). Available for download on nga.gov object pages. Courtesy of the National Gallery of Art, Washington, DC, USA.

Succeeding conservationists in the United States shared the ideal of a national park to preserve nature, but not Indians. They retained the vision of natural beauty and pristine wilderness, but also the conviction of the inferiority of Indigenous Peoples who therefore could not contribute to such "conservation" effort. The first such park was Yosemite, established by the state of California in 1864 at the urging of America's most prominent conservationist John Muir,[23] which became a US national park in 1890. Muir considered that the Indians in Yosemite "walk softly and hurt the landscape hardly more than the birds and squirrels,"[24] but advocated their removal. Many had already been killed in armed conflicts with gold prospectors since the 1840s.[25] Some of those who remained after 1864, mostly members of the Miwok Tribe, were forcibly relocated, while others stayed on as laborers and as attractions for tourists (there was a "Miwok Village" in Yosemite until the 1960s). Over time their numbers dwindled. The last Miwok employee left in 1996.[26]

The first US National Park (established in 1872) was Yellowstone, an area of 2.2 million acres in northwest Wyoming. The area was inhabited by the Tukarika Shoshone year round, and used in summer by the Bannock and other Shoshone tribes to hunt and fish in its highlands.[27] Evidence suggests that the region had been occupied for at least 11,000 years and its landscape modified by extensive and intentional burning by Native Americans including the Shoshone, who were still setting fires in Yellowstone as late as 1886, when the last of the Indians were evicted.[23] Such removals were not an expression of Park directives but of the Army's policy to resettle Indigenous Peoples on reservations with defined boundaries to aid the western expansion of white settlers.[25] In the United States, Native Americans were not removed from their homelands to create national parks, but their removals allowed national parks to be created in their former homelands.

Exporting Yellowstone – Defining Relations of Power

The "Yellowstone Model" was attractive to colonist governments throughout the world and widely emulated. Canada was among the first to follow the example in forming its national park system (today Parks Canada). Canada's first park, Rocky Mountains National Park (now Banff National Park), created in 1885, displaced the resident Nakoda People and denied them access to the Park for hunting, fishing, gathering, and other subsistence practices because they were perceived as a threat to the mission of protecting the Park's resources, especially the wildlife.[28] In contrast, sport hunting and fishing by Canadian and European tourists were encouraged in this and later in other Canadian parks. Hunting and fishing were acceptable as recreation *for society* (humans living without nature), but not as subsistence practices in areas protected *for nature* (nature without humans). The displacement of the Nakoda also ensured they would not interfere with the Park's tourism industry and its growing revenue. Instead, the Nakoda were placed proximate to church and government residential schools which would assimilate them into colonist culture.[29]

In sub-Saharan Africa, national parks began to be formed in East Africa in the 1930s. Here the Indigenous Peoples included the Maasai, who were nomadic cattle, goat, and sheep herders. For thousands of years, the Maasai had learned that the semi-arid grasslands of East Africa could, with the use of careful movements of their herds and skillful application of prescribed burning, sustain a pastoral lifestyle and economy, but that "fixed site" agriculture and grazing were not sustainable. What the British settlers failed to perceive upon seeing these landscapes was that they were looking at managed grasslands reflecting thousands of years of human intervention.[30] What they did see was expressed in the words of the East Africa British Commissioner, Sir Charles Elliot. "I cannot admit," Elliot declared, "that wandering tribes have a right to keep other and superior races out of large tracts of land merely because they have acquired a habit of straggling over far more land than they can utilize."[31]

The Maasai began to be excluded from their traditional pasture lands and pasturing routes that they had formerly "straggled over" even before national parks were created in East Africa. Prior to the creation of such parks, British settlers took the best Maasai pasture for their own farms and cattle ranches. Later, as national parks were established, more land was taken as colonial government officials claimed that Maasai herding practices were destructive to the land's ecology and a threat to its wildlife. With these changes, much of the Maasai's shared cultural knowledge necessary to successfully maintain the grazing system and repeatedly rejuvenate its grasslands was lost.[32] As the exclusions and forced relocations of the Maasai increased, a small group of wealthy Europeans was able to extend their control over large areas. And so later African national parks set aside large areas of the East African landscape for the enjoyment of European elites based on a narrow definition of nature as wilderness – a landscape free of human beings (at least certain classes of human beings) and any evidence of their activity.

As implemented and practiced by colonial settlers in Africa, conservation was the rationale and means for displacing Maasai herders and taking their grazing lands. In

Tanzania, Maasai herders in Simanjiro, an area historically favored for livestock grazing, saw their grazing lands and water resources enclosed with the creation of Tarangire National Park in 1971. Such "conservation," as seen by the Maasai, amounted to the taking of traditional grazing lands by other people, the majority of whom were white and not even from Tanzania. More enclosures of grazing lands were created, followed by the arrival of commercial farms in the 1980s, and herd mobility became even more restricted. Herders could no longer move their animals in search of pasture and water as they had done in the past. Deprived of fallow periods in which it could recover, pasture in the area began to decline. A proposal to further expand the Park and create more forced relocations led Indigenous residents to create the "Committee to Stop Conservation."[32] Such a response was predictable. Conservation was not a method of "saving nature," but a justification for strengthening existing relationships of power and further depriving the Maasai of the national sovereignty they had formerly enjoyed. As the Maasai themselves became increasingly desperate under these conditions, they engaged in unsustainable agricultural practices confined to smaller land allotments, with further detriment to the environment. Poverty and environmental decline began to reinforce one another, resulting in food insecurity and increasing levels of malnutrition.[32]

Some forced resettlements came at gunpoint. As one member of the Twa people, expelled from the Kahuzi-Biega National Park in the Democratic Republic of Congo in the 1960s, related,

> It was early in the morning. I heard people around my house. I looked through the door and saw people in uniforms with guns. Then one of them forced the door of our house and started shouting that we had to leave immediately because the park is not our land. I first did not understand what he was talking about because all my ancestors have lived on these lands. They were so violent that I left with my children.[33]

Denied their traditional lands and livelihoods, these former hunters and gatherers were forced into squatter camps on the fringes of their once extensive forest territory. Here they suffered malnutrition, landlessness, demoralization, and despair. As another Twa explained, "Since we were expelled from our lands, death is following us. The village is becoming empty. We are heading toward extinction. Now the old people have died. Our culture is dying too."[34] This testimony is not unique, but describes practices characteristic throughout Central Africa, where the creation of PAs has continued to include the forced relocation of Indigenous Peoples, often without plans for resettlement or compensation.[32] By the beginning of the current millennium, approximately 1 million km^2 of forests, pasture, and farmlands have been expropriated in Africa to facilitate "conservation."[34]

The South American Experience

Indigenous Peoples in South America, particularly in Amazonia, were not so much subjected to overt, government-led "Indian wars," but found themselves pressured to surrender their lands, and their resources, in well-coordinated efforts by government and foreign industries. In the Peruvian Amazon in the late eighteenth century, European explorers described the plant resources of the region as so great that "all of them could scarcely be examined in one hundred years by a whole succession of botanists."[35] Depending heavily on the help and knowledge of current Indigenous residents, especially the Yánesha people, the explorers returned with stories of their "discoveries" of vast amounts of plant resources and their potential value. The process followed the familiar pattern of Western scientific investigation and value creation characteristic of the capitalist framework of interaction with nature – the transformation of natural entities into inventories of market-based commodities.

Resource inventory was followed by road construction to facilitate resource removal, and the creation of "fields" (deforested lands) was used by the Peruvian government as the basis to define private property, a pathway for foreign companies and individuals to gain legally recognized ownership of the land, and a means to transform the resources of such land into commodities that could be exported to generate wealth in global markets.[35] By the late nineteenth century, the government of Peru offered a land concession of approximately 500,000 hectares (about 1.25 million acres) to the British-controlled Peruvian Corporation Company. As a result, Yánesha settlements were reduced to small enclaves within increasingly colonized landscapes.[35] The Yánesha were now described as an "economic problem." One of the first International Labour Organization (ILO) reports stated that the Yánesha produced little, consumed less, and were a detriment to the national economy. The perceived solution was to assimilate the Yánesha and other Indigenous Peoples of the region and unite them with surrounding "society" in a unified conquest of nature.[35]

Such efforts were aided by the perception of Western conservationists that "nature" in the Peruvian Amazon, like nature in all other places, was to be conserved as areas in their natural state without human habitation. As recently as 1969, the IUCN at its 10th General Assembly in New Delhi (India) formally defined a national park as "a relatively large area where one or several ecosystems are not materially altered by human exploitation and occupation."[36] The government of Peru held the same view. As anthropologist Peter Bille Larsen noted, "The first protected areas in Peru reflected a preservationist ethos present since the 1940s. In 1968, Salomon Vilchez Murga thus presented Peruvian national parks as 'works of God that the state reserves and protects for posterity.'"[37] This confluence of capitalism, national government, and "conservation" defined a terminal trajectory for the Yánesha people and their culture. Ironically, however, it was only a year later (1969) that a "congress" of the Amuesha (the Indigenous Peoples in Peru, including the Yánesha) entered the political arena as a collective voice for Indigenous Peruvians. By 1974, Peru's "native community" legislation was one of the first legal actions in South America, and perhaps the world, to recognize the collective land rights of Indigenous Peoples in the Amazon. The new law, among other things, replaced the term *tribus selvicolas* (jungle tribes) with *comunidades nativas* (native communities).[35]

The Recognition and Rights of Indigenous Communities

What Does It Mean to Be "Indigenous?"

The word "Indigenous" has been applied to those who are the inhabitants of a particular place predating a colonial period and who have retained their traditional ways of life despite the modernization and social change occurring around them. Indigenous Peoples possess a self-conscious determination to preserve their own identity, and to transmit that identity to their own future generations within their traditional homeland and culture. Although they are resident on every inhabited continent, there have long been and yet remain barriers and prejudices against their recognition. Anthropologist Jim Igoe summarized the problem this way:

> Being a consumer ... is something that Westerners take for granted. However, it is not a natural way of "being" for many peoples. Being a consumer means having money and accumulating things. It also assumes that people are naturally self-interested and acquisitive ... A large part of the world's people grow their own food, build their own houses from materials they find in their environment, and require little in the way of consumer goods ... acquisitiveness and competition are frowned on. People share and cooperate. Those who do not are viewed with suspicion. Such cultural proclivities are often seen as a barrier to globalization and progress; people who hold them are said to have the wrong "mind-set."[38]

In these words, we can begin to understand why the emergence of political organizations representing and advocating for Indigenous Peoples was not only met with suspicion, but active opposition. Perhaps an even greater divide is the difference in the perception of "place." "Modern" people are often fundamentally *placeless*. They move at will to other locations, continuing to earn income and enjoying their same life and livelihood. Similarly, modern economic structures use resources in one location, and when these have been extracted move to another that promises a new supply. In so doing, they are simply imitating the older precedents of colonist cultures.

In contrast, "place" is central to the identity of Indigenous Peoples. In many regions, such as North America, Indigenous Peoples have always been *recognized* as sovereign nations, but *treated* as inferior races by colonist cultures and subjected to their sovereignty. What Indigenous Peoples seek, and are increasingly articulating today, is that their sovereignty be recognized and respected, foundationally through their right of self-determination intrinsically connected to their ancestral "place" or homeland. They desire full jurisdiction of their traditional territory and political control within it, the authority to manage its resources as they see fit, and the right to bar entrance to outsiders and block intrusions of infrastructure (such as dams and highways) which would degrade their resources or culture. They seek the freedom to maintain a self-sufficient, communal (non-capitalist) socioeconomic system that operates according to their own traditions and subject to their own mechanisms of social control.[39]

Indigenous Peoples have long sought not only national but international recognition of their sovereignty, but have been frustrated in the past. In 1923, Cayuga Chief Deskaheh, representing the Iroquois Nation of North America, traveled to Geneva, Switzerland seeking an audience with the League of Nations, the precursor to the United Nations. He argued for recognition of the Iroquois as a sovereign state, asserting their rights to live on their own lands, follow their own laws, and practice their own religion. His petition was denied, and met with what Deskaheh described as "cruel indifference."[40] Beginning in the 1950s, the international community's increasing concern for human rights brought attention to the conditions of Indigenous Peoples worldwide and, by the 1990s, networks of political and social activism created by Indigenous Peoples themselves led to the establishment and growing cohesion of a global Indigenous community.[40] Today the rights and recognition of Indigenous Peoples have become explicit and formally declared. Some of the most important of these declarations are now stated in the Convention on Biological Diversity's Nagoya Protocol, the United Nations Declaration on the Rights of Indigenous Peoples (UNDRIP), the United Nations Charter, the International Covenant on Economic, Social, and Cultural Rights, the International Covenant on Civil and Political Rights, and the Vienna Declaration and Program of Action.[41]

By 2007, a more comprehensive statement on Indigenous Peoples and their rights, with significant input from Indigenous Peoples themselves, was developed and expressed in UNDRIP.[42] UNDRIP's foundational premise was that, henceforth in international law, Indigenous rights were not conditional privileges granted by states, but universal, inherent, indivisible, and independent rights belonging to Indigenous Peoples themselves.[42] Specifically, Article 31 of UNDRIP declared that "Indigenous peoples have the right to maintain, control, protect and develop their ... traditional knowledge as well as the manifestations of their sciences, technologies and cultures ... They also have the right to maintain, control, protect and develop their intellectual property" and further that "the recognition of the rights of Indigenous peoples in this Declaration will enhance harmonious and cooperative relations between the State and Indigenous Peoples, based on principles of justice, democracy, respect for human rights, non-discrimination, and good faith."[42] As Janis Alcorn, an expert on the culture and history of Indigenous Peoples, notes, "The United Nations Declaration of the Rights of Indigenous Peoples (UNDRIP) is now the key defining reference document guiding engagements with Indigenous Peoples."[43]

Indigenous Peoples and Conservation

A World of Difference

Because Indigenous Peoples and their cultures are intrinsically land-based, it is impossible to talk about their identities or rights without reference to land use. Paradoxically, it is just as difficult to talk about Indigenous Peoples and "conservation." Indigenous Peoples do not practice "conservation" in ways that Western culture has historically understood it. As environmentalist Azlan Abas and his colleagues explain regarding the treatment of nature by people indigenous to the forests of Borneo,

> ... conservation and the use of natural resources amount to the same thing: to care for the forest as a source of livelihood, food, and good health, as well as cultural identity ... conservation is inclusive of those positive human components and the traditional knowledge and values of the communities that depend on those resources for a living and for their cultural survival.[44]

It does not help to replace "conservation" with "sustainable use." Indigenous cultures do not practice "sustainable use" as an *objective* of their interaction with nature but as an *outcome* of their practices in nature. If we use the words and thought forms of Indigenous Peoples themselves, what is common to most is a relationship to nature of *mutual nurture and benefit*. They desire to gain well-being (what Westerners sometimes condescendingly refer to as "subsistence") from the land and what it gives. They modify landscapes where changes benefit their needs, not for *extracting* resources, but to make the land more productive at *producing* resources. In fact, they generally do not think of what they gather as "resources" at all, but as gifts of the land and of the spirits that inhabit it, which gives the land its unique identity.[45]

The First Indigenous "Preserves"

From 1900 to 1957, the government of Brazil encountered approximately 80 different Indigenous People Groups within Brazilian Amazonia. In 1900, their population was estimated to be about 1 million. By 1957 it had declined to approximately 200,000, with remaining groups having been decimated or "assimilated." As early as 1910, the government had established an agency called the Indian Protection Service, whose function was to shield Indigenous Brazilians from persecution and oppression. It accomplished little in advancing this mission over the next 50 years, but in 1962, the Brazilian government established and demarcated 7.4 million acres of Brazilian Amazonia as Xingu National Park (more than three times the size of Yellowstone). The objectives were to preserve the flora and fauna of Xingu so that they would be safeguarded for the country's future and preserve what Brazil was like at the time of its "discovery" by European colonizers. The Park also extended protection to the Indigenous tribes of the region within Xingu, offering them assistance, and defending them from "premature and harmful contacts" with the expanding frontiers of Brazilian society. Over the next ten years, three additional parks were created with the same intent.[30]

Following the Western model of the national park, the state was the land's governing body, drew the park's boundaries, and guarded the wild flora and fauna within the designated conservation "fortress" through its administration and legal enforcement. Indigenous Peoples were thus "protected" like any other species within the park, and were among the most endangered. They could remain on some of their land and continue, at least in part, their traditional way of life, but they had neither the rights of land ownership nor the privilege of land management.[30]

One of the largest and most powerful tribes in Xingu was the Kayapo, who believed that the land belonged to them, not the Brazilian government. In the 1970s, when the government attempted to build a road through Xingu to connect isolated mining efforts (established before the Park's creation) to a larger network of mining settlements further north, open conflict erupted with the Kayapo, who instigated armed attacks on anyone who attempted to settle in the region.[30] Eventually, the Kayapo succeeded in making their homeland free of settlers, but Kayapo leaders then began to augment their armed resistance with political engagement by traveling to Brasilia, the capital of Brazil, and meeting with political leaders. The Brazilian government responded by demarcating a special section of Xingu as the Kayapo Indigenous Area (KIA) and accepted legal obligation to protect it. In the KIA, the Kayapo acquired gold and timber concessions and associated financial revenues, recognizing their authority over the use of these resources. Empowered by these funds, the Kayapo took their lobbying to international levels to further protect the area from development, even persuading the World Bank to withdraw loans it had extended to Brazilian contractors to build dams on rivers within or affecting Xingu.[30] Although the conception of Indigenous Peoples like the Kayapo has now moved toward a more accurate, less stereotyped picture of them as "people," the limitations and problems associated with the early "Indigenous Preserves" have led to the development of other ways to include and collaborate with Indigenous Peoples in conservation.

Broken Treaties, Lasting Legal Protection

In Australia the government began establishing national parks almost immediately after the establishment of Yellowstone. Royal National Park, located just south of Sydney, was in place by 1879, making it the world's second national park. More parks followed, often located near cities to relieve the crowding and other pressures of city life. What differed from the US experience, however, was that, as the government began to expand its national park system to areas with fewer non-indigenous Australians, it was creating national parks in areas where Aboriginals were not only greater in number, but still legal owners of large land areas, even, in some cases, almost the entire region or state.[46]

Protesting the loss of their ancestral lands and their exclusion from areas that had provided their livelihood for generations, Aboriginals sought legal redress in court. By the 1970s, they had forced the government to create new laws that protected their historic land rights and permitted them to continue living in the parks where they held legal title to the park's designated area. Australia's National Parks and Wildlife Conservation Act of 1974 and its Aboriginal Land Rights Act of 1976, for example, allowed Aboriginals to establish "land trusts," organizations that were effectively corporations with the power to claim land not already leased or otherwise encumbered by the federal government. These and other statutes, as well as growing alliances between Aboriginals and other conservation groups at both national and international levels, gave Aboriginals legal leverage to have an increasing role in park management and policy, not only permitting their residence in and use of some sections of the parks, but restricting or excluding activities deemed harmful to the Aboriginals or their culture.[30]

By 1992, the Australian Supreme Court ruled in *Eddie Mabo and Others v. The State of Queensland (1992) 66 ALJR 408* that the Indigenous Merriam people of the Murray Islands off the Australian coast held native title to these offshore waters. Further, the Court stated that "the mere circumscribing of land as national parks does not exclude native title."[47] Therefore, in the judgment of the Court, land inside national parks anywhere in Australia might be available to Aboriginal communities that could prove, in the Court's words, a "continuity of traditional ownership."[47] This decision was followed by the passage of the Native Title Act of 1993 which not only gave Aboriginals the right to claim land, but declared that *terra nullius* was a deception, and land could be granted to Indigenous Australians on the basis of native title.[47] But it was not legal support alone that engaged the Aboriginal community in Australian conservation.

Co-management in Australia, Canada, and the United States

Common Burdens – Distinct Responses

Unlike Native Americans in the conterminous United States, some Indigenous Peoples in Australia, Canada, and the US state of Alaska did not experience forced physical relocation to new areas and were able to remain on all or part of their historic homelands with much of their traditional culture intact. In these instances, Indigenous Peoples have not only been able to assert legal ownership of land, but have pursued an active and growing role in managing national parks, and even creating new ones, within their homelands.[48] Such conditions have provided the soil for the growth of "co-management" structures for parks and preserved areas on lands composed of all or part of the historic homelands of Indigenous Peoples.

Co-management is a form of shared decision making between Indigenous Peoples, who often directly use the land's resources, and formally trained government resource managers who have traditionally set policies guiding the use of parks and PAs. For a co-management structure to "work," it must create effective interaction between some sort of management "committee" of Indigenous residents and traditional park administrators, addressing: 1) *Jurisdiction* – Can the co-management committee make binding decisions? 2) *Institutional structure* – Is the local committee of Indigenous Peoples integrated into the actual administration of the park? 3) *Autonomy* – Are committee members able to decide their meeting times and agendas? 4) *Community connections* – Are Indigenous committee members chosen according to local traditions? 5) *Capacity* – Have Indigenous committee members been trained to undertake required activities? 6) *NGOs* – Are there NGOs that provide funding and other support, especially to facilitate indigenous participation and management? 7) *Benefits* – Is co-management of benefit to indigenous residents and effective at involving them in the area's conservation? [30, 46]

Australia

Aboriginals in Australia had managed their landscapes for millennia, including those later designated as national parks, and the effects of removing them from such lands resulted in uncontrolled wildfires and degradation of the ecological qualities for which this landscape was originally valued.[49, 50] Some Aboriginals in Australia viewed what colonist Australians called "wild country" (wilderness areas and national parks) as "sick country," land that had been degraded through a lack of care expressed in appropriate use.[51] Employing the traditional thinking of Western conservation, Australian managers of national parks had followed a policy of "non-intervention" in natural processes, with few or no intrusive management activities.

Aboriginal societies in the same areas had long been active land managers, maintaining a complex system integrating fire and native plant life cycles to ensure food abundance throughout the year.[52] These indigenous fire regimes and traditional knowledge are being used today to recover native biodiversity and ecosystem functions in restoration efforts.[53, 54] Conservation alliances with Aboriginal communities as well as new and existing laws led to the establishment of the Kakadu National Park, protecting one Aboriginal community, the Gagadju, from land claims arising from competing interests (such as mining) and helping the Gagadju determine the management regime most aligned with their aspirations.[30] Even more recently the government recognized the importance of Aboriginal communities to safeguard other lands, returning 160,000 hectares of rainforest within the Daintree National Park to the Eastern Kuku Yalaji community.[54, 55]

Changes in policy in these and other national parks have stimulated the development of co-management models of developing park policy, with Aboriginals having an expanding role in actual park management. One co-management model developed in Australia has been in the Uluru-Katatjuta National Park through its "hand-back, lease-back" model created by the Gagadju. In this arrangement, the National Parks and Wildlife Service (NPWS) has recognized Aboriginal claims ("hand-back") under the condition that the Aboriginals would lease the park back to the NPWS for 99 years ("lease-back"). The lease agreement contains a "detriment clause" that allows the Gagadju to determine if the NPWS has done anything detrimental to their interests, forcing the Service to negotiate binding agreements with the Gagadju, not merely "consult" with them. The lease stipulates that Aboriginals can live and move within the park, and the Service will create training and employment programs for them in the NPWS.[30]

United States

The Alaskan Experience

Of the 84 million acres of US national parks as well as the majority of land in the US Wilderness Preservation System, 54 million acres (64%) are found in Alaska. In the conterminous United States, areas could be designated as parks and wilderness areas because Native Americans had been removed, but this was not the case in Alaska. Instead, Alaskan parks had been mostly superimposed on the traditional homelands of Indigenous Peoples while they were still residing in them and pursuing their traditional livelihoods. Such Indigenous communities were able to retain residential and cultural integrity, fostered through articulation of their own identities, and through what anthropologists would describe as *topophilia* (their love of place) and *historia* (their "atlas of eternity") which conveyed historical meaning and purpose to their present place, and the historical "story" of origin in it and love for it.[56] Grounded in these convictions, Indigenous Alaskans resisted removal, assimilation, regulation, and even representation of their interests by the US National Park Service (NPS). They intended to remain in place (no removal), maintain their culture (no assimilation), continue their traditional way of life (no regulation), and articulate their interests in their own words (no surrogate representation).

Secure in their own identity, Indigenous Alaskans petitioned the government to permit them to continue their traditional hunting, gathering, and harvesting activities on their ancestral homelands, even if these were within a national park. Their efforts were often effective because they composed the numerical majority of the rural Alaska population, which made their interests matter to lawmakers in the Alaska legislature or to those representing the interests of Alaskans in the US Congress. Reflecting this reality, the Alaska National Interest Land Conservation Act of 1980 specifically allowed "traditional and customary" resource users to pursue subsistence activities inside Alaskan national parks.[30]

Such existing laws and political realities provoked the development of co-management approaches in Alaska, now often described as "joint management," or "Co-productive Conservation" – a framework that encompasses both the co-production of knowledge and public services to pursue ethically conscious, culturally relevant, and fully knowledge-based approaches to biodiversity concerns (see Figure 8.2). Six iterative and reflexive co-production processes – co-planning, co-prioritizing, co-learning, co-managing, co-delivering, and co-assessing – focus on the human dimensions and interactions that can allow research, management, and conservation to affect change.[56]

The desire and incentive for Co-productive Conservation has affected the management of existing national parks and created new ones. In 1973, a group of native Inuit Alaskans, the Nunamuits, initiated the creation of a national park in which they would reside. Taking this initiative to protect their subsistence base from economic development, they proposed cooperatively managed governance shared by the Nunamuit and the NPS. In 1980, their vision became a reality in the Gates of the Arctic National Park. Gates of the Arctic was only the beginning of numerous initiatives in Co-productive Conservation that have begun to emerge in Alaska's unique social, political, and physical environment.[30]

The Conterminous United States

In the 48 conterminous US states, the singular example of formal co-management between Native Americans and the NPS is found in Badlands National Park in South Dakota, where the South Unit of Badlands is co-managed with the Oglala Sioux Nation as part of their ancestral homeland.[46] But although the national parks of other US states have not developed formal co-management models, they have increasingly incorporated

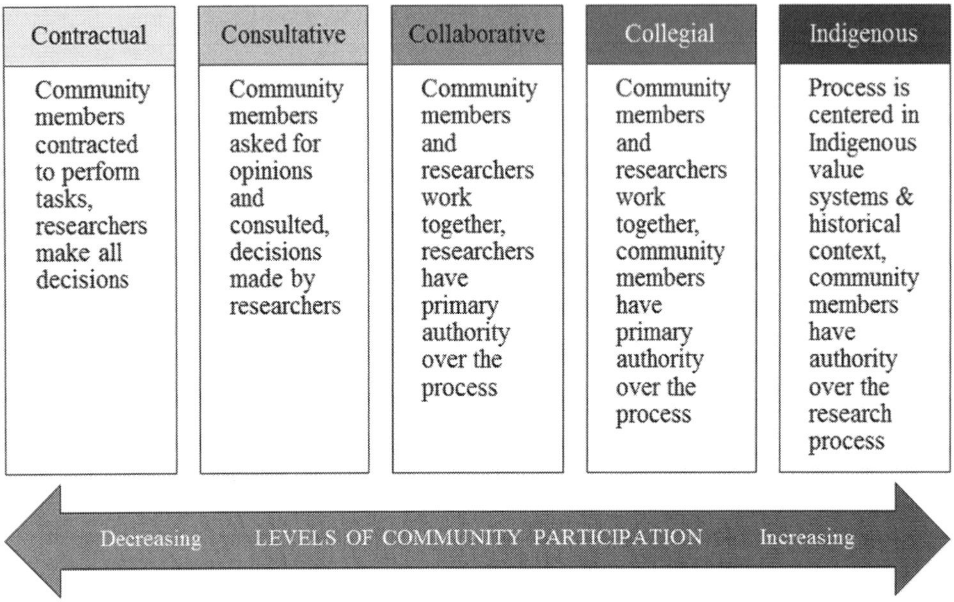

Figure 8.2 Levels and categories of Co-productive Conservation (co-management) between Indigenous Peoples and government conservation agencies, ranging from the Contractual Level (least involvement) to fully Indigenous Management in which Indigenous community members have influence over the research and management process.

Native American management of national parks and monuments using traditional bureaucratic channels in imaginative and non-traditional ways. The engagement of Native Americans in national park management has also increased through a growing number of US court decisions and Acts of Congress that have enforced neglected treaty agreements, giving Native Americans jurisdiction of land and the rights of residence in several national parks, and providing an increasing level of Indigenous autonomy. Examples are multiplying.

Death Valley National Park in California today includes a second national park, the Timbisha Shoshone Natural and Cultural Preservation Area – an area that has been restituted to Timbasha Shoshone. Other tribal reservations also overlap with national park spaces, like the Canyon de Chelly National Monument within the Navajo Nation reservation in New Mexico, Grand Portage National Monument within the boundaries of the Grand Portage Indian Reservation in northern Minnesota, and Big Cypress National Preserve in Florida, surrounded by lands of the Miccosukee Tribe. In the last case, the Miccosukee have lived and governed their affairs in the Big Cypress Preserve on the northern edge of Everglades National Park through NPS permits since 1964. In 1998, the special use permit granting these rights was replaced with a formal legislative act and a legal framework under which the Tribe could live permanently within the Park's boundaries and govern their own affairs. During the tenure of NPS Director John Jarvis from 2009 to 2017, the NPS restored the rights of several Indigenous Tribes to collect plants within a park's boundaries associated with their ancestral or reservation lands. Such provisions have been implemented by the Nisqually Tribe in Mount Rainier National Park

(Washington) as well as the Wabanaki Nations in Acadia National Park (Maine). There are a number of federal agreements between NPS and specific tribes to ensure tribal rights to enter parks to collect foods and medicines, as well as conduct prayer and ceremonies within the parks, although not all such agreements are currently being exercised.[46]

More recently, the US government has gone further, placing Native Americans in positions of federal agency leadership. In December 2020, the Biden Administration nominated Deb Haaland, a member of the Laguna Pueblo Tribe (New Mexico), to become the first Native American to serve in a cabinet-level position as Secretary of the Department of the Interior, a nomination confirmed by the US Senate in August 2021 (see Figure 8.3). In this position, Haaland has administrative oversight of more than 70,000 employees responsible for 20% of US lands, including national parks, monuments, and wildlife refuges, as well as the enforcement of treaty-mandated trust responsibilities to 574 federally recognized American Indian Tribes and Alaska Natives.[57] One of Haaland's first official acts was her issuance of *Secretary's Order 3403: Joint Secretarial Order on Fulfilling the Trust Responsibility to Indian Tribes in the Stewardship of Federal Lands and Waters*, which recognized that federal lands were previously owned and managed by Indian Tribes, contain cultural and religious resources important to them, and lie within areas where Indian Tribes have the reserved right to hunt, fish, gather, and pray pursuant to ratified treaties and agreements with the United States.[57] In August 2022, the Senate

Figure 8.3 Deb Haaland, US Secretary of the Interior, appointed by the Biden Administration in 2021. Haaland, a member of the Laguna Pueblo Native American Tribe in New Mexico, is the first Native American appointed to a US federal cabinet position. As Secretary of the Interior, she is responsible for the administration of all federal lands, national parks, monuments, and wildlife refuges, as well as the enforcement of treaty-mandated trust responsibilities and obligations to 574 federally recognized Indian Tribes and Native Alaskans. Photo US Department of the Interior. <https://www.doi.gov/secretary-deb-haaland>. Accessed 7 May 2024. Public domain.

unanimously confirmed Biden nominee Charles "Chuck" Sams, a member of the Cayuse and Walla Walla Tribes of northeastern Oregon, as the first Native American to serve as Director of the NPS (see Figure 8.4). Since his appointment, Sams has been working with NPS park superintendents nationwide to expand the narrative of Native Americans in national parks, improving how each park's educational program relates the story of Native American history and its relationship to the park. Sams has labored to ensure increased recognition and use of approximately 80 existing collaborative and cooperative agreements and opportunities already in place with various tribes,[58] which will increase the ability of Native Americans to use national parks and to cooperate with the NPS in the management of individual parks.

Figure 8.4 Charles "Chuck" Sams, a member of the Cayuse and Walla Walla Native American Tribes of Oregon (US) and the first US Native American to be appointed, in 2022, to serve as the Director of the US National Park Service.

Canada

Displacement, Residence, and Management

With the 1885 creation of Canada's first park, Banff National Park (formerly Rocky Mountains National Park), First Nations Peoples were denied access to the region because their subsistence practices (hunting, fishing, gathering) conflicted with both late nineteenth-century perspectives of conservation and the objectives of an emerging tourism industry.[59, 60, 61] Consistent with these actions, subsequent government policies were imposed with little regard for the material needs and political priorities of First Nation Peoples.

Such practices began to shift in the 1970s as Indigenous Dene and Inuit Tribes gained more political influence, especially through their opposition to the development of the Mackenzie Valley Pipeline and their insistence on satisfactory resolution of outstanding land claims. Indigenous cultures also proved surprisingly resilient and, when necessary, self-assertive.[46] The Dene, in particular, not only successfully mobilized their own people for political influence but public opinion as well, leading to active and effective negotiation with both provincial and federal government agencies and the settlement of multiple land claims in their favor. Likewise, the Nakoda People, driven from their lands in the creation of Banff National Park, refused to accept colonial policies excluding them from their homeland and continued to access the park for cultural or spiritual purposes, as well as to hunt, fish, and gather.[46] More recently, Nakoda have slowly but persistently increased their presence in the park, first by returning to sacred locations, subsequently gaining plant and medicinal harvesting rights, and then becoming involved with elk culls and grizzly bear monitoring.[46]

From the 1970s onward, the Canadian government began to abandon expropriation and forced expulsion as a mechanism for creating more parks. The testimony of Indigenous representatives before Canada's House of Commons and Senate Committees convinced government representatives that the establishment of national parks beyond the provisions of unresolved land claims amounted to unauthorized taking of Indigenous land.[46] Persuaded by such testimony, three parks were established as National Park Reserves, a new category that did not formally extinguish First Nations land ownership, but made the parks subject to the ongoing unresolved land claims negotiation process. By 1994, Parks Canada had established the Guiding Principles Operational Policies (GPOP) in cooperation with First Nations, formalizing and expanding legislative changes permitting Indigenous hunters to harvest wildlife in select national parks, especially in northern Canada.[46]

The Co-management Model in Canada

In Canada, formal adoption of GPOP along with a growing number of laws and policies intended to implement them led to the formation of park co-management boards comprised of First Nations representatives and expected to provide a means for First Nations members to have a more meaningful role in park management. Unfortunately, most early co-management boards remained largely advisory to park managers, with final power in the hands of the Park Superintendent or the Federal Cabinet. Traditional knowledge of Indigenous Peoples was not consistently incorporated into management policy, a testimony to the still-prevailing view that Indigenous communities were less capable than non-native outsiders of managing parks.[46]

Such attitudes are changing in Canada, and the changes are becoming apparent in park governance. First Nations people still exercise their traditional rights to hunt, fish, and gather wild food on their lands, but are also becoming more integrated into the overall process of park stewardship and history. At Canada's Gwaii Haanas National Park Reserve and Haida Heritage Site, a Coastal Watchman and Forest Guardian program employs people of the Haida Nation as protectors and stewards of the land, and an evolving heritage management program seeks to implement a Haida Land Use Vision based on ethno-ecological values defined in the Haida language.[46] The Kluane National Park Management Board, comprised of First Nations representatives, can make recommendations regarding the development and management of the park direction to the

Canadian Minister of Environment. The Board also has authority to coordinate, with the Fish and Wildlife Management Board, the management of fish and wildlife populations which cross the park's boundaries, with authority over existing and proposed legislation relating to the park and the designation or modification of no-harvesting zones within the park.[21] An employment program to hire First Nations people in park positions is also increasing the role of Indigenous Peoples in Kluane's planning and management.[21]

Indigenous Conservation – Current Results, Future Promise

Recognition and Agency

The increasing recognition of the rights on Indigenous Peoples is just, but is conservation by Indigenous Peoples effective? It had better be. Recent reports by the UN and IUCN assert that 43.5 million km^2 (32%) of the world's land area, across 132 nations and territories, is owned or governed by Indigenous Peoples,[62] including 36% of all Key Biodiversity Areas (KBAs), 40% of all terrestrial PAs and ecologically intact landscapes, and 36% of intact forest landscapes mapped through 2021.[62, 63] Half of the world's 14 terrestrial biomes are more than 30% covered by Indigenous lands. Of 847 global terrestrial ecoregions (including rock and ice), 75% have some overlap with Indigenous lands; 91 ecoregions are more than 75% covered by Indigenous lands, and 179 are more than 50% covered.[62] For some ecoregions, Indigenous governance is the system's only form of protection.[62]

Loss of tropical forests was declared the Problem of the Decade by IUCN and the World Wildlife Fund (WWF) in 1974.[54] Today Indigenous communities are one of the principal agents that prevent it. Lands managed by Indigenous communities have lower deforestation rates than comparable forests in the same regions.[54, 64] In fact, with the exception of the Americas, Indigenous lands have ~26% lower deforestation rates than *protected forests*.[55] In Nepal, deforested lowland regions have improved since the assignment of state forests to local community management, which has slowed deforestation and restored communal forests and watersheds.[64, 65] The lands of Indigenous Peoples in Brazilian Amazonia have contributed a measurable impact in reducing forest carbon emissions and mitigating climate change. From 2010 to 2019, the Amazon Forest could potentially absorb an average of 26,376.66 tons particulate matter of small size ($PM_{2.5}$) per year, with the territories of Indigenous Peoples contributing 27% of this potential absorption. By maintaining the protection of these territories, over 7,000 tons of $PM_{2.5}$ could be absorbed annually, avoiding 15 million cases of respiratory and cardiovascular diseases and infections and annually saving US$2,079,000,000 in health costs to the Brazilian government.[66]

The scientific community has been slow to acknowledge, respect, and publish the traditional and local knowledge of Indigenous Peoples. The first peer-reviewed paper that made use of such knowledge did not appear until 1991.[44] Since then, a number of publications have elaborated the traditional wisdom of Indigenous Peoples, with the most prevalent being focused on forest management (21%), flora and fauna conservation (18%), food security (15%), water management (13%), and land management (12%). Unfortunately, most publications involving traditional knowledge are descriptions of its content or examples of its practice in the daily lives of Indigenous Peoples, not practical applications of its principles in actual management practices.[44] Nevertheless, interest in traditional knowledge of Indigenous Peoples is growing. A formal movement toward greater respect for and accessibility to such traditional Indigenous knowledge has been initiated through the journal *AlterNative: An International Journal of Indigenous Peoples*, founded by New Zealand's Māori Centre of Research Excellence. "AlterNative" requires that authors have approval from Indigenous knowledge holders through research agreements and from ethics committees. They must design a study with participatory research and give the community and individual participants a say in the use of data.[67]

Conservation efforts that seek to separate Indigenous Peoples from nature, or separate their livelihoods from their surrounding biodiversity, have often produced negative results, not only for Indigenous cultures but for conservation.[68] Imported Western market- and commodity-oriented programs to achieve conservation based on capitalist economic assumptions, such as Payment for Ecosystem Services (PES), ecotourism, or revenue-sharing schemes involving extracting natural resources from local lands, although sometimes useful with limited and circumspect applications, have not proved to be the best means of engagement with Indigenous cultures. Such efforts have frequently created perverse incentives or social divisions based on differential income benefits. They have often corrupted Indigenous value systems, trivialized or degraded Indigenous customs and practices into tourist-oriented entertainment, or removed essential resources and landscape elements needed for Indigenous communities to continue their traditional livelihoods. These forms of imposed conservation have often generated local hostility toward the conservation effort itself. It is, rather, fully empowered Indigenous communities, in which local knowledge, local practices, and local governance define conservation and social objectives, that produce the most substantial and enduring gains in conservation efforts.[68]

Defined Rights – The Key to Conservation Achievement

After an exhaustive review of conservation efforts using different approaches and incentives with Indigenous communities, Neil Dawson et al. note that "material benefit alone is rarely sufficient as a pathway to compensate or motivate Indigenous Peoples to align with conservation regulations."[68] It is rather the provision of enhanced participation and increasing autonomy of Indigenous Peoples that permit them to form their governance of conservation agenda which is most likely to foster cooperative conservation efforts with others. Such recognition of the importance of Indigenous government is apparent in the IUCN's current six categories of PAs: 1) strict nature reserves or wilderness areas; 2) national parks; 3) national monuments; 4) habitat/species management areas; 5) protected landscapes or seascapes; 6) PAs with sustainable use of natural resources.[69] Although the IUCN does not recognize the first category (strict nature reserves) as one including the residence of Indigenous Peoples, Indigenous Peoples worldwide routinely establish so-called "no-go zones," including but not limited to sacred burial sites, which are considered the habitations of ancestral or non-human spirits and guarded by them, and which local Indigenous People do not enter.[70] In terms of demonstrated conservation value, such sites have been shown to hold higher densities of studied species than non-protected sites.[71] When Indigenous no-go zones are considered, it is correct to assert that Indigenous Peoples are involved in the governance of all IUCN PA categories.[69, 70]

In 2003, for the first time in modern conservation history, many Indigenous Peoples participated in the World Parks Congress (WPC) in Durban, South Africa. Taghi Farvar, chair of the Congress, noted that such attendance "put a tremendous impact on conservation, probably forever."[72] At that Congress, representatives of Indigenous Peoples led the writing of two important statements: *The Indigenous Peoples' Declaration to the World Parks Congress* and the Congress's closing plenary statement *The Declaration of Protected Areas on Indigenous Territories without Consent and Engagement*. In these statements, representatives of Indigenous Peoples declared that they were not and would not be treated as mere "stakeholders" affected by conservation actions in a specific place, but "rightsholders," having defined rights that determined the use of the land.

In their declarations, Indigenous Peoples stressed seven core principles.[42] First, the free, prior, and informed consent of Indigenous Peoples must be obtained and an appropriate social and cultural assessment carried out. Indigenous Peoples must reserve the right to say "no" regarding the establishment of any new PAs. Second, existing PAs on Indigenous Peoples' lands should now have the rapid establishment of a legal framework to ensure culturally appropriate, full, and effective participation of the Indigenous Peoples concerned regarding the administration and management of the PAs. Third, the WPC must recognize the cultural integrity of Indigenous Peoples and ensure the integration of traditional collective management systems as a basis for the management of PAs. Fourth, the WPC should uphold civil, political, economic, social, and cultural rights in all PA policies, programs, projects, and activities. Fifth, there must be recognition and respect of the rights of Indigenous Peoples in existing and proposed PAs. Sixth, there must be prioritization of recognition of Indigenous territories and areas as a sound basis for conservation. And seventh, Indigenous Peoples and their lands and territories are not objectives of tourism development – tourism must be under the control of the Indigenous Peoples. When these principles are fully integrated into an effective political economy with supporting legal and policy frameworks, Indigenous communities can maintain their own cultures with integrity and provide effective conservation management of their lands (see Figure 8.5).

From these statements and further discussions, the IUCN identified three targets with regard to Indigenous Peoples for all of its future conservation efforts: 1) all established and future PAs would be established and managed in full compliance with the rights of Indigenous Peoples; 2) the management of all relevant PAs would involve representatives chosen by Indigenous Peoples in numbers proportionate to their rights and interests; and 3) participatory mechanisms for restitution of Indigenous Peoples' traditional lands and territories that were incorporated into PAs without their free and informed consent would be established and implemented by the next IUCN WPC.[42]

Such targets are positive steps. They are not uniformly enforced worldwide, but they have established a clear global standard that is increasingly the rule in conservation planning. However, it is not merely the traditional knowledge of Indigenous Peoples that must inform the current conservation effort, but also their radically different valuations of nature, intentions toward nature, and interactions with nature. Today areas under Indigenous management are among the most biodiverse areas remaining on the planet.[73] Landscapes under their traditional patterns of low-intensity use are generally much more biodiverse than those governed by high-intensity agricultural and industrial economies.[73, 74] All global societies are not Indigenous, but all non-Indigenous societies can learn from Indigenous societies. The things that conservationists want to achieve in conservation, Indigenous Peoples are doing.

Indigenous communities practice conservation where they live as a normal part of daily life. Their motive is not to interact with the land to gain wealth from it, but to better understand it (Indigenous knowledge) so that they may apprehend and nurture the

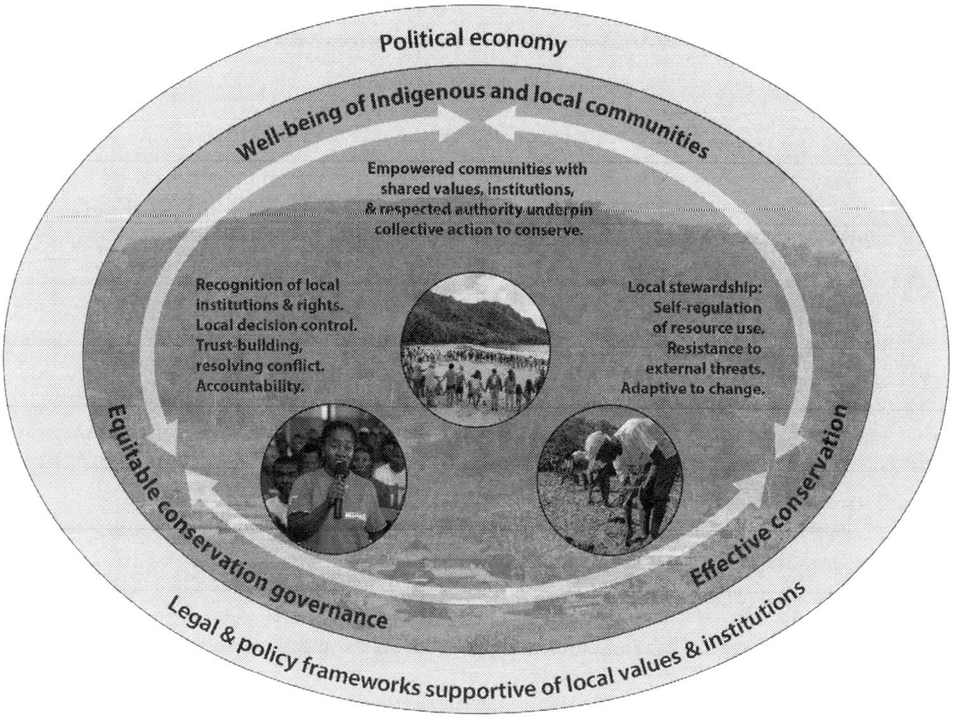

Figure 8.5 The central and inseparable role of Indigenous Peoples in equitable and effective biodiversity conservation in a context of legal and policy frameworks supportive of local values and institutions of Indigenous Peoples. Dawson et al. 2021. The role of Indigenous Peoples and local communities in effective and equitable conservation. Ecology and Society 26(3):19. DOI: 10.5751/ES-12625–260319. <https://www.ecologyandsociety.org/vol26/iss3/art19/>.

land's value, from which they derive their livelihood (Indigenous culture). They value the land through emotional bonds of place and history, not merely through concepts of functions and processes. Such attachment to the land is not abstract. Indigenous communities think of prominent landscape features (mountains, rivers, valleys) as personages which have interests of their own and are the habitation of resident spirits, and to which people are connected in mutual relationships of shared concern. In some countries, like New Zealand, natural features are beginning to receive the legal status of "persons" by the government through the increasing influence of Indigenous Peoples.[75]

Let conservationists learn from the practices of Indigenous communities. The economist Marc Sagoff perceived this truth poignantly.

> If you want to understand what makes the economic use of environmental resources sustainable, … then look to the relationships, cultural and political, of the people in them. Look for affection not for efficiency as the trait with which people treat their surroundings. Where family and community ties are strong, where shared memories and commitments root citizens to places, people can manage changing conditions, and they will do so in ways that respect nature and conserve the environment.[76]

References

1. The World Bank: Indigenous Peoples. 2023. ttps://www.worldbank.org/en/topic/indigenous-peoples#:~:text=There%20are%20an%20estimated%20476,of%20non%2DIndigenous%20Peoples%20worldwide>. Accessed 7 January 2024.
2. World Wildlife Fund, United Nations Environmental Programme – World Conservation Monitoring Centre, GEF Small Grants Program/ICCA-Global Support Initiative, LandMark Global Platform of Indigenous and Community Lands, The Nature Conservancy, Conservation International, et al. 2021. The date of Indigenous Peoples' and Local Communities' lands and territories: A technical review of the state of Indigenous Peoples' and Local Communities' lands, their contributions to global biodiversity conservation and ecosystem services, the pressures they face, and recommendations for actions. IUCN, Gland, Switzerland.
3. Mann, C. C. 2005. *1491 – New revelations of the Americas before Columbus*. Knopf, New York, USA.
4. Ellis, E., N. Gauthierb, K. K. Goldewijkd, et al. 2021. People have shaped most of terrestrial nature for at least 12,000 years. *Proceedings of the National Academy of Sciences* 18(17):e2023483118.
5. Hicks, F. 1982. Tetzcoco in the early 16th century: The state, the city and the Calpolli. *American Ethnologist* 9:230–249. <https://doi.org/10.1073/pnas.2023483118>.
6. Kratz, J. 2022. Geronimo, Apache chief.ttps://prologue.blogs.archives.gov/2022/11/22/geronimo-apache-chief/>. Accessed 20 July 2024.
7. Betz, C. M., editor. 2005. *Diba Jimooyung – Telling our story. A history of the Saginaw Ojibwe Anishinabek*. Saginaw Chippewa Indian Tribe of Michigan and the Ziibiwing Cultural Society, Mount Pleasant, Michigan, USA, 1.
8. De Lacy, T., and B. Lawson. 1997. The Uluru/Kakado Model: Joint management of Aboriginal-owned national parks in Australia. Pages 155–192 *in* S. Stevens, editor. *Conservation through cultural survival: Indigenous Peoples and protected areas*. Island Press, Washington DC, USA.
9. Hendlin, Y. H. 2014. From *terra nullius* to *terra communis*: Reconsidering wild land in an era of conservation and Indigenous rights. *Environmental Philosophy* 11:141–174.
10. Gilder Lehrman Institute of American History. The Doctrine of Discovery, 1493. 2024. <https://www.gilderlehrman.org/history-resources/spotlight-primary-source/doctrine-discovery-1493>. Accessed 6 January 2024.
11. Justia US Supreme Court. 2024. *Johnson & Graham's Lessee v. McIntosh, 21 U.S. 543 (1823)*. <https://supreme.justia.com/cases/federal/us/21/543/>. Accessed 6 January 2024.
12. Boule, K., J. Vayro, and C. W. Mason. 2021. Conservation, hunting policy, and rural livelihoods in British Columbia. *Journal of Rural and Community Development* 6:108–132.
13. Candid. 2024. U.S. pays missionaries to "civilize" Native Americans. <https://nativephilanthropy.candid.org/events/u-s-pays-missionaries-to-civilize-native-americans/>. Accessed 6 January 2024.
14. Nitah, S. 2021. Indigenous Peoples proven to sustain biodiversity and address climate change: Now it's time to recognize and support this leadership. *One Earth* 4:907–909.
15. Brown, M. 2024. Investigation finds at least 973 Native American children died in US government boarding schools. Associated Press. <https://apnews.com/article/indian-boarding-schools-deaths-investigation-82645234fe9d7ce689e8375a51d7e161>. Accessed 30 July 2024.
16. Hanson, E., D. P. Games, and A. Manuel. 2009. The residential school system. Indigenous Foundations. <https://indigenousfoundations.arts.ubc.ca/residential-school-system-2020/>. Accessed 6 January 2024.
17. United Church of Canada. 2024. The apologies. United Church of Canada. <https://united-church.ca/social-action/justice-initiatives/reconciliation-and-indigenous-justice/apologies>. Accessed 19 July 2024.
18. Parliament of Australia. 2010. Apology to Australia's Indigenous Peoples. House of Representatives, Parliament House, Canberra. <https://www.aph.gov.au/Visit_Parliament/Art/Icons/Apology_to_Australias_Indigenous_Peoples>. Accessed 18 July 2024.

19. Capriccioso, R. 2010. A sorry saga: Obama signs Native American apology resolution; fails to draw attention to it. Indian Law Resource Center. <https://indianlaw.org/node/529>. Accessed 6 January 2024.
20. Watson, B. 2002. George Catlin's obsession. Smithsonian Magazine. <https://www.smithsonianmag.com/arts-culture/george-catlins-obsession-72840046/>. Accessed 6 January 2024.
21. Stevens, S., editor. 1997. *Conservation through cultural survival: Indigenous Peoples and protected areas*. Island Press, Washington, DC, USA, 30.
22. Catlin, G. 1841. *Letters and notes on the manners, customs, and condition of the North American Indians*. Toswell and Myers, Piccadilly, UK, 241.
23. Petulla, J. M. 1977. *American environmental history*. Boyd and Fraser, San Francisco, California, USA.
24. Muir, J. 1911. My first summer in the Sierra. *Atlantic Monthly*. January.
25. Keller, R., and M. Turek. 1998. *American Indians and national parks*. University of Arizona Press, Tucson, USA.
26. Patterson, A. 2019. Indian removal from Yosemite National Park. Intermountain Histories. <https://www.intermountainhistories.org/items/show/339>. Accessed 6 January 2024.
27. Bartlett, R. A. 1974. *Nature's Yellowstone*. University of New Mexico Press, Albuquerque, USA.
28. Mason, C. W. 2020. Colonial encounters, conservation, and sport hunting in Banff National Park. Pages 77–100 *in* C. Adams, editor. *Sport and recreation in Canadian history*. Human Kinetics, Champaign, Illinois, USA.
29. Cruikshank, J. 2005. *Do glaciers listen? Local knowledge, colonial encounters, and social imagination*. University of British Columbia Press, Vancouver, Canada.
30. Igoe, J. 2004. *Conservation and globalization: A study of national parks and Indigenous communities from East Africa to South Dakota*. Wadsworth/Thomson, Belmont, California, USA.
31. Elliot, C. 1933. Kenya land commission report. Nairobi: Government Printer, 642.
32. Nelson, J., and L. Hossack, editors. 2003. *From principles to practice: Indigenous Peoples and protected areas in Africa*. Forest Peoples Programme, Moreton-in-Marsh, UK.
33. Kwokwo Barume, A. 2000. *Heading towards extinction: Indigenous rights in Africa – the case of the Twa of the Kahuzi-Biega National Park, Democratic Republic of Congo*. Forest Peoples Programme and International Work Group for Indigenous Affairs, Copenhagen.
34. Colchester, M. 2004. Conservation policy and Indigenous Peoples. *Environmental Science and Policy* 7:145–153. <https://doi.org/10.1016/j.envsci.2004.02.004>, 2.
35. Larsen, P. B. 2015. *Post-frontier resource governance*. Palgrave Macmillan, Houndmills, UK, 38.
36. Dudley, N., and S. Stolton, editors. 2008. *Defining protected areas: An international conference in Almeria, Spain*. IUCN, Gland, Switzerland.
37. Larsen, P. B. 2015. *Post-frontier resource governance*. Palgrave Macmillan, Houndmills, UK, 69.
38. Igoe, J. 2004. *Conservation and globalization: A study of national parks and Indigenous communities from East Africa to South Dakota*. Wadsworth/Thomson, Belmont, California, USA, 2.
39. Bodley, J. H., editor. 1988. *Tribal peoples and development issues: A global overview*. Mayfield, Mountain View, California, USA, 5.
40. Stamatopoulou, E. 2024. *Origins of the movement: Indigenous Peoples' rights*. Institute for the Study of Human Rights, Columbia University, New York, USA. <https://www.youtube.com/watch?v=CracfCTDbE0&list=PLSuwqsAnJMtw1YgJIjSMLN3EXcPj8p8gw&index=2>. Accessed 18 July 2024.
41. McCartney, A. M., M. A. Head, K. S. Tsosie, et al. 2023. Indigenous Peoples and local communities as partners in the sequencing of global eukaryotic biodiversity. *npj Biodiversity* 2:8. <https://doi.org/10.1038/s44185-023-00013-7>.
42. United Nations. 2007. United Nations Declaration on the rights of Indigenous People. <https://www.un.org/development/desa/indigenouspeoples/wp-content/uploads/sites/19/2018/11/UNDRIP_E_web.pdf>. Accessed 29 December 2023.
43. Alcorn, J. 2010. Indigenous Peoples and conservation. MacArthur Foundation White Paper Series. <https://www.macfound.org/media/files/csd_indigenous_peoples_white_paper.pdf>. Accessed 6 January 2024.

44. Abas, A., A. Aziz, and A. Awang. 2022. A systematic review on the local wisdom of Indigenous People in nature conservation. *Sustainability* 14(6):3415. <https://doi.org/10.3390/su14063415>.
45. Htoo, M. H. C., B. Steenhuisen, and B. Verschuuren. 2022. Different natures of reality inform different realities of nature: What Karen perceptions of forest reveal about nature conservation in Indigenous contexts. *Conservation and Society AOP*:1–10.
46. Stevens, S., editor. 2014. *Indigenous Peoples, national parks, and protected areas*. University of Arizona Press, Tucson, USA.
47. The High Court of Australia. *Mabo v Queensland (No 2) ("Mabo case") [1992] HCA 23; (1992) 175 CLR 1 (3 June 1992)*. <https://www.austlii.edu.au/cgi-bin/viewdoc/au/cases/cth/HCA/1992/23.html?stem=0%253D0%253DMabo%2520175%2520CLR>. Accessed 6 January 2024.
48. Fletchera, M.-S., R. Hamilton, W. Dressler, et al. 2021. Indigenous knowledge and the shackles of wilderness. *Proceedings of the National Academy of Sciences USA* 118(40):e2022218118.
49. Crabtree, S. A., D. W. Bird, and R. B. Bird. 2019. Subsistence transitions and the simplification of ecological networks in the Western Desert of Australia. *Human Ecology* 47:165–177.
50. Bird, R. B., D. W. Bird, B. F. Codding, et al. 2008. The "fire stick farming" hypothesis: Australian Aboriginal foraging strategies, biodiversity, and anthropogenic fire mosaics. *Proceedings of the National Academy of Sciences USA* 105:14796–14801.
51. Rose, D. B. 2004. *Reports from a wild country: Ethics for decolonization*. University of New South Wales Press, Sydney, Australia.
52. Bardsley, D. K., T. A. A. Prowse, and C. Siegfriedt. 2019. Seeking knowledge of traditional Indigenous burning practices to inform regional bushfire management. *Local Environment* 24:727–745. DOI: 10.1080/13549839.2019.1640667.
53. Russell-Smith, J., C. P. Yates, A. C. Edwards, et al. 2015. Deriving multiple benefits from carbon market-based savanna fire management: An Australian example. *PLOS ONE* 10(12):e0143426. DOI: 10.1371/journal.pone.0143426.
54. Santini, N. S., and Y. Miquelajauregui. 2022. The restoration of degraded lands by local communities and Indigenous Peoples. *Frontiers in Conservation Science* 3:873659. DOI: 10.3389/fcosc.2022.873659.
55. Langton, M., L. Palmer, and Z. Ma Rhea. 2014. Community-oriented protected areas for Indigenous Peoples and Local Communities: Indigenous protected areas in Australia. Pages 84–107 *in* S. Stevens, editor. *Indigenous Peoples, national parks, and protected areas*. University of Arizona Press, Tucson, USA.
56. Buschman, V. Q. 2022. Framing co-productive conservation in partnership with Arctic Indigenous Peoples. *Conservation Biology* 2(36):e13972. DOI: 10.1111/cobi.13972.
57. US Secretary of the Interior, Secretary Deb Haaland. 2024. US Department of the Interior. <https://www.doi.gov/secretary-deb-haaland>. Accessed 20 July 2024.
58. US National Park Service Office of Communications. 2021. Charles F. Sams III sworn in as National Park Service Director. US National Park Service. <https://www.nps.gov/orgs/1207/director-chuck-sams-sworn-in.htm>. Accessed 20 July 2024.
59. Mason, C. W., A. Carr, E. Vandermale, et al. 2022. Rethinking the role of Indigenous knowledge in sustainable mountain development and protected area management in Canada and Aotearoa/New Zealand. *Mountain Research and Development* 42(4):A1–A9. <https://doi.org/10.1659/mrd.2022.00016>.
60. Snow, J. 2005. *These mountains are our sacred place: The story of the Stoney People*. Fifth House, Calgary, Alberta, Canada.
61. Mason, C. W. 2015. The Banff Indian Days tourism festivals. *Annals of Tourism Research* 53:77–95. <https://doi.org/10.1016/j.annals.2015.04.008>.
62. International Union for the Conservation of Nature. 2021. The state of Indigenous Peoples' and Local Communities' lands and territories: A technical review of the state of Indigenous Peoples' and Local Communities' lands, their contributions to global biodiversity conservation

and ecosystem services, the pressures they face, and recommendations for actions. IUCN, Gland, Switzerland.

63. Sze, J. S., L. R. Carrasco, D. Childs, et al. 2022. Reduced deforestation and degradation in Indigenous Lands pan-tropically. *Nature Sustainability* 5:123–130. DOI: 10.1038/s41893-021-00815-2.
64. Garnett, S., N. D. Burgess, J. E. Fa, et al. 2018. A spatial overview of the global importance of Indigenous lands for conservation. *Nature Sustainability* 1:369–374.
65. Nagendra, H. 2007. Drivers of reforestation in human-dominated forests. *Proceedings of the National Academy of Sciences* 104:15218–15223. DOI: 10.1073/pnas.0702319104.
66. Prist, P. R., F. Sangermano, A. Bailey, et al. 2023. Protecting Brazilian Amazon Indigenous territories reduces atmospheric particulates and avoids associated health impacts and costs. *Earth and Environment* 4:34. <https://doi.org/10.1038/s43247-023-00704-w>.
67. *AlterNative: An international journal of Indigenous Peoples*. 2021. Māramatanga: New Zealand's Māori Centre of Research Excellence. <https://www.journal.mai.ac.nz/content/alternative-international-journal-indigenous-peoples>. Accessed 20 July 2024.
68. Dawson, N., B. Coolsaet, E. Sterling, et al. 2021. The role of Indigenous Peoples and Local Communities in effective and equitable conservation. *Ecology and Society* 26(3):19. DOI: 10.5751/ES-12625–260319. <https://www.ecologyandsociety.org/vol26/iss3/art19/>.
69. International Union for the Conservation of Nature. 2024. Effective protected areas. <https://www.iucn.org/our-work/topic/effective-protected-areas>. Accessed 11 January 2024.
70. Amazon Watch. 2016. Defend the sacred: Making Indigenous sacred sites "No-Go" Zones for extractive industries. <https://amazonwatch.org/news/2016/0908-defend-the-sacred-making-indigenous-sacred-sites-no-go-zones-for-extractive-industries>. Accessed 22 July 2024.
71. Ntiamoa-Baidu, Y. 2008. Indigenous beliefs and biodiversity conservation: The effectiveness of sacred groves, taboos and totems in Ghana for wildlife and species conservation. *Journal for the Study of Religion, Nature and Culture* 2:309–326.
72. Quoted in Stevens, S., editor. 2014. *Indigenous Peoples, national parks, and protected areas*. University of Arizona Press, Tucson, USA, 47.
73. Chapin, M. 2004. A challenge to conservationists. World Watch. <https://www.scribd.com/doc/104065669/A-Challenge-to-Conservationists-Mac-Chapin>. Accessed 5 January 2024.
74. O'Bryan, C. J., S. T. Garnett, J. E. Fa, et al. 2021. The importance of Indigenous Peoples' lands for the conservation of terrestrial mammals. *Conservation Biology* 35:102–108. DOI: 10.1111/cobi.13620.
75. Mason, C. W., A. Carr, E. Vandermale, et al. 2022. Rethinking the role of Indigenous knowledge in sustainable mountain development and protected area management in Canada and Aotearoa/New Zealand. *Mountain Research and Development* 42(4):A1–A9. <https://doi.org/10.1659/mrd.2022.00016>.
76. Sagoff, M. 2008. *The economy of the Earth*. Cambridge University Press, Cambridge, UK, 165.

9 Protected Areas, Parks, and Wilderness

What Is a Protected Area?

The United Nation's 1992 Convention on Biological Diversity, the first global agreement to address all aspects of the conservation of biodiversity, defines a protected area (PA) as "a geographically defined area which is designated or regulated and managed to achieve specific conservation objectives."[1] Similarly, the IUCN defines a PA as "a clearly defined geographical space, recognized, dedicated, and managed, through legal or other effective means, to achieve the long-term conservation of nature with associated ecosystem services and cultural values."[2] PAs have a long history in conservation, but one that has not always been grounded in conservation intention.

The earliest PAs were established as expressions of privilege by royalty or aristocracy for the purpose of exclusive use.[3] Hunting was one such use, so game species were protected from excessive harvests and their preferred habitats increased and enhanced by hired gamekeepers, but their "conservation" was not the goal. The modern concept of a PA – from the nineteenth century onward – was conceived under the assumption that conservation was best accomplished by setting apart ("protecting") certain areas from human land-use transformation.[4] This was understandable, because, when the first PAs were being established, people were beginning to see the destructive effects of human impacts on natural ecosystems.[5]

Such PAs were not informed by scientific investigation, but more often created by public sentiment. For example, despite the size (2.2 million acres) of Yellowstone National Park in the United States, most of its land lies within a high-elevation plateau. The plateau provides excellent summer range for herbivores but is snow-covered in winter. Larger, migratory species like bison (*Bison bison*) and elk (*Cervus canadensis*) leave the park in autumn to forage in lower-elevation winter ranges beyond Yellowstone's boundaries. When they do, the animals are subjected to a "firing line" (or perhaps, "firing squad") of hunters assembled en masse at some boundary lines, killing animals in large numbers. In many African parks, normally migratory herbivores have prudently remained with park boundaries, but damaged the park's vegetation through excessive, year-round grazing. In others, herbivores such as elephants destroyed much of their edible forage, then left the park in search of food and suffered the absence of its protection. Interactions between humans and elephants subsequently increased, and all were negative, usually for the animals.

In the United States, it was not until 1934 that a national park, the Everglades, would be established specifically to protect an *ecosystem*. But from this point on, the designation of PAs shifted from conserving landscapes to conserving wildlife and their habitats,

DOI: 10.4324/9781003404415-9

with area designations informed, at least in part, by biologists and environmental organizations with extensive and area-specific knowledge about biodiversity and ecology. Despite an infamous history coupled with a host of ongoing problems and controversies, PAs remain essential to modern conservation. In the words of the Secretariat of the Convention on Biological Diversity, "Protected areas are a vital contribution to the conservation of the world's natural and cultural resources … [they are] the cornerstones for in situ conservation of biological diversity."[6] But the Secretariat admits, not many pages later, that "current global systems of protected areas are not sufficiently large, sufficiently well-planned, nor sufficiently well-managed to maximize their contribution to biodiversity conservation."[6]

Currently about one-sixth (17%) of the Earth's terrestrial surface is designated as PAs.[7] Among these, the IUCN recognizes seven categories of PAs: strict protection (two categories – strict nature reserves and wilderness areas); ecosystem conservation and protection (national parks); conservation of natural features (natural monuments); conservation through active management (habitat/species management area such as wildlife refuges in the US National Wildlife Refuge System); landscape/seascape conservation and recreation (protected landscapes and seascapes); and areas dedicated to sustainable use of natural resources (managed resource PAs).[8] Despite their large areas and diverse categories, current PAs are insufficient to fully preserve the Earth's biodiversity. And because this is so, environmental scientist Erle Ellis has warned that "there is an urgent need to take action to improve the coverage, representativeness and management of protected areas nationally, regionally and globally."[9]

Proponents of the "half Earth" strategy believe that the solution to the problem Ellis describes is to create more PAs – so many that they would eventually cover at least 50% of the Earth's terrestrial surface![10] Reaching this level of coverage is theoretically possible, but practically difficult, as it would require reallocation of existing land use, including agricultural lands, across a wide spectrum of political, economic, and social stakeholders in scores of countries.[11] Despite the ongoing difficulties associated with both current and proposed PAs, their long history and enduring use testify to the belief among conservationists that PAs are better at protecting ecosystems and wildlife than other methods, and certainly better than no effort at all. As conservationist Nigel Dudley expressed it,

> Protected areas don't work but they are the only things that do work. Convention on Biological Diversity recognizes that its Programme of Work on Protected Areas was its most successful activity in terms of meeting targets and making real progress.[12]

But do PAs *really* work? Only recently has this conviction been put to the test in carefully designed experiments and analyses.

Do PAs Work?

Today some PAs are the only places where certain rare species exist. And some rare species have become extinct in PAs. But only recently has there been careful analysis and counterfactual evaluation of the effectiveness of PAs for conservation compared to similar unprotected areas. In tropical bird species, Victor Cazalis et al. found that a PA

> generally has positive effects of protection on the diversity of bird species that are forest-dependent, endemic to the hotspots, or threatened or Near Threatened, but

> not on overall bird species richness … [and] this benefit is driven by protected areas preventing both forest loss and degradation.[13]

In plant conservation, Yuhao Feng et al. determined, based on an extensive review of global plant conservation literature, that slightly over half (56%) of PAs were "effective in maintaining plant productivity, and PAs in forests were more effective than those in non-forests.... [Overall] global PAs had been playing a positive and persistent role in maintaining plant productivity."[14] In a global review of the effectiveness of PAs in maintaining ecological stability, He and Wei found that "Despite the increasing impacts of climate change and human disturbances on biodiversity, changes in land cover/use and landscape patterns within PAs are generally less than outside PAs … reflecting the stability of PAs in response to external disturbances,"[15] a finding that had been confirmed by earlier studies.[16, 17]

Other investigations have produced a less favorable assessment. Jonas Geldmann and his colleagues compiled data from 12,315 PAs across 152 countries to evaluate the ability of PAs to reduce human pressure compared to its reduction in unprotected areas. "Pressure" was indexed by changes in human population density, night-visible infrastructure, and the proportion of area under cropland. For PAs, Geldmann et al. evaluated "performance" of the PA as being equivalent to changes in these variables and related these changes to site-specific assessments of PA management to determine the role of management inputs in mitigating pressure, then compared the results from the PA to matched unprotected areas with similar environmental conditions in the same region over the same period.[8] PAs in Holarctic and Nearctic Regions (northern Europe and North America) experienced less pressure than their counterfactuals, but those in tropical areas experienced more. Pressure changes were most expressed in agricultural expansion, especially in Indomalaya and the Afrotropics.[8] The authors also note that

> PAs located in areas of lower initial human pressure and limited human access experienced the highest increase in pressure compared with their counterfactual. This suggests that the most remote PAs that had low human pressure in 1995 have suffered more from increased human pressure than PAs under greater initial pressure.[18]

In other words, PAs that had the most to lose (in remoteness and absence of human influence) lost the most compared to their initial states.

PAs have also suffered from changes to their size and status *after* they have been established. One problem is PA downgrading (decreases in legal restrictions regarding protection or use of the PA), downsizing (decreasing the size of the PA), and degazettement (removal of the legal protection or status of the entire PA) (PADDD). Michael Mascia and Sharon Paille of the WWF found 89 instances of PADDD across 27 countries since 1900, and many cases have occurred more recently or are now being considered.[19] One might think that PADDD would be more likely to occur in smaller and more obscure PAs that lack sufficient visibility, notoriety, and concern to prevent such degradations. This is not the case. Large iconic PAs such as the Arctic National Wildlife Refuge (ANWR) in the United States and the Coral Sea Marine Park in Australia have experienced downgrading (see Figure 9.1). Overall, Mascia and Paille found that the problem of PADDD occurred because PAs were "socially defined and socially constructed governance regimes, responsive to social pressures—including conservation demands—at local to global scale."[20] This condition makes the effectiveness of many PAs not only insufficient for conservation

Figure 9.1 Coral Sea Marine Park in Australia was recently downgraded through "degazettement," a change in the legal status of the protected area (PA) that reduces a PA's legal standing. The downgrade opened large portions of formerly restricted areas to commercial fishing, affecting 75% of the total area of the PA. This action currently constitutes the largest downgrade to a single (marine or terrestrial) PA in history worldwide.

but unstable in their restrictions, size, and legal status. PAs work – until they don't. What factors make a PA truly "protected?"

What Makes a PA "Protected?"

In 2004, the Secretariat of the Convention on Biological Diversity noted that 11% of the Earth's land surface has been reserved in conservation PAs.[7] By 2019, the percentage had grown to 17%.[8] Such increase is an achievement of a policy goal (more PAs), but not an achievement of a conservation objective (more biodiversity conserved). As Jonas Geldmann et al. note, "merely designating PAs does not ensure protection of biodiversity. PAs must deliver real conservation benefits by buffering the wild populations and habitats they contain from human pressures on the environment."[21] In some countries, the establishment of PAs can weaken the tenure rights of Indigenous communities, and erode their authority to deter trespass or prevent individuals or companies from extracting resources from the PA.[7] But if the same community, or the government itself, can "sell" extractive rights to resources within the PA, resource extraction activities actually increase and the PA may be degraded. If the prohibitions of activities imposed in the PA reduce subsistence or economic opportunities for local residents, and surrounding lands outside the PA cannot replace them, then illegal activities increase in the PA, not only to gain lost subsistence and economic benefit, but as an act of defiance against the regulations themselves. In a study in and around Rajiv Gandhi National Park (RGNP) in India, for example, the proportion of local people expressing negative attitudes toward the Park and fostering adversarial relationships between the Park staff and local residents increased after the Park imposed bans on livestock, collecting non-timber products, and excluding local communities from participating in the decision-making process governing management actions.[22]

There are four essential qualities (ICIS) in the governance of the PA that make it "work" to preserve biodiversity.

Intention – It might seem obvious that the *intention* of a PA is to protect biodiversity. Regrettably this is not always the case. The first national parks, for example, were explicitly designated to preserve scenery, not species. Such intention usually fostered success in preserving the first, but failure in protecting the second. A stated intention is, at one level, only words on paper. But clarity of intention – thoughtfully expressed, strenuously repeated, and consistently applied – creates a standard by which managers of a PA judge actions, asking at each point whether they are fulfilling the *intention* for which this PA was created.

Capacity – *Capacity* is the ability to carry out intention. If a PA requires the prohibition of certain activities, managers must have the means to enforce those prohibitions. If there is to be informed management action consistent with the PA's intention, then the PA staff must have the capacity to conduct regular monitoring of key variables to determine changes in environmental conditions. And if there is to be any management at all, managers must have the appropriate skills and tools to do it. It is no good saying you are going to remove invasive fish species from the lake if you don't even have a net. Capacity is often fundamentally a matter of having the right tools, the right legitimacy (legal status), the right people, and the right expertise to carry out the intention. Many PAs fail because their protective authority (usually the government) lacks the capacity to keep the species in the PA from being killed in it or removed from it. If poachers and trappers are armed with better weapons than wardens, the poachers and trappers have greater capacity to carry out their intentions than the wardens do to carry out theirs. This is why no PA should be designated without, first, a firm commitment of money, people, and law that provides the capacity to protect the PA, including popular support from the local community.

Integrity – One definition of *integrity* is the state of being honest, fair, and good. Etymologically, it is the state of being "whole" (from the Latin *integer*, "whole, complete, or intact"). Integrity is the consistently applied synthesis of professed intention with completed action. It is nurtured and guarded through purposeful avoidance of conflict of interest. Integrity is the consistent determination to perform the intention even when such effort is difficult or inconvenient. And integrity is to consider the appropriate reward of that effort as the accomplishment of the intention, rejecting *any other reward*, no matter how beneficial or lucrative, that fails to advance the purpose of the PA. If consistently and faithfully practiced, integrity transforms intention into mission.

Not long ago, reflecting on over 40 years of my own work in conservation, I was surprised, even a little shocked, to realize that in all of those efforts, collaborations with dozens of agencies and hundreds of people, including a fair share of disagreements, differences of opinion, and personality conflicts, I was unable to recall a single example of any person I had worked with who lacked *integrity*. I can see now that such integrity, the unified commitment to the "rightness" of the conservation goal, was invariably a major contributor to achieving it. Integrity, in fact, was the essential and irreplaceable component of sustained effort, without which no amount of money, expertise, or influence could have gained the objective.

Based on interactions with people who are conducting conservation efforts in other parts of the world, I have learned of their integrity too. But they have confessed that sometimes they are not always dealing with other people of integrity. In some places, a culture of corruption, bribery, inappropriate influence, and conflict of interest is "normal." Where this is the case in the governance structures or in the people entrusted to the care of the PA, the PA will fail to conserve biodiversity. Worse, such governance will destroy credibility with others, and in that loss forfeit opportunities for collaboration essential to further success.

Stability – Because conservation must have a long-term perspective to frame meaningful goals, the governance structure of a PA requires *stability*. People must be able to count on it as "being there" over the long haul. Such confidence creates credibility, and credibility generates investment. Those who would begin to contribute to and collaborate with the intention and mission of the PA must believe that its governance structure is secure. Even one example is sufficient to show just how important stability is to all facets of conservation.

The okapi (*Okapia johnstonia*) (see Figure 9.2), a large, charismatic mammal of Central Africa, historically inhabited mature, tropical moist forests. Reduced in numbers and range over time, all wild populations of okapi eventually resided in a single African nation, the Democratic Republic of Congo (DRC). Although several PAs were established in the DRC, including four World Heritage sites, for the conservation of the okapi and other species, enforcement of reserve boundaries and protection of species within the reserves was never secure. Conditions degraded further due to high levels of human population growth and immigration of refugees into the PAs and surrounding areas – creating a problem that was beyond the capacity of both the reserve staff and the DRC government to manage.[23, 24, 25] Okapi populations declined as a result of slash-and-burn cultivation and subsequent forest succession associated with human population growth.[23] To make matters worse, the government was destabilized by civil war and the incursions of neighboring countries, which backed the rebel forces. For several years, rebels held the eastern portion of the DRC, where all four of the World Heritage Site PAs were located. The headquarters of three of these four were looted and all technical equipment and vehicles were stolen. Some park guards were killed.[26] In the face of such chaos, conservation funding from western organizations declined.[27] Open hostilities in the civil war came to an end in 2003 and the DRC has moved toward political stability, but civil unrest remains and conservation outcomes are uncertain, especially regarding the protection of existing PAs and okapi populations within them. In 2012, an armed militia group opened fire on the headquarters of the Okapi Conservation Project in eastern DRC, killing 6 people and 13 of the 14 "ambassador" okapi that lived on public display at the center.[28]

These outcomes represent the cost of instability – at multiple levels – and the DRC is not the world's only example of it. The *designation* of PAs does not protect *anything*. Only the effective *protection* of PAs conserves their *biodiversity*. When considering the designation of a new PA, all four components of ICIS – intention, capacity, integrity, and stability – are critical, and all four must be in place for the PA to succeed.

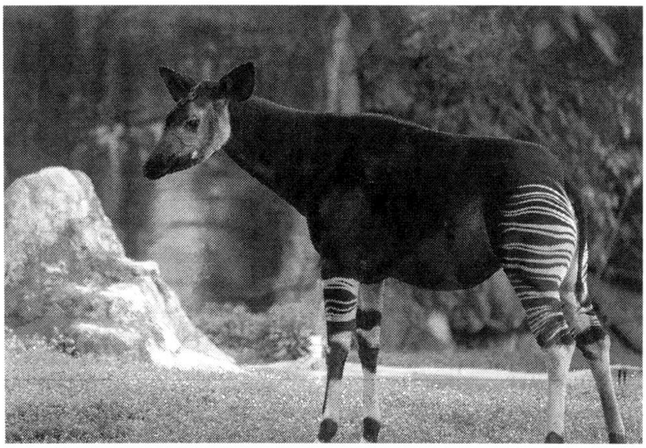

Figure 9.2 The okapi (*Okapia johnstoni*), a forest-dwelling relative of the giraffe, has seen its range reduced to a single African nation, the Democratic Republic of Congo (DRC). Eric Kelby. CC-BY-SA 2.0. <https://creativecommons.org/licenses/by-sa/2.0/deed.en>.

National Parks

Definition and History

The IUCN defines a "national park" as an area to be managed for ecosystem protection and recreation (IUCN Protected Areas Management Category II).[29] Current estimates now place the number of national parks at over 6,000 worldwide.[30] It is difficult to be more precise because category assignments do not always match category realities. Some PAs called "national parks" are not "national" (they are not owned or managed by the nation's government). Some do not allow human recreation (they are not "parks" that people can use). And some have failed to protect the ecosystems they were created to save. Nevertheless, national parks are, in many nations, the crown jewels of conservation – the premier example of its PAs.

The national park concept has long endured a mission of inherent contradiction. Yellowstone, the first US national park, was dedicated by the US Congress in 1872 to provide "a public park or pleasuring-ground for the benefit and enjoyment of the people."[31] Such a statement is not far from the definition of an amusement park. It does not prioritize conservation, but it does offer means and incentive for profit through tourism. A different definition would not arrive until the formation of the US National Park Service (NPS) in 1916 (the US Army administered the US National Parks from 1872 until this date). With the passage of the NPS Organic Act, the mission of the Park Service was declared to

> conserve the scenery and the natural and historic objects and the wild life therein, and to provide for the enjoyment of the same in such manner and by such means as will leave them unimpaired for the enjoyment of future generations.[32]

Here was a full-throated expression of "fortress conservation," the belief that biodiversity is best conserved by creating PAs where ecosystems function in isolation from human disturbance. As a fortress, the park was to ensure that the "scenery," natural objects, and wildlife would continue "unimpaired" happily ever after, "protected" within the park's boundaries from disturbances that would frustrate this mission. Except! – to provide for the "enjoyment" of such scenery, natural and historical objects, and wildlife by park visitors (tourists).

Such declaration of mission sowed conflicts of interest from the beginning. Roads and railroads had to be built to, in, and through the park so that tourists could reach Yellowstone and then travel within it, otherwise there would be few means to "enjoy" what the park had preserved. Campgrounds, hotels, food services, and stores also had to be constructed. Boardwalks had to be placed around geysers, trails routed beside hot springs, and viewpoints on roads enhanced by car pullout areas to help visitors fully appreciate the "scenery." For decades, park rangers killed predators such as wolves, mountain lions, and grizzly bears to increase populations of elk (which were more "viewable" wildlife), allowed tourists to watch bears feed at garbage dumps, and removed "nuisance" wildlife that posed a threat to visitors.

Practices like these no longer occur in Yellowstone or any other US national park, or in well-managed national parks anywhere on Earth. In Yellowstone, predator populations have recovered (or been actively restored), garbage dumps closed, and visitors counseled not to be a "nuisance" to wildlife, rather than vice versa. But as the number of annual visitors grows, national parks often still exist in two separate worlds. In the realm of park "attractions" like geysers, hot springs, and boiling mud pots, roads and trails make access easy. Crowds on peak days can rival those at Disney World. Road and trail traffic is dense and congested. But travel as little as a quarter mile away from these wonders and you will find yourself in a landscape that seems to have never been disturbed at all, and contains no people in it to enjoy the "scenery."

A Walk in the Wildlands (of a National Park)

Informed by many visits to Yellowstone, I was aware of this dichotomy of crowds at attractions and solitude everywhere else, but I wanted to experience more of the latter. One autumn I came to the park to attend the Yellowstone Science Symposium, an annual gathering of conservation scientists working in and around Yellowstone. The conference ended on a Thursday. I was a college professor at that time, and I had previously determined that if I arranged my return trip just right I could be back in time for my classes on Friday. Instead, I made an intentionally irresponsible decision – to leave the conference on Thursday afternoon and spend the next three days hiking around Shoshone Lake, Yellowstone's second largest lake and the largest in its "backcountry." Armed with my well-prepared backpack and ranger-approved backcountry camping permit, I set out, following the trail along the Firehole River from the Kepler Cascade southwest toward the Lone Star Geyser and one of the park's "backcountry" geyser basins just beyond it. After the first few minutes I did not pass another living soul until meeting a father and son exploring the Basin. This geyser basin had no boardwalks, one trail, and a single interpretive sign warning hikers not to leave the path. These geysers were unpredictable, and, if careless, you could find yourself standing on one when it erupted.

Not much further west the trail gave me a view of a long meadow, surrounded by tall spruce and subalpine fir. About 30 bison, well spread across this open space, were grazing. Staying behind some low vegetation, I worked my way to the meadow's edge, with no agenda but to watch. I would find a virtual companion, years later, in the writings of George Catlin, the American portrait painter of Native Americans and western lands, who referred to the things he saw on his journeys as "soul-melting scenery." Perhaps that is the best that words can do, if in vain, to capture that connection between the observer and the observed in such places, and the depth of almost painful penetration that comes in the "receiving" of what you behold. I wondered if explorers like Jim Bridger or Lewis and Clark had witnessed scenes like this, albeit in other places. And in that instant, I was glad that the mission of the Park Service had been accomplished. It had "preserved the scenery" of Yellowstone, although, at that moment, I seemed to be the only visitor who was enjoying it.

To prevent crowding and properly disperse the backpackers the Park Service assigns a specific campsite, by number, to each person or party on their backcountry hikes – one to a customer. They give you a map and mark where it is. Somehow I managed to miss mine, and realized, studying the map as the sun was setting, that I was now at least several miles beyond it. For a fleeting instant an ingrained, rule-following citizen of my conscience urged me to go back and camp at my assigned place. After a reflection of at least three or four seconds, I dismissed the voice, traveled on, and found a good off-trail site in a pleasant grove of pines. There I made camp for the night.

It was late September and, for elk, the beginning of the season of the "rut." At this time, the bulls, calling as loudly as possible, exert themselves to gather groups of females ("harems") and then guard them to ensure from these cows exclusive breeding rights to father the next generation. That effort was now being pursued in force. It was not overwhelming during the day, but as the sun made its exit, the coming darkness was but a rising curtain for a symphony of the bulls and their performance. A bull elk's mating cry is traditionally referred to as "bugling," but I always would have described it as more like "squealing" or "shrieking," and the closer you are, the more piercing it becomes. Warming up their throat harps in the twilight, the bulls' concert soon burst upon the night. More and more musicians, with louder and louder volume – the cacophony made "fortissimo" seem like an understatement. As I passed the advancing hours in the tent after dinner, vainly attempting to sleep, I began to suspect that I must have concealed my tent well, because there were moments when it sounded like the biggest, loudest bull in Yellowstone was right at the door. Slumber was delayed, but being trampled by a bull in an ecstasy of erotic frenzy was avoided. I woke the next morning and, finding my body without injury, counted myself fortunate and declared the experience a good night's sleep.

The next two days brought me round the waters of Shoshone Lake. By the second day I began to realize that I should have packed more food, and so cut back to half rations. On the third day I followed DeLacy Creek away from the lake and back to the road, then west to the Kepler Cascades along the Firehole where I had parked my car. Hunger was trying to start a conversation at this point, but he was no help in bearing the pack, so there was nothing for it but to go on. As I traveled toward the Cascades, some bison had gathered near the highway, and a crowd of tourists had parked their cars and advanced on foot to greet them. One of these was making an attempt at hand feeding, and getting much too close to a large bull. I stopped and advised him to draw back, but, holding some sort of pastry in his left hand, he replied, in a clear British accent, "I want to give this one a bun." After a moment's reflection, I made the choice not to intervene with physical force to deter the tourist (the bison was too large to persuade) and went on. I hope the bison showed the same restraint. Hunger was now increasing its commentary and helped me to quicken my pace. Returning to the Old Faithful Inn, I silenced my unwelcome companion with a chocolate malt at the nearby Hamilton Store. To this day, I have never consumed any food or drink that provided such satisfaction.

I look back on those three days of irresponsibility as one of the best decisions I ever made, and one of the most instructive. The experience also personally illustrated, for me, the dilemma of national parks worldwide. Their attractions swarmed by admiring crowds, their wildlife provoked by annoying tourists (at Yellowstone there are always regrettable injuries), and much of their true life concealed from view. This is the paradox of the national park.

National Parks – Then and Now

The history of national parks has often been a story of displacement and impoverishment of Indigenous Peoples (Chapter 8). In some countries, Indigenous Peoples have been repatriated to continue their traditional life on their historic homelands. In others they work as park entertainers, providing songs, dances, and ceremonies to tourist audiences. Some parks have *lost* species they were established to preserve. And in this present age, national parks, because of their fixed boundaries, have become vulnerable to climate change.[33]

For all their shortcomings, national parks remain essential contributors to conservation. They have been the centers for the reintroduction of species and the rewilding of landscapes, as was the case in Yellowstone with the reintroduction of wolves (*Canis lupus*) to the US Intermountain West and in Grand Canyon to aid the return of the California condor (*Gymnogyps californianus*). They are active centers of conservation research. Yet the persistent tension that remains is how the parks should guide and affect human interaction with nature – given the abundance of visitors in these natural environments.

In the United States, the core employees of the NPS are the *rangers*. Some are trained in law enforcement, others are environmental experts or park historians. But perhaps the most visible to tourists are interpretive naturalists. By *informing* park visitors about the ecological processes, environmental history, and plant and animal residents *of the park*, they guide the *interaction* of people with nature *in the park*.

Not far from my home is one of three US National Lakeshores – the Sleeping Bear Dunes (see Figure 9.3). National Lakeshores are essentially national parks with beaches – all of them adjoining the US Great Lakes. In 2022, Sleeping Bear welcomed over 1.5 million visitors, making it the 56th most visited site among 387 administered by the NPS. Here the NPS's "Junior Ranger" program engages youth in the skills needed to both interact with nature and interpret natural surroundings. In winter, rangers lead tourists on snowshoe hikes (with snowshoes provided) to show visitors how to interact with nature in circumstances that might otherwise appear inhospitable. And in an experience unique to Sleeping Bear's place and history, rangers lead visitors through a reenactment of the rescue of passengers from a Lake Michigan shipwreck in their program "Heroes of the Storm." This experience can literally "immerse" people in nature (and even in the waters of Lake Michigan), as well as connect them to the role of nature in history.[34]

For all their limitations and flawed histories, national parks provide opportunity for informed interaction with the natural world for their millions of visitors, many of whom may be victims of "Nature Deficit Syndrome"[35] or "Nature-deficit Disorder"[36] But by shaping such interactions in ways that value and respect nature, national parks change human intentions toward nature. Perhaps, in the end, the greatest contribution of the national park will not be the species or ecosystems it saved, but the people it taught to better understand the reality of the natural world.

Figure 9.3 Sleeping Bear Dunes, a US National Lakeshore which engages visitors directly in nature through participation in interpretation programs as "junior rangers," winter snowshoe hikes, and reenactments of rescues of passengers from sinking ships.

The Indigenous PA

Indigenous Peoples and Existing National Parks

Throughout the world, most Indigenous cultures have experienced a history of social, cultural, historical, and subsistence interaction with nature that has been beneficial to both. Many national parks have been created through their forced relocations. Today some Indigenous Peoples have seen these lands returned to them. Others have become increasingly involved with government agencies in co-management roles in specific parks (Chapter 8). But even in the best of co-management relationships, Indigenous community members can rightfully ask the question, "Why should we have to co-manage land with you that really belongs to us?" Why indeed?

Early interactions between national parks and Indigenous Peoples were sometimes framed as the creation of a park to "preserve the natives," as a park might preserve an endangered species. Governments worldwide have repented of this approach, and today have even created national parks at the behest of Indigenous Peoples, with their involvement in the park's management. But given the shortcomings of traditional national park structures, why not make Indigenous Peoples *the administrators* of national parks? Or, putting it more directly, why not permit Indigenous Peoples to manage their own homelands in traditional ways and retain the conservation of the landscape and its biodiversity as they historically did?

The Indigenous Conservation Area

There are now a growing number of conservation initiatives which answer this question affirmatively – let's do it! In the United States, some of this affirmation has taken the form of putting Native Americans *in charge* of the NPS or even the cabinet-level Department of the Interior (Chapter 8). More globally and systemically, that affirmation is expressing itself in the creation of Indigenous and Community Conserved Areas (ICCAs) also designated as Indigenous Protected and Conserved Areas (IPCAs). ICCAs and IPCAs are managed not by government agencies or national or international conservation organizations, but by local Indigenous cultures who govern such areas as their own home. The IUCN now formally defines ICCAs as "natural and/or modified ecosystems containing significant biodiversity values and ecological services, voluntarily conserved by (sedentary and mobile) indigenous and local communities, through customary laws or other effective means."[37] In an ICCA, Indigenous residents are the primary decision makers and Indigenous governments have the primary role in protecting and conserving ecosystems through their own laws and traditional knowledge. As anthropologist Peter Bille Larsen put it, "ICCAs are grounded on the principle of recognizing protected areas run by indigenous and local communities as a distinct 'governance' modality, distinguishing them from state-, co-management- and privately-run areas."[38]

One example in Canada is Edéhzhíe (eh-day-shae). Established as a Dehcho PA in their traditional homelands in the southwestern part of the Northwest Territories in 2018, the Dehcho First Nations Assembly designated Edéhzhíe as a National Wildlife Area in 2022, making it Canada's first IPCA. Edéhzhíe is an example of co-management (Chapter 8) and now formally administered as the Edéhzhíe National Wildlife Area (Edéhzhíe NWA) and Edéhzhíe Dehcho PA by a management board composed of representatives of the Dehcho First Nations and the Government of Canada. The board will make decisions by consensus and encourage Indigenous presence on the land,[39] making Edéhzhíe vitally important to the Dehcho culture, language, and way of life. Edéhzhíe is no small area. At 14,218 km^2 in size, Edéhzhíe is over 5,000 km^2 larger than Yellowstone and more than twice the size of Canada's Banff National Park (see Figure 9.4). Similarly the Lutsël K'e Dene First Nation has succeeded in protecting an area known as Thaidene Nëné € ("land of the ancestors"), one of the largest PAs in North America (26,376 km^2). Preserved according to Dene laws and with the cooperation of the Canadian government, Thaidene Nëné € has been designated as an Indigenous PA, with portions of it variously designated as a national park, territorial park, and wildlife conservation area.[40]

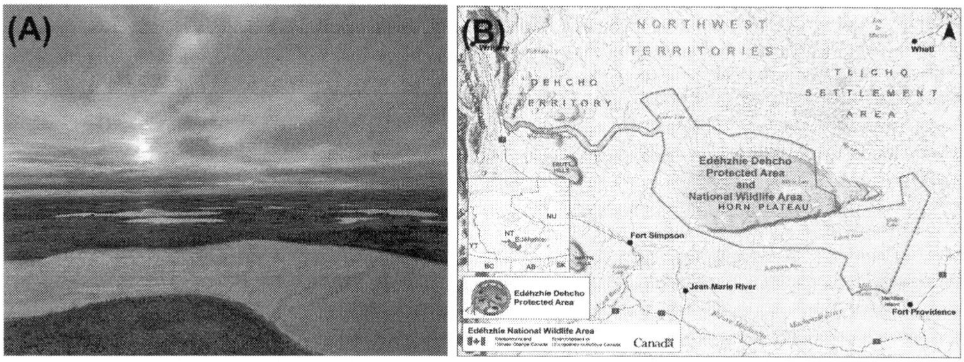

Figure 9.4 (A) The Edéhzhíe (eh-day-shae) National Wildlife Area and Edéhzhíe Dehcho Protected Area in Canada is an Indigenous Protected and Conserved Area (IPCA) jointly administered by the Dehcho First Nations and the Government of Canada. (B) Located in Canada's Northwest Territories, the combined area encompasses over 14,000 km^2, making it one of the world's largest Indigenous Community Conservation Areas (ICCAs) and twice the size of Canada's Banff National Park.

The importance of Indigenous PAs is increasing in conservation because evidence supports their effectiveness, as well as their geographic relevance. The global distribution of "high biodiversity zones" corresponds closely with areas that have experienced the most intense Indigenous cultural impact. To formally establish ICCAs worldwide, nations must: 1) legally recognize and enable relatively autonomous Indigenous and locally led and managed territories; 2) prioritize Indigenous and local knowledge as the basis for the area's management; and 3) support Indigenous rights to land, resources, diverse livelihoods, and traditional ways of living.[41]

Within its category of ICCAs, the IUCN also recognizes a subcategory called Indigenous Conservation Territories (ICTs). ICTs are comprised of all or parts of Indigenous Peoples' territories or lands that they conserve through their own administration of traditional and sustainable ways of life. Such Indigenous control can take the form of an Indigenous or "Tribal" park more or less the equivalent of a national park, but administered by Indigenous residents instead of the government. Anthropologist Jim Igoe explains that, in Nepal,

> Parks controlled by Indigenous communities exist primarily in places where they have treaties that give them control of their own natural resources and the legal authority to exclude outsiders. ... In terms of creating a park, such communities usually have the authority to create some sort of management bureaucracy. If not, they have the

authority to negotiate with conservation NGOs and government agencies to provide the support necessary to create one.[42]

An important element of the success of ICCAs in Canada is the government's explicit recognition of the essential role of reconciliation with First Nations, expressed in multiple areas of Canadian life, including the establishment of new conservation areas. Parks Canada's definition of reconciliation has three components: 1) strengthening Indigenous connections with traditionally used lands and waters; 2) expanding and ensuring presentation and commemoration of Indigenous histories and cultures; and 3) increasing economic opportunities related to Indigenous tourism.[40] When these elements are in place at the national level in any country, Indigenous Peoples are given legitimacy and support in their efforts to assert sovereignty over their traditional homelands, establish permanent residence in such lands if they have not already done so, and practice livelihoods and subsistence activities fostering the conservation and nurture of the land's biodiversity and ecological systems, as well as the integrity of their own culture.

Indigenous National Parks

In the United States, the first declared tribal park was the Monument Valley Tribal Park, or, in Navajo, *Tse'Bii'Ndzisgaii*. Spanning lands in the states of both Utah and Arizona, the park is approximately 92,000 acres in size. It is dedicated to the preservation of Navajo culture, but open to visitors during daylight hours via a scenic drive.[43] Yet establishing a tribal park can generate opposition or conflict with the government's own parks' management agency, especially if it appears to compete with it.

The Ute Indians, a Native American Community living in southwestern Colorado and northwestern New Mexico, reside on lands that adjoin an established US national park, Mesa Verde. Mesa Verde is famous for its preservation of the residences and art of the Anasazi Indians, a cliff-dwelling people historically resident to the US southwest, arising as early as 1500 BC, then disappearing suddenly around AD 1200. Today the descendants of the Anasazi are the Pueblo Indians, now living further south. Given the 300,000 to 500,000 people that visit Mesa Verde annually and, in keeping with its mission to manage the park "for the enjoyment" of park visitors, the ruins and artifacts of the cliff dwellings of the Anasazi have been made easier for visitors to access but suffered deterioration as a result.

Today, adjoining the boundary of Mesa Verde National Park, the Utes have created their own protected lands, the Ute Mountain Tribal Park, an area of about 125,000 acres dedicated to preserve many of the same ruins and artifacts as Mesa Verde, as well as those of past Pueblo and Ute tribal cultures. In the tribal park, tourists must make reservations for a tour of the park and be accompanied by a Ute guide.[44] On the guided tours, visitors follow their guide along ancient Anasazi paths into the ruins rather than simply exploring on their own. Not dependent on tourist revenues, the Utes have gained the legal right to pursue other kinds of economic development within the park, but entirely under their own governance. Under their current management, they have maintained their park's artifacts without damage or deterioration.[45]

ICCAs, now given the support of international recognition and legal empowerment, can provide Indigenous Peoples with greater agency, autonomy, and sovereign control over how, when, where, and why natural and cultural resources are used and managed on their own territories – territories that often overlap locally with critical conservation

PAs and, globally, with Key Biodiversity Areas.[46, 47] Around the world in multiple and varied forms, Indigenous Peoples working with governments, international agencies, and conservation NGOs are envisioning and developing new models of conservation in just and creative ways that begin with a right relationship between people and nature.[48]

Wilderness

Origin, Fallacy, and Reality

The word "wilderness" is derived from the Old English *wildeornes*, meaning "land inhabited only by wild animals," in turn derived from *wild dēor* (*noun* "wild animal, wild beast" + "ness" (land)). The IUCN offers a more contemporary context by placing "wilderness" in its conservation classification of Category IB – *wilderness area*. US law defines wilderness as a legally recognized and designated PA. In the words of the US Wilderness Act (US Public Law 88–577, §2c), wilderness is "an area where the earth and its community of life are untrammeled by man, where man himself is a visitor who does not remain," and an area "which is protected and managed so as to preserve its natural conditions." As a legal entity it is of US origin. Its primary voice, The Wilderness Society, is an American organization founded in 1935 by conservationists Aldo Leopold, Robert Sterling Yard, Benton MacKaye, Ernest Oberholtzer, Harvey Broome, Bernard Frank, Harold C. Anderson, and Robert (Bob) Marshall, the Society's leader. Marshall did not equivocate regarding the Society's intention. "There is just one hope," he wrote, "of repulsing the tyrannical ambition of civilization to conquer every niche on the whole earth. That hope is the organization of spirited people who will fight for the freedom of the wilderness."[49]

As a Forest Service administrator, Marshall was all too familiar with the prevailing view that national forests and other public lands were repositories of "resources" for logging, mining, and other kinds of extractive "development," which effectively reduced nature to a kind of private property for those who had the means to exploit it. Under Marshall and succeeding leaders, The Wilderness Society grew from an organization that initially worked with other conservation groups to *stop such development* to one that eventually succeeded in designating wild areas for *complete protection from development*, beginning with the Arctic National Wildlife Refuge in northeast Alaska, a landscape of 19 million acres (about the size of the US state of South Carolina). The first efforts to legislate wilderness as a category of its own began in 1956 when Howard Zahniser, then one of the Society's leaders, wrote the first draft of legislation to protect wilderness areas in the United States. Over the next 8 years, Zahniser's bill would be rewritten or resubmitted 66 times, subjected to 18 public hearings, and debated in 16,000 pages of testimony. But in 1964, it became law as *The Wilderness Act*, protecting 9.1 million acres of land and placing these in the newly created National Wilderness Preservation System (NWPS), a bureaucracy created to be co-managed by the NPS, Bureau of Land Management (BLM), US FWS, and US Forest Service (FS). By 2024, the NWPS included nearly 112 million acres of wilderness in 803 designated wilderness areas.

Given its legal constraints, there were no "new" lands that could be acquired specifically for the NWPS. Wilderness lands had to be selected from existing holdings of one of the four management agencies. Neither state, nor private, nor Indian reservation lands could be appropriated into the System, but instead had to be designated from among the most remote (roadless) areas in national parks, BLM holdings, FWS refuges, and FS lands. Wilderness areas were prescribed to be at least 5,000 acres in size (most were much larger) and permit no motorized travel within the wilderness by land, water, or air.

Some designated wilderness areas did contain lands of significance to Native Americans. In the case of *Pueblo of Sandia v. Babbitt*, the Pueblo Indians of the Sandia Tribal Council claimed that the area around Sandia Mountain, part of a designated wilderness area in the Cibola National Forest in the state of New Mexico, belonged to the Tribe according to a 1748 grant from the King of Spain. The land was religiously significant to the Tribe, who worshiped at shrines within the Sandia Mountain area. The Pueblo sued the FS for limiting access to worship at these sites because of wilderness restrictions on modes of travel and trail access. The district court ruled in favor of the Pueblo, ordering the FS to transfer the land to the Tribe. The FS appealed the verdict, which was ultimately resolved after eight years of litigation, in favor of the Pueblo.[50]

Although legally designated wilderness created controversy at Sandia Mountain, only one led to relocation of some Native Americans from their tribal lands. That exception was the Arctic National Wildlife Refuge (ANWR), which included traditional lands, residences, and use areas of the Indigenous Gwich'in people. Although some relocations took place, residence in and use by ANWR by the Gwich'in continued in ANWR after its wilderness classification. Such designation served to protect the Gwich'in from outside interference, leaving them free to pursue their traditional culture and livelihood. Conflicts have arisen more recently because of permissions given by the US government to oil and gas companies that allowed them to operate in ANWR,[51] but such permissions were not part of, and in fact violate, the original terms of the ANWR designation and its intent.

It is not merely the designation or administration of wilderness that has generated controversy – it is the concept itself. In recent and modern conservation literature, no idea has been so vilified as that of "wilderness." The notion has been derided by philosophers as a Western romantic illusion,[52] by historians as a human creation to preserve a mythical condition that never existed,[53, 54] by conservation scientists as a state of nature that does not now remain,[55, 56] by sociologists as a tool for dispossessing Indigenous Peoples,[41] and by conservation activists as a false description of natural areas that should instead be managed by people for human benefit.[57]

Controversy is fed by ignorance and nurtured by misunderstanding. If the term "wilderness" is used to mean "uninhabited land," as it was by British explorers when they declared Australia to be *terra nullius*, or by the Brazilian government when it described Brazilian Amazonia as "a land without men," such statements reflect this ignorance and the condemnation of "wilderness" in these contexts is just. If the word is used in its legal and statutory definition as "an area where the earth and its community of life are untrammeled by man," the problem shifts from ignorance to misunderstanding. Many critics of wilderness assume that "untrammeled" is intended to mean "untouched," "unaffected," "pristine," or "never inhabited." The word "trammel," a verb, means none of these things. It is rather employed to describe actions *that restrict or limit another's freedom of action*. "Trammel," in the Wilderness Act, refers to a condition of nature that is not intentionally and directly manipulated by human activity, not to any previous condition or interaction with humans.[58] "Wilderness," in the minds of the Act's framers, is a descriptor applied to a land where the freedom of nature to shape the forces and processes of the landscape is not significantly restricted or limited. Public lands research scientist George Wuenther put it this way: "Because humans have lived in all landscapes except Antarctica does not mean that human influence is uniformly distributed. Wilderness should be viewed as those places largely influenced by natural forces, rather than dominated by human manipulation and presence."[59]

Wilderness is not to be valued because it is uninhabited, but because our inhabitation of it and interaction with it require unique responses from us necessitated by our habitation in it. To reside for any length of time in a "wilderness" requires skills, practices, and attitudes different from those common to human-dominated environments. In wilderness, nature is the prevailing governance. Humans must face the reality of that governance and conform to it, or they will not survive. This knowledge quiets us – the hubris of ego, the demands of desire, and the insistence that things must go "our way." In wilderness, "our way" no longer exists, or if it does, in some attenuated form, you had better get nature to agree with it. The environmental philosopher Paul Keeling explains,

> We value the wildness of our wilderness areas, not only because restraint is a human virtue, but because that restraint results in unique conditions that cannot be gotten any other way. Untrammeled conditions are necessary for preserving both the process of wildness and its products.[60]

Wilderness is more than an arena for human transformation, it is a place that gives life to its residents. As FS scientists Patrick Kelly and Peter Landres explain, "Wilderness plays a crucial role in wildlife conservation, with recent studies showing that these protected wild areas are important for the long-term well-being of species."[61] In the mid-1970s there were less than 200 grizzly bears in Yellowstone Park and its surroundings. Today the number in Yellowstone is still estimated at 150–200, but, according to the Interagency Grizzly Bear Study Team, grizzlies in the Greater Yellowstone Ecosystem numbered more than 1,000 in 2021 (see Figure 9.5).[62] The population's ability to expand has been aided by the movement of bears into surrounding wilderness areas – the Gros Ventre to the south, the North Absaroka to the east, the Lee Metcalf to the west, and the Absaroka Beartooth to the north. These wildlands also have limits. Beyond them grizzlies cannot have long careers. But without room in the wilderness, their numbers would not have grown.

Figure 9.5 Grizzly bears (*Ursus arctos horribilis*) in the Greater Yellowstone Ecosystem (GYE) of the United States were mainly confined to Yellowstone National Park through the 1970s. Since then, wilderness areas surrounding the Park have contributed to a population growth of more than 400%, with grizzly numbers now estimated at more than 1,000 individuals in the GYE in 2021.

The Indigenous Wilderness

While the global proportion of terrestrial land surface designated as PAs has grown, the increase in wilderness protection has diminished. In the past two decades, 2.5 million km^2 of wilderness areas (including 2.1 million km^2 considered globally significant) have been added into protective status and 3.3 million km^2 (including 2.7 million km^2 considered globally significant) have been lost.[63] And wilderness designation has consistently been omitted from multilateral international conservation agreements like the United Nations Framework on Climate Change, the World Heritage Convention, and the Convention on Biological Diversity.[63]

Some omissions reflect a failure to clarify appropriate definitions of wilderness and distinguish it from other kinds of PAs, but much of the neglect is a direct result of current statements from members of the conservation community itself that wilderness is only an intellectual construct, a historical and contemporary myth, or a non-existent state of nature, and that its designation results in injustice toward Indigenous Peoples. Wilderness designations have been used in the past to dispossess or forcibly relocate Indigenous Peoples from their traditional homelands, but the idea that Indigenous cultures have no concept of wilderness is false. As philosopher Holmes Rolston III has noted, "Every culture can, to some extent, see beyond itself to a spontaneous nature, unaffected by human agency."[64] Indigenous Peoples worldwide have a clear conception of and appreciation of wilderness, although sometimes by other names. But their vision is that of an *inhabited* wilderness, not an *uninhabited* wilderness.[65] Those who work in conservation must always be careful enough, and wise enough, to distinguish the difference between nature *unaffected* by humans and nature *uninhabited* by humans. Indigenous cultures understand this distinction, and because they do, many today are establishing their own wilderness areas – they are defining *tribal wilderness*.

In the late 1930s, Salish and Kootenai tribal lands in the Mission Mountains in Montana were designated for consideration as a future wilderness area. Because the government did this without the consent of the Native American Salish and Kootenai tribes, who had land claims that included the Mission Mountains, the tribes protested, and, after extended conflicts and discussions, the area was declassified in 1959 and made available for various other uses, including logging, mining, and oil exploration. After another 25 years of conflict with the US government over these and other proposed activities, which the tribes opposed, the Salish and Kootenai people had gained decision-making influence over the use of the area. In 1982, the Tribal Council created the Missions Mountain Tribal Wilderness (MMTW) (see Figure 9.6), an area of approximately 70,000 acres, and ended further consideration of logging, road building, mining, or other activities that involved removing resources from the Mission Mountains.[66]

The MMTW now borders a portion of the Flathead National Forest also designated as wilderness – the Mission Mountain Wilderness (MMW), managed by the FS. Confusion between the two areas is understandable, but management philosophies and strategies are different. The FS manages their wilderness for recreation. Hikers, backpackers, and, in season, hunters frequent the MMW, and professional outfitting guide services are permitted to operate in it. No hunting or outfitting services are permitted in the tribal wilderness, and hikers and backpackers must obtain a permit from the tribes to enter. For the benefit of their wilderness users, the FS conducts regular trail maintenance in the MMW. The tribe does not maintain trails, so many of these former pathways in the MMTW have faded back into the landscape.[66]

Figure 9.6 The Mission Mountains Tribal Wilderness in the US state of Montana, a wilderness area designated and dedicated by the Salish and Kootenai Native American tribes for the preservation of tribal culture.

Within the tribal wilderness area, 10,000 acres have been set aside as a Grizzly Conservation Zone which delimits an area of high use by grizzlies in summer months from which human visitors are prohibited. The concern is not so much to protect the people as to not disturb the bears. On lands to the west, also under tribal jurisdiction, the tribes have created a "buffer zone" where livestock grazing, timber harvest, and private development are permitted, but restricted and monitored by the tribes. The FS has no buffer zone around its wilderness.[66]

These differences reflect contrasts of intention. Although much of the language the tribes composed in their own definition of wilderness matches the language of the Wilderness Act, the purpose of the MMTW is not to simply preserve *land*, but to preserve *tribal culture* for which the wilderness is essential.[66] Today there are other tribal wilderness areas in the United States, such as the Inter-Tribal Sinkyone Wilderness in northern California and the Timbisha Shoshone Natural and Cultural Preservation Area in southern California, which shares land with Death Valley National Park's backcountry and wilderness areas (like the MMTW). Management of the latter is dedicated to preserving the traditional cultural and religious activities of the Timbisha Shoshone, an effort not only consistent with the Wilderness Act, but now legally supported by the American Indian Religious Freedom Act (42 USC 1996 et seq.). Some Indigenous writers have become advocates for specific wilderness areas, and for the value of wilderness itself. Bernadette Dimientieff, a citizen of the Alaskan–Canadian Gwich'in people, explains,

> We advocate for Wilderness because it will provide the strongest protection for the birthing grounds of the Porcupine caribou herd. We are the caribou people; a piece of the caribou heart lies within us and a piece of the Gwich'in heart lies within the caribou, and it has been so since time immemorial. They are a part of who we are as a people, our way of life, our food security, and our identity.[67]

PAs, Parks, and Wilderness in Conservation

We have long confused delineation with protection. This confusion reflects a false understanding of "boundary." The *boundary* of a PA does not "protect" anything, nor does it stop, or even restrict, the flow of processes and species that pass through it to and from the surrounding landscape. PAs need boundaries, but boundaries do not protect areas, and boundaries need connections. The practice of conservation within the boundary cannot be accomplished without a shared intention of conservation in lands beyond the boundary. Practices inside and outside the boundary may be different, but they must not be adversarial or counterproductive.

Conservationists have continued to make a second error in their neglect of governance. In fact, *governance* of PAs is more important than *boundaries* of PAs. Wherever and whenever the governance of a PA lacks commitment to the intention of its conservation mission, capacity to carry out that intention, integrity to resist abandoning the intention to competing interests, and stability to endure over the time needed to save endangered populations, the PA will fail, and would have been better to have never been "designated" at all. The Anthropocene deserves its reputation as the age of human domination, and some such domination is expressed in empty designations that attempt to impersonate conservation. To correct such hypocrisy, PAs *designated in the Anthropocene must be placed in governance systems that provide genuine protection.*

In the specific category of national parks, we find a dark and bloody history. Such parks have often been an instrument of oppression and injustice, actively or passively, in the expression of conservation worldwide. Today national parks must make their Indigenous history – all of it – part of the story of the park, and reconciliation with those harmed by the park's creation must now be part of the park's ongoing work. The future of national parks lies in management that is shared with or returned to the former, rightful residents of the park's land. Indigenous history must become the core of the park's interpretive effort and Indigenous traditional knowledge must take a meaningful role in the park's management. And the park must create pathways through which Indigenous Peoples can enter into and participate in the park's operation and management. In the 2023 US Federal Budget for the NPS, all of these concerns are addressed in its total allocation of $3.8 billion and designated to "better serve Tribal Nations and Indigenous people by hiring Tribal liaisons and subsistence managers, supporting Tribal co-stewardship efforts, and funding the Tribal heritage grant program." The US NPS program is now devoted to protecting elements of national parks that are of greatest importance to Native Americans – oral history, plant and animal species important in cultural traditions, sacred and historic places, and the establishment of Tribal historic preservation offices. Since 1990, more than $17 million has been awarded to over 460 Indian and Alaskan Native communities toward these goals.[68] While the future of such parks is slowly reformed in these and a myriad of other ways worldwide, national parks will remain essential to conservation in the Anthropocene, not only because of the creatures and ecosystems within them, but through the interactions between humans and nature that they are uniquely equipped to guide, and the millions of people who can be changed through such guidance.

Wilderness, in many forms and expressions, has been used, like parks, as an instrument of dispossession of Indigenous Peoples. Such dispossessions were wrong, but, in its rightful expression, the reality of wilderness is part of Indigenous culture – an inhabited wilderness that nurtures human character as an intended cultural outcome. Indigenous Peoples have historically recognized and set aside portions of their own lands in which they permit little or no human use or permanent residence. These places were devoted to the welfare of non-human creatures as well as to express reverence for spiritual beings. Modern people may have regard for neither, but they still need wilderness. In it they have the opportunity to learn the reality of nature's governance and, with humility, gain the wisdom to conform to it.

References

1. Convention on Biological Diversity. 2006. United Nations Environment Programme. <https://www.cbd.int/convention/text/>. Accessed 9 January 2024.
2. Dudley, N., editor. 2008. *Guidelines for applying protected areas management categories.* International Union for Conservation of Nature, Gland, Switzerland.
3. Van Dyke, F., and R. L. Lamb. 2020. *Conservation biology: Foundations, concepts, applications.* 3rd edition. Springer, Cham, Switzerland.
4. Chape, S., J. Harrison, M. Spalding, et al. 2005. Measuring the extent and effectiveness of protected areas as an indicator for meeting global biodiversity targets. *Philosophical Transactions of the Royal Society B* 360:443–455.
5. Palomo, I., C. Montes, B. Martín-López, et al. 2014. Incorporating the social–ecological approach in protected areas in the Anthropocene. *BioScience* 64:181–191. DOI: 10.1093/biosci/bit033.
6. Secretariat of the Convention on Biological Diversity. 2004. Programme of Work on Protected Areas. CBD Programmes of Work, Montreal, Canada.
7. Geldmann, J., A. Manica, N. D. Burgess, et al. 2019. A global-level assessment of the effectiveness of protected areas at resisting anthropogenic pressures. *Proceedings of the National Academy of Sciences* 116:23209–23215.
8. International Union for the Conservation of Nature. 2024. Effective protected areas. <https://www.iucn.org/our-work/topic/effective-protected-areas>. Accessed 11 January 2024.
9. Ellis, E. C. 2019. To conserve nature in the Anthropocene, Half Earth is not nearly enough. *One Earth* 1:163–167, 164. <https://doi.org/10.1016/j.oneear.2019.10.009>.
10. Wilson, E. O. 2016. *Half-Earth: Our planet's fight for life.* Liveright/W. W. Norton, New York, USA.
11. Ellis, E. C. 2019. To conserve nature in the Anthropocene, Half Earth is not nearly enough. *One Earth* 1:163–167, 164. <https://doi.org/10.1016/j.oneear.2019.10.009>.
12. Dudley, N. 2011. *Authenticity in nature: Making choices about the naturalness of ecosystems.* Earthscan, London, UK.
13. Cazalis, V., K. Princé, J. B. Mihoub, et al. 2020. Effectiveness of protected areas in conserving tropical forest birds. *Nature Communications* 11:4461. <https://doi.org/10.1038/s41467-020-18230-0>.
14. Feng, Y., Y. Wang, H. Su, et al. 2021. Assessing the effectiveness of global protected areas based on the difference in differences model. *Ecological Indicators* 130:108078. <https://doi.org/10.1016/j.ecolind.2021.108078>.
15. He, X., and H. Wei. 2023. Biodiversity conservation and ecological value of protected areas: A review of current situation and future prospects. *Frontiers in Ecology and Evolution* 11. <https://doi.org/10.3389/fevo.2023.1261265>.
16. Nagendra, H. 2008. Do parks work? Impact of protected areas on land cover clearing. *Ambio* 37:330–337. DOI: 10.1579/06-R-184.1.
17. Rodriguez-Rodriguez, D., J. Martínez-Vega, and P. Echavarría. 2019. A twenty-year GIS-based assessment of environmental sustainability of land use changes in and around protected areas of a fast developing country: Spain. *International Journal of Applied Earth Observation Geoinformation* 74:169–179. DOI: 10.1016/j.jag.2018.08.006.
18. Geldman, J., A. Manica, N. D. Burgess, et al. 2019. A global-level assessment of the effectiveness of protected areas at resisting anthropogenic pressures. *Proceedings of the National Academy of Sciences* 116:23212.

19. Mascia, M. B., and S. Paille. 2011. Protected area downgrading, downsizing, and degazettement (PADDD) and its conservation implications. *Conservation Letters* 4:9–20. DOI: 10.1111/j.1755-263X.2010.00147.x.
20. Mascia, M. B., and S. Paille. 2011. Protected area downgrading, downsizing, and degazettement (PADDD) and its conservation implications. *Conservation Letters* 4:16. DOI: 10.1111/j. 1755-263X.2010.00147.x.
21. Geldman, J., A. Manica, N. D. Burgess, et al. 2019. A global-level assessment of the effectiveness of protected areas at resisting anthropogenic pressures. *Proceedings of the National Academy of Sciences* 116:23209.
22. Nautiyal, S., and R. R. Nidamanuri. 2012. Ecological and sociological impacts of conservation policies in biodiversity hotspots: A case from Rajiv Gandhi National Park, India. *Journal of Environmental Studies and Sciences* 2:165–177.
23. Wilkie, D. S., and J. T. Finn. 1990. Slash-burn conservation and mammal abundance in the Ituri Forest, Zaire. *Biotropica* 22:90–99.
24. Hart, T. B., and J. S. Hall. 1996. Status of eastern Zaire's forest parks and reserves. *Conservation Biology* 10:316–327.
25. Wilkie, D. S., B. Curran, R. Tshcombe, et al. 1997. Modeling for sustainability of subsistence farming and hunting in the Ituri forest of Zaire. *Conservation Biology* 12:137–147.
26. Hart, T. B., and J. A. Hart. 1997. Zaire: New models for an emerging state. *Conservation Biology* 11:308–309.
27. Hart, T. B., J. A. Hart, and J. S. Hall. 1996. Conservation in the declining nation state: A view from eastern Zaire. *Conservation Biology* 10:685–686.
28. Platt, J. R. 2012. Okapi Conservation Center recovering after militia attack that killed 6 people and 14 animals. Scientific American Blog Network. <https://www.zoochat.com/community/threads/okapi-conservation-project-attacked-by-militia.284266/>. Accessed 17 October 2024.
29. International Union for the Conservation of Nature. 1994. *Guidelines for protected areas management categories*. IUCN, Cambridge, UK and Gland, Switzerland.
30. World National Parks. 2024. <https://worldnationalparks.com/>. Accessed 12 January 2024.
31. US Congress. 1872. An act to set apart a certain tract of land lying near the headwaters of the Yellowstone River as a public park. Forty-second Congress of the United States. <https://www.archives.gov/milestone-documents/act-establishing-yellowstone-national-park>. Accessed 6 July 2024.
32. National Park Service Organic Act. 1916. 39 Statute 535, 16 U. S. C. 1.
33. Coldrey, K. M., J. K. Turpie, G. Midgley, et al. 2022. Assessing protected area vulnerability to climate change in a case study of South African national parks. *Conservation Biology* 36(5):e13941. DOI: 10.1111/cobi.13941. ePub 2022 Jun 1. PMID:35648687; PMCID:PMC9796953.
34. US National Park Service. 2024. Sleeping Bear Dunes National Lakeshore Michigan Ranger-led programs. <https://www.nps.gov/slbe/planyourvisit/calendar.htm>. Accessed 6 July. 2024.
35. Warber, S. L., A. A. DeHudy, M. F. Bialko, et al. 2015. Addressing "Nature-Deficit Disorder": A mixed methods pilot study of young adults attending a wilderness camp. *Evidence-Based Complementary and Alternative Medicine*. https://doi.org/10.1155/2015/651827>.
36. Louv, R. 2005. *Last child in the woods: Saving our children from Nature-Deficit Disorder*. Allenand Unwin, New South Wales, Australia.
37. Borrini, G., A. Kothari, and G. Oviedo. 2004. *Indigenous and local communities and protected areas: Towards equity and enhanced conservation*. IUCN, Gland, Switzerland. <https://doi.org/10.2305/IUCN.CH.2004.PAG.11.en>. Accessed 17 October 2024.
38. Larsen, P. B. 2015. *Post-frontier resource governance*. Palgrave Macmillan, Houndmills, UK, 79.
39. Mason, C. W., A. Carr, E. Vandermale, et al. 2022. Rethinking the role of Indigenous knowledge in sustainable mountain development and protected area management in Canada and Aotearoa/New Zealand. *Mountain Research and Development* 42(4):A1–A9. <https://doi.org/10.1659/mrd.2022.00016>.
40. Nitah, S. 2021. Indigenous peoples proven to sustain biodiversity and address climate change : Now it's time to recognize and support this leadership. *One Earth* 4:907–909.

41. Fletcher, M.-S., R. Hamilton, W. Dressler, et al. 2021. Indigenous knowledge and the shackles of wilderness. *Proceedings of the National Academy of Sciences* 118(40):e2022218118. <https://doi.org/10.1073/pnas.2022218118>.
42. Igoe, J. 2004. *Conservation and globalization: A study of national parks and Indigenous communities from East Africa to South Dakota*. Wadsworth/Thomson, Belmont, California, USA, 158.
43. Navajo Nation Park and Recreation. 2024. Monument Valley Navajo Tribal Park. <https://navajonationparks.org/navajo-tribal-parks/monument-valley/>. Accessed 7 July 2024.
44. Mesa Verde Country. 2024. Ute Mountain Tribal Park. <https://mesaverdecountry.com/ute-mountain-tribal-park/>. Accessed 14 January 2024.
45. Igoe, J. 2004. *Conservation and globalization: A study of national parks and Indigenous communities from East Africa to South Dakota*. Wadsworth/Thomson, Belmont, California, USA, 162–165.
46. Johnston, J. W., and C. W. Mason. 2020. The paths to realizing reconciliation: Indigenous consultation in Jasper National Park. *International Indigenous Policy Journal* 11(4):1–27. <https://doi.org/10.18584/iipj.2020.11.4.9348>.
47. Tran, T. C., N. C. Ban, and J. Bhattacharyya. 2020. A review of successes, challenges, and lessons from Indigenous protected and conserved areas. *Biological Conservation* 241:108271.
48. Moola, F., and R. Roth. 2018. Moving beyond colonial conservation models: Indigenous protected and conserved areas offer hope for biodiversity and advancing reconciliation in the Canadian boreal forest. *Environmental Review* 27(2). DOI: 10.1139/er-2018-0091.
49. Marshall, R. 1930. The problem of the wilderness. *Scientific Monthly* 30(2):141–148, 148.
50. Miranda, S. 2022. The real meaning of wilderness: Impacts of the Wilderness Act on Native Americans. American Bar Association. <https://www.americanbar.org/groups/environment_energy_resources/publications/nar/the-real-meaning-of-wilderness/>. Accessed 13 January 2024.
51. Hagar, J. 2023. Conflict in Alaska's Arctic: Better conservation, less oil and gas production. *Polar Journal*. <https://polarjournal.ch/en/2023/09/11/conflict-in-alaskas-arctic-better-conservation-less-oil-and-gas-production/>. Accessed 6 July 2024.
52. Nash, R. F. 2014. *Wilderness and the American mind*. 5th edition. Yale University Press, New Haven, Connecticut, USA.
53. Cronon, W. 1995. The trouble with wilderness; or, getting back to the wrong nature. Pages 69–90 *in* W. Cronon, editor. *Uncommon ground: Rethinking the human place in nature*. W. W. Norton & Co., New York, USA.
54. Mann, C. C. 2005. *1491 – New revelations of the Americas before Columbus*. Knopf, New York, USA.
55. Kareiva, P., R. Lalasz, and M. Marvier. 2012. Conservation in the Anthropocene: Beyond solitude and fragility. <https://is.gd/Lvn7uA>. Accessed 13 January 2024.
56. Ellis, E. C., N. Gauthier, K. K. Goldewijk, et al. 2021. People have shaped most of terrestrial nature for at least 12,000 years. *Proceedings of the US National Academy of Sciences* 118(17):e2023483118.
57. Marris, E. 2011. *Rambunctious garden: Saving nature in a post-wild world*. Bloomsbury, New York, USA.
58. Landres, P., C. Barns, S. Boutcher, et al. 2015. Keeping it wild 2: An updated interagency strategy to monitor trends in wilderness character across the national wilderness preservation system. General Technical Report RMRS-GTR-3.
59. Wuerthner, G. 2018. Anthropocene boosters and the attack on wilderness conservation. *Ecological Citizen* 1:161–166, 163.
60. Keeling, P. M. 2007. Beyond the symbolic value of wildness. *International Journal of Wilderness* 13:19–23, 22–23.
61. Kelly, P., and P. Landres. 2022. Does wilderness matter in the Anthropocene? Resolving a fundamental dilemma about the role of wilderness in 21st century conservation. *Ethics, Policy & Environment*. <https://doi.org/10.1080/21550085.2022.2133942>.
62. Kudelska, K. 2021. New calculation method results in the largest grizzly bear population. Wyoming Public Radio. Accessed 14 May 2024.

63. Watson, J. E. M., D. F. Shanahan, M. Di Marco, et al. 2016. Catastrophic declines in wilderness areas undermine global environment targets. *Current Biology* 26:2929–2934. <http://dx.doi.org/10.1016/j.cub.2016.08.049>.
64. Rolston, H. III. 2001. Natural and unnatural: Wild and cultural. *Western North American Naturalist* 61:267–276, 274.
65. Stevens, S., editor. 1997. *Conservation through cultural survival: Indigenous Peoples and protected areas*. Island Press, Washington, DC, USA.
66. Delhomme, P. J. The Mission Mountain Divide. 2023. <https://crazycanyonmedia.com/mission-mountains-first-tribal-wilderness-in-the-united-states>. Accessed 15 January 2024.
67. Himrod, C. 2021. Wilderness and traditional Indigenous beliefs: Conflicting or intersecting perspectives on the human–nature relationship. Alaska Wilderness League. <https://alaskawild.org/blog/wilderness-and-traditional-indigenous-beliefs/>. Accessed 15 January 2024.
68. Tribal Heritage. 2024. National Park Service. <https://www.nps.gov/subjects/historicpreservationfund/tribal-heritage-grants.htm>. Accessed 18 January 2024.

10 Conservation Interventions – Saving Species Like We Mean It

Restoring the Sturgeon in the Manistee River

In 2015, I received a grant from the Little River Band of Ottawa Indians (LRBOI), a Native American Tribal Nation in Michigan whose lands are in the Manistee River Valley. The grant would fund my efforts to determine if the mottled sculpin (*Cottus bairdii*) and slimy sculpin (*C. cognatus*), two small bottom-dwelling stream fish, could serve as "indicator species" to assess the quality of freshwater rivers.[1] I was glad to have the Tribe's support, but, with the funding, I also agreed to assist the Tribe in their work of restoring the lake sturgeon (*Acipenser fulvescens*) to the Manistee River.

Once abundant in US Great Lakes, the lake sturgeon was decimated by fishing efforts in the late nineteenth and early twentieth centuries. One couldn't call the problem "over-harvest" because the fishermen were trying to catch lake trout (*Salvelinus namaycush*) and whitefish (*Coregonus clupeaformis*), not "harvest" lake sturgeon. Sturgeon were abundant then. Adults lived long (up to 100 years!) and grew large (up to 8 feet and 300 pounds). Often caught and entangled in the same nets as the trout and whitefish, the sturgeon thrashed and twisted their powerful bodies attempting to escape, often tearing the netting. In angry response, fishermen beat the sturgeon to death with clubs and then discarded their bodies into the lake as "trash." Over time, this technique solved the "sturgeon problem," and as the years progressed, sturgeon became increasingly rare. They also ceased to return to the Manistee and other rivers to spawn.

To the Ottawa, the sturgeon is *nmé*, a "clan spirit" – the "grandfather fish" of the Tribe's Sturgeon Clan. Before the European colonization of the upper Great Lakes region, many Ottawa lived in settlements on or near the eastern shore of Lake Michigan where they harvested sturgeon and other fish from the lake and adjoining rivers, celebrating the sturgeons' return to the streams each spring in tribal ceremonies.[2, 3] Today the LRBOI is committed to restoring the sturgeon. Their effort begins by taking infant sturgeon ("fry") from the Manistee River and placing them in a rearing facility where, under controlled conditions, they grow, and then are released a few months later when they are larger and less vulnerable to predators.

The contribution my research team and I make is in the catching. Entering the river near a point called Rainbow Bend, we carry large D-frame nets with us, orienting them into the current, each anchored by a heavy iron brace. Sturgeon are more active at night. Therefore, we begin at sundown and the light is passing by the time our nets are arrayed across the channel. A sandhill crane (*Grus canadensis*) passes on the opposite shore, slow wingbeats keeping the crane just above the trees as it flies upstream. More interested in its own affairs than ours, the crane offers no comment on what we contrive. As the last

DOI: 10.4324/9781003404415-10

light fades, a woodcock (*Scolopax minor*) flashes past, only a few feet over our heads, then drops like a feathered stone into an alder thicket.

We avoid darkness because it robs us of sight. But if one lets the darkness come unhindered, meeting it on its own terms, there is no theft. We begin to "see" the dark. Our eyes open wider, and we watch the stars come forth, one by one, not in garish parade, but orderly procession, each one as if summoned, arriving by appointed invitation. The moon is nearly full, and its light is not meagre. The water is illuminated, and we enter, seeing. Riffle and current glide past us like shining silk. We wade deeper to inspect the catch. Each net glows brilliantly, adorned in its place by starlight and water. They sweep back from their frames, luminous kites in the river's watery wind, each one aligned with the next, head to head and tail to tail, a dazzling array of white mesh under silver moonlight. Never before have I seen scientific methodology bear such beauty in the weight of its work. The river has transformed our tools into art. Its rushing brush strokes press against me. I stand upon its canvas, and gaze on the masterpiece it has made of our rude cloth and iron.

At 1 in the morning we retrieve the nets. The evening brought us many fry, but most were suckers, only a few sturgeon. My younger colleagues quickly become adept at distinguishing the two. I make many errors, but their skills correct these and keep them from becoming permanently recorded "mistakes" in identification. We take the sturgeon fry upstream to the "Sturgeon Start Center," a small trailer carefully placed along the banks of the Manistee. Here water from the river itself is pumped into holding tanks using a "flow through" plumbing system. Nurtured with water from a river that has been the home of their ancestors, the sturgeon in the trailer can grow from "fry size" (about 0.5 to 1 inch) to "fingerling size" (7 to 9 inches) (see Figure 10.1), from June to September. Passing their most vulnerable life stage in the safety of the Center's rearing tanks the larger fingerlings are less vulnerable to predators after release. Research results so far are mixed as to whether wild or captive-raised fry are better at avoiding predators, but those studies were done in the controlled environment of captivity.[4, 5] What is certain is that the fry in the Center are not *exposed to* predators like those in the river until they are no longer fry. More important is the effort's clarity of outcome. In the Manistee, sturgeon numbers are growing. From a population of 20 adults found in 2002, the surveys of 2023 found 400.[6]

Have Conservation Actions Prevented Extinction Outcomes?

Capturing the Last of the Wild to Return the Last of the Wild

This restoration of one species – the lake sturgeon – in one stream – the Manistee River – is one example of *conservation intervention*, an initiative intended to save a declining population, an intrusion into its life and habitat, and an interruption in the natural ecological processes in which it participates. It is a deliberate human disruption of natural events to save a species that, left to those events alone, might come to a natural end of local extinction. It is an intrusion in which human beings take on a primary role in preventing the extinction of a species through their direct efforts on its behalf. As philosopher John Visander has said, "We need to understand both the 'natural' and the 'wild' in such a way that we can imagine giving more to the world around us than the gift of our mere absence."[7] In the Anthropocene, conservation efforts must get conservation results. If they do, it will not be through preservation, but intervention. It will not measure success by knowledge gained, but by outcomes achieved. It will happen because

Figure 10.1 (A) A juvenile lake sturgeon (*Acipenser fulvescens*), a species once common in the US Great Lakes but reduced in abundance by overharvesting and habitat degradation. Photo courtesy of M. Holtgren. (B) An adult lake sturgeon, which requires three people to lift it. US FWS.

of intentional human action, intervention, and care, especially as they are directed to endangered species. So, the first question to ask is – does conservation action do endangered species any good?

In 2024, the IUCN estimated that, of 157,000 species assessed to date, 44,000 (28%) are endangered, including 26% of mammals, 12% of birds, 21% of reptiles, 41% of amphibians, 25% of fishes, and 40% of plant species.[8] In an assessment of all vertebrates from 1988 to 2008, the average Red List Index value of listed species declined (the species became more prone to extinction) by 0.49%. Less than one half of 1% over a 20-year period seems small, but it is equivalent to 52 species of vertebrates moving closer to extinction.[9] Ten species of birds and five species of mammals did become extinct (or are strongly suspected to have done so) since 1993, but estimated extinction rates would have been 2.9–4.2 times greater without conservation effort.[10] Just what does it take to save a species in the Anthropocene?

The California condor (*Gymnogyps californianus*) is one of the largest birds in North America – up to 26 pounds, 3 feet tall, with a 9.5-foot wingspan. Historically it ranged from Baja California (Mexico) to British Columbia (Canada), but had disappeared across nearly all of its range by the 1940s. A remnant of perhaps 150 birds survived in arid canyons in the United States in southern California, but even this population was declining, decimated by habitat loss, poaching, and poisoning ingested from eating carcasses of game killed with lead gunshot and livestock laced with cyanide left to kill predators.[11] With the condor already rare, the US Forest Service (FS) established two sanctuaries for their preservation, the Sespe Condor Sanctuary and the Sisquoc Condor Sanctuary. These areas provided nesting habitat, but adults still foraged for food on private lands where they were unprotected, and unsurprisingly suffered ongoing mortality from poison and poaching.[12]

By 1984–1985, only nine wild birds remained, and but a single breeding pair. The US FWS, carrying out their recovery plan to save the condor, prepared to capture the survivors and place these in captivity to enhance an existing captive breeding program. The Audubon Society was appalled at the idea and sued the FWS. But the female of the last breeding pair died, and, under the circumstances, a US federal court ruled that the FWS had the right to capture remaining birds. They did, beginning in 1985.[13] With the capture of the last individual in 1987, the California condor was extinct in the wild.

The breeding effort had success, and by 1991 the captive population had surpassed 50 birds. Releases began in 1992, but many of the birds had maladaptive behavior (gained in their captive environment) and were neither good nest builders nor good parents. It was not until 2000 that a breeding "group" of one male and two females produced the first healthy egg, but upon hatching one of the females killed the chick.[11] The first successful nesting was not documented until 2006.[14] More breeding pairs formed, more nests were built, and more chicks hatched, yet numbers were slow to increase. But by 2010, there were an estimated 177 wild condors in California and the neighboring state of Arizona. Counts put the wild population at 210 by 2011, 235 by 2012, and 276 by 2016.[15] By the end of 2022, the US National Park Service numbered 347 birds in 5 wild populations, with the 2 largest concentrated in Grand Canyon National Park (Arizona) and Zion National Park (Utah).[16]

The efforts to save the California condor reach such heights of organizational dedication, government support, and individual heroism that it might be assumed to be a singular event. It is not. Pére David's deer (*Elaphurus davidianus*), once common in China's lower Yangtze River Basin, was extinct in the wild by 1939. Through dedicated conservation efforts and captive breeding programs in Europe, the species was re-established in the wild by the 1990s.[11] In the US, the last colony of black-footed ferrets (*Mustela nigripes*), surviving in Wyoming, was all but exterminated after being struck by sylvatic plague and distemper (communicated to it by a dog) in 1981.[17] Now, like the condor, FWS scientists captured all remaining ferrets and brought them into captivity. An unprecedented collaboration of wildlife biologists, geneticists, veterinary scientists, and others spawned innovations in artificial insemination,[18] conservation of genetic diversity,[19] behavioral training for release to the wild (including the amazingly creative development of "Robobadger," perhaps the first robotic predator [a road-killed badger (*Taxidea taxus*) mounted on the chassis of a toy truck!] (see Figure 10.2) used to train a species in predator avoidance), and detailed habitat evaluation and monitoring to find the best release sites.[20] All these actions were essential to begin the reintroduction effort, which commenced in 1991. By 2010, the ferret had established a wild population of 700–800 individuals at 19 sites in Canada, Mexico, and 8 US states.[20] Sylvatic plague, spread to ferrets from their primary prey, the prairie dog (*Cynomys* spp.), has continued to ravage populations in the last decade, reducing the total number of ferrets to fewer than 500. Yet black-footed ferret colonies, albeit reduced in numbers, remain active at 30 sites in the United States, Canada, and Mexico.[21]

These efforts and others were "edge of the cliff" rescues of species that, on their own, should have disappeared. The intention, intervention, and intrusion of human effort to save them cannot be exaggerated. That is what it took to save endangered species in the Anthropocene. What conservation interventions are needed to save populations whose straits are not yet so desperate, but still declining, before they reach the need for such urgent care?

Figure 10.2 "Robobadger," the carcass of a road-killed badger (*Taxidea taxus*) mounted on the chassis of a remote-controlled toy truck, used to teach captive-raised black-footed ferrets (*Mustela nigripes*) how to avoid this predator when they were released into the wild. <https://cameratrapcodger.blogspot.com/2008/01/adventures-with-robobadger.html>. Creative Commons Attribution-Noncommercial-NoDerivatives 3.0 Unported License.

Augmenting Wild Populations

Adding More Individuals

The obvious way to increase the number of individuals in a population is to add more. The newcomers can be obtained from other populations (if the target population is not the only one left) or from those raised in captivity and released into the wild population. The "addition" method is not only one of the most obvious approaches, but one of the oldest. Its most basic (and, in many ways, least reputable) form is "put and take," in which managers add to a population of animals that is hunted or fished, usually just before the hunting or fishing season opens. The added individuals might come from a game farm or fish hatchery. If done with the right numbers and timing, hunters and anglers go home happy. This approach requires investment in raising or capturing the animals to be added, but no investment in improving the habitat into which they are released.

Effective and efficient as this practice may be, it has no conservation value. Aside from the experience in the gained hunting or fishing, it is no different from buying the animal at the grocery store. The foundation of natural population growth is habitat quality – does the habitat provide the species with the right resources in the right amounts at the right time? If the habitat must be supplemented with additions, the obvious question is – why can't this habitat produce a sustainable yield on its own? Either something is deficient about the habitat or something is excessive about the harvest – or both. Increasingly state and federal fish and wildlife management agencies are moving away from "stocking" populations and investing in acquiring and developing better habitat and making better decisions about how many hunting and fishing licenses to sell.

Assisted Natural Recovery

Assisted Natural Recovery, or ANR, a technique with a long history in ecosystem restoration, is now being used more directly with plant and animal populations. In ANR, managers change *structures* or *resources* needed to enhance habitat development or ecosystem recovery, assuming that if ecosystem processes or population growth are resource-limited, adding more of that resource should improve the function of ecosystem processes and, in consequence, the survival or reproduction of the targeted plants or animals. ANR is most effective when applied to a successional sequence or a colonization effort already in progress.

A recent application has been with the long-spined sea urchin (*Diadema antillarum*). Many species of sea urchins have suffered massive die-offs, a loss which in some areas can depress the establishment and recovery of corals. It takes decades for urchin populations to recover once they have declined or been locally exterminated, but their revival can be aided by adding a suitable substrate to which larval sea urchins can attach. Scientists working to both aid the recovery of sea urchins and establish new corals developed one using strings of plastic "bio balls" deployed at mid-depths in ocean waters and physically anchored to coral reefs. The bio balls were polypropylene spheres 3 cm in diameter attached to stainless steel rings which were epoxied (glued) into the reef matrix and kept upright using a small buoy on top (see Figure 10.3).[22] Over time, reefs with streamers averaged 6.3 urchin recruits per reef, *21 times higher* than reefs without streamers (0.3 recruits per reef)! This difference was, in the words of the investigators, "highly significant." The scientists concluded that "bio ball streamers were able to provide a suitable settlement substrate for *D. antillarum* [urchin] larvae and thereby likely able to help circumvent one of the key potential barriers to natural recovery."[22]

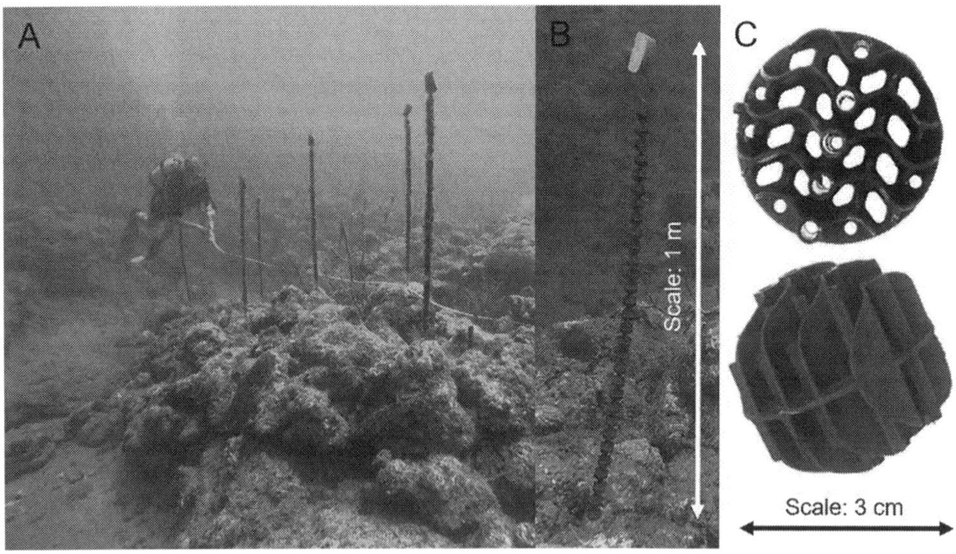

Figure 10.3 "Bio ball streamers" provide effective substrates for attachment of larval sea urchins (*Diadema antillarum*) which may enhance recovery and new establishment of coral reefs. (A) A coral reef with "bio ball streamers." (B) An individual streamer with its flotation buoy on top. (C) A close-up view of the bio balls. From Hylkema et al. 2022. Assisted natural recovery: A novel approach to enhance *Diadema antillarum* recruitment. Frontiers in Marine Science 9:929355.

Head-starting Populations with Juveniles Raised in Captivity

The Ottawa Indians' effort to restore the lake sturgeon in the Manistee River is an example of a "head start" intervention for population recovery – a technique most useful in cases in which the infant or juvenile cohort of the population is more vulnerable to mortality than the adults. If the younger cohorts can be protected from such mortality, population growth will follow. Head start efforts are common in fish conservation, habitat management, forest restoration, and other conservation practices. Today millions of salmon (*Onchorynchus* spp.) are raised in hatcheries before release to the wild. Young trees are usually grown for at least one or two years in a nursery before being planted in a natural environment. If the young trees, like oaks (*Quercus* spp.), are palatable to herbivores, such as deer (*Odocoileus* spp.), they may be caged or tubed to keep them from being eaten, outgrowing their protection (or having it removed) by the time they are large enough to withstand browsing. In ecological restoration, plants that have been reduced or lost in the disturbed ecosystem may be restored, after first being grown in a greenhouse, by inserting individual plants ("plugs") of the desired species, often one by one (it is at this point that restoration ecology perhaps most resembles gardening, but gardening works).

Head start programs risk creating "adaptation to captivity," which may be counterproductive to survival in the wild. For example, lifetime reproductive success and overall fitness of steelhead trout (*Oncorhynchus mykiss*) declined sharply in just two generations of domestication.[23] Rearing in a hatchery environment reduced their reproductive capabilities by ~40% per captive-reared generation when fish moved to natural environments.[23] This problem of "genetic adaptation" (GA) in captivity can be reduced by shortening the time in captivity, reducing the strength of selection in captivity (by making the captive environment more like the wild environment), or adding more individuals from wild populations to slow the rate of GA. The sturgeon head start program addresses GA at two levels: first, by keeping the sturgeon in the Start Center (the captive environment) for a relatively short time (about three months); and, second, by recirculating water into the tanks from the river (making tank water more like "wild" water). Nevertheless, head start efforts always face the problem of minimizing adaptation to the captive environment so that the "protection" of captivity does not reduce the ability to thrive in the wild.

Raising New Individuals in Captivity from Gametes – The Art of Cryptopreservation

Environmental change – like global warming (Chapter 5) – affects reproductive capacity. The overall effect is referred to by geneticists as "allostatic load" – the cumulative burden of chronic stress, especially as induced by environmental change. When such environmental challenges exceed the individual's ability to cope with them, "allostatic overload" sets in, and reproductive capacities may be reduced or destroyed.[24] When this happens, especially in rare species, conservationists may have to take drastic measures to preserve reproductive capacities. One is to raise new individuals in captivity from gametes. To do this, spermatogonia (sperm-producing) and oogonia (egg-producing) cells are extracted from ovaries or testes, frozen and stored ("cryopreserved"), then used later by implanting them into the ovaries and testes of recipient animals. For example, when spermatogonia from the testes of rainbow trout (*Oncorhynchus mykiss*) were injected into the peritoneal cavity (a membrane-lined cavity surrounding the abdominal organs) of recently hatched masu salmon (*Oncorhynchus masou*), they invaded the gonads of the salmon and were able to generate either spermatozoa or oocytes.[25] And these procedures worked with spermatogonia that had been in cold storage for *five years*. The same technique has been used in the conservation of the endangered Manchurian trout (*Brachymystax lenok*)[26] and has been successful in producing elkhorn coral (*Acropora palmata*) using gametes from geographically separated and genetically isolated populations.[27]

Assisted Migration and Colonization

Every species has a "climate niche," a constellation of climate variables within which it can survive and beyond which it will perish (Chapter 5). As climate change is present and accelerating, species must either find ways to adapt to the new climate or "track" their climate niche across latitude, longitude, and elevation to remain within it. The "protected areas" (PAs) they may inhabit do not protect against climate change. If the climate changes sufficiently, species must abandon the protection of their preserve and find new (perhaps otherwise unprotected) areas that conform to their own climate tolerances. Conservation under climate change often will not depend on *surrounding* species with PAs, but on *moving* species to areas of tolerable climate. Conservation in the Anthropocene must include the intervention of assisted migration (AM) and assisted colonization (AC).

AM is the movement of animals to facilitate or mimic range expansion, or enhance depleted populations by their human-assisted transport to areas more suitable for their persistence.[28, 29] AM is also a way to replenish extinct or endangered populations, to create safe havens for threatened species, and to mitigate habitat loss from climate change. Plants and animals can move on their own, but AM must be used when there are barriers to movement that the species cannot, by itself, overcome. In a review of 443 cases describing the response of individual species' populations or assemblages to particular interventions, Katharine Bowgen et al. found that 30% demonstrated at least one positive impact on a population affected by a climatic variable, and 73% did so in interventions focused on a single species. Interventions spanned a range of responses associated with drought, flooding, changes in rainfall, and increases in mean temperature or reductions in cold extremes.[30]

Not all interventions are about endangered species. One recent test of AM focused on the southern rock lobster (*Jasus edwardsii*), an abundant, commercially valuable species sustaining an important fishery in the oceans around Tasmania and South Australia. These oceans are among the fastest-warming waters on Earth, and there is already evidence of declines in lobster recruitment consistent with changes predicted by their warming.[31]

Bridget Green and her colleagues tested the effect of moving lobsters from deep water to inshore sites with climates that the lobsters are predicted to encounter under future climate change scenarios.[31, 32] About 30% of relocated female lobsters delayed reproduction in their new locations in the first year, but 93% of these resumed reproduction in the second year. Despite first-year losses, egg production over the combined two-year period *improved* by approximately 35,000 eggs/year for translocated individuals compared to lobsters that remained at the source site because lobsters that deferred reproduction for one year increased egg production by 65,000 eggs in the second year.[32]

The same lobsters also demonstrated increased growth "to more than double that of the residents in the area to which they were moved and four times that of the lobsters in their original location."[32] These changes would be appreciated by commercial fishers, as lobsters are normally sold by the pound. And such changes showed "plasticity" in the lobsters that permitted them to change both their growth rates and their reproductive output when subjected to changes in temperature.[32] Lobsters are resilient (at least within limits), and such results also suggest that the lobster fishery could be improved, in the words of Green et al., "by moving slow growing lobsters to sites that are more favourable to growth, and therefore egg production."[32]

AM and AC can enhance conservation efforts, but also pose dangers. As conservation scientist Ben Minteer notes, such efforts

> will prove meaningless – and maybe perversely counterproductive – if they only convince us that we've solved the underlying conservation problem by moving populations to more hospitable environments ... if proposed strategies such as AC end up taking the place of more serious attempts to control and mitigate our environmental destructiveness.[33]

Risk assessment is an essential component in AM and AC, but low-risk uses of these strategies may in some cases be the best (sometimes the only) chance for climate-vulnerable species to survive in the short term, especially if their traditional range has become fragmented by barriers (natural or manmade) they cannot cross and altered in climate beyond their adaptive capacities.

Rehabilitation and Reintroduction

Prolonged contact of species with humans can occur through the pet trade, illegal private or domestic exploitation, recovery from injury in a rehabilitation center, or other kinds of contact that can change the animal's behavior, reduce its ability to function in its natural environment, or impair its interaction with members of its own species. In these cases, effective intervention takes the form of rehabilitation and reintroduction. *Rehabilitation* is the process by which captive animals are treated for medical and physical disabilities until they regain health, enabled to acquire or regain natural social and ecological skills, and finally weaned from human contact and dependence to survive independently in the wild. In contrast, *reintroduction* attempts to establish a species in an area in its historic range, but from which it has been extirpated or become extinct.[34]

The problems associated with behavioral, social, and intellectual changes that can occur through extended contact with humans are especially prominent in primates, particularly the great apes, because their survival competencies depend on learning and flexible behavior, coupled with slow development and reproduction. One species facing serious conservation needs, yet offering perhaps the greatest potential for conservation effect, has stimulated some of the most ingenious responses in conservation effort – the rehabilitation and reintroduction of the orangutan.

The Bornean orangutan (*Pongo pygmaeus*), Sumatran orangutan (*P. abelii*), and Tapanuli orangutan (*P. tapanuliensis*) are arboreal, forest-dwelling apes whose populations are now restricted to the islands of Borneo and Sumatra, mainly in Indonesia. IUCN lists the Bornean as Endangered (population 104,700), the Sumatran and Tapanuli as Critically Endangered (populations 13,846 and fewer than 800, respectively) (see Figure 10.4).[35] All three are declining, primarily due to forest habitat loss from logging, agriculture, plantation forestry, and fires. Orangutans are also victims of illegal trade. Infants bring high prices as pets, motivating traders to capture them by killing their mothers and selling the infants at a favorable, albeit illegal, profit.

Some traders are arrested, but the captured infants lack skills to survive on their own. Because the Indonesian government lacks funds, and political will, to care for, let alone

Figure 10.4 The Tapanuli orangutan (*Pongo tapanuliensis*), one of three endangered orangutan species now restricted to the islands of Borneo and Sumatra, mainly in Indonesia. All three species are exploited by the pet trade, raising the need for rehabilitation of formerly captive orangutans before they can return to the wild. Tim Laman, CCBY.

rehabilitate, such orphans, NGOs have created centers to receive orphaned orangutans in need of medical attention and rehabilitation. After receiving medical care at the centers, orangutans are progressively trained in how to interact with other orangutans in specially designed "socialization cages" and then progress to learning forest survival skills in a "forest school."[35] When such training is completed, the final step (reintroduction) is to release them in a protected forest. The rehabilitation process takes years, but overcrowded conditions at the centers, government pressure, and public (often donor) demand persistently call for the release of rehabilitated individuals, ready or not.

The motives for these efforts are praised, but their effectiveness isn't. Critics accuse such work of being no more than conservation propaganda and fund-raising promotion,[34] failing to succeed with most ex-captives, and exploiting those at the centers as commodities for tourist revenue.[36, 37, 38] These criticisms have merit. Uniform standards for rehabilitation training, with defined social and behavioral competencies, have only recently been developed,[34, 39] and rigorous assessments of rehabilitation and reintroduction success are just beginning. Dominick Fechter and his colleagues monitored 32 rehabilitated and radio-fitted ex-captive Sumatran orangutans in a study developed by the Frankfurt (Germany) Zoological Society (FZS). They compared the behaviors of the rehabilitated orangutans with behaviors expected for and observed in wild orangutans using rehabilitated orangutans which were released in an area around the FZS field station near the border of the Bukit Tigapuluh National Park in Sumatra.

In diet selection, experienced (older) rehabilitants made food choices more like wild orangutans than those of novices (recently released orangutans). Younger orangutans selected areas closer to the FZS station while older rehabilitants selected sites further from it, usually closer to rivers and including lower-elevation areas with optimal food availability.[40] Regardless of the duration of their rehabilitation, all orangutans decreased their selection for areas closer to the FZS station over time and strengthened their selection for elevations associated with greater food availability. Addressing the call to release orangutans to the wild as soon as possible, Fechter et al. declared, "The current release practice favors releasing orangutans at a young age ... because this stage of semi-independence is believed to be the best period for learning and integration ... Based on our findings, however, we suggest releasing orangutans at an older age."[40]

Fechter et al.'s results show that the rehabilitation of ex-captive orangutans can be successful, but what proportions of ex-captives actually survive reintroduction? Rescue facilities have reported post-release survival rates ranging from 20% to 80% in past years, but other, more recent evidence suggests long-term survival rates actually range from 6% to 27%.[41] If rehabilitation and reintroduction interventions are going to matter for orangutans, survival rates must increase because the influx of new, orphaned, and ex-captive orangutans to the centers continues unabated. The number of orangutans held in the centers was estimated to be around 1,500 in 2013. That figure may have since increased, not only because orangutans are being brought to the centers faster than they can be responsibly released, but also because of orangutan births at the centers themselves. The arrival of new "orangutan orphans" continues because conservation law is not enforced. As human behavior created the problem, so it sustains the same dilemma.

Just as PAs and national parks (Chapter 9) require governance structures possessed of intention, capacity, integrity, and stability, so interventions in rehabilitation and reintroduction need the same. The work of these interventions – to heal a creature from injury and restore it to a life of its own – are necessary but insufficient conditions for the goals of restoration and reintroduction to be completed. They must be supported by governance that provides the animal with protection and by cultural practices of people in its environment that value its own, self-willed life. In Indonesia, orangutan rehabilitation and reintroduction continue and are increasingly informed by scientific study and practiced technique, but they can only be completed by changes in governance and culture. In the end, rehabilitation requires relationship. The animals are hurt and need care. They should receive it from a human hand.

Rewilding

Rewilding as Concept and Intervention

Rehabilitation and reintroduction require the preparation of captive animals to return to a wild environment. But what if wild animals were to colonize a "human" (or human-altered) environment from which they had been previously removed? That is the genesis of re-wilding. Formally rewilding has been defined (or, perhaps better, described) as the process of protecting an environment and returning it to its natural state, especially by bringing back wild animals that were formally resident in it.[42] For some, rewilding runs counter to human attempts to control and cultivate nature. To others, the vision of "planned rewilding," even to the extent of bringing back mastodons and saber-toothed tigers, only strengthens their vision of the Earth as a human-controlled "garden" which they can reshape at will. It is best from the beginning to differentiate between "human-initiated rewilding," which, like other strategies covered in this chapter, is a human effort, and "nature-initiated rewilding," a return of species not according to human planning but through animal resilience and adaptive capacity.

In the United States, the terms *rewild* and *rewilding* were becoming commonly, if informally, used by some environmental activists by the 1980s. The idea emerged as a more defined concept in the writings of conservation biologist Reed Noss in 1990, in which Noss stressed the importance of combining composition, structure, and function in preserving forest ecosystems, and proposed a series of steps in biodiversity monitoring at a regional scale which would provide feedback for adaptive management over large areas.[43] As Noss and others began to provide an intellectual and scientific foundation for rewilding, the term itself arrived in the literature in 1991 through the "Wildlands Project," a collaborative effort which brought together, in the words of environmentalist Dave Foreman, "citizen conservationists and conservation biologists to formulate a new idea of conservation and to apply science to the design and management of protected areas."[44] Their 1991 report was the first explicit expression of "rewilding" as a conservation strategy. Perceiving national parks and wilderness areas as "islands" of "relatively unproductive habitat" that had already lost species and were certain to lose more, Noss proposed a corrective strategy expressed in four principles: 1) represent, in a system of PAs, all native ecosystem types and seral stages across their natural range of variation; 2) maintain viable populations of all native species in natural patterns of abundance and distribution; 3) maintain ecological and evolutionary processes, such as disturbance regimes, hydrological processes, nutrient cycles, and biotic interactions, including predation; 4) design and manage the system to be responsive to short-term and long-term environmental change and to maintain the evolutionary potential of lineages.[45] Using this strategy, Foreman asserted that core wilderness areas could be managed to preserve and "where necessary" restore native biological diversity and natural processes. The rewilding would be concentrated on specific "focal species" including those identified as keystone species (which enrich ecological function by their activities), umbrella species (whose protection, with their large home areas, would offer protection for other species), flagship species (charismatic species that would generate popular support), and indicator species (that would provide early warning of environmental change).[44]

Most segments of the conservation community were initially reticent to legitimate rewilding as an official "slogan" for conservation, but sometimes ideas become more pervasive than programs. The concept of and attraction to rewilding as a conservation strategy has grown over time, expressing itself worldwide in unique ways in different places. In the Republic of Korea (South Korea), the Korea National Park Service has been working to restore multiple endangered species, including the Asiatic black bear (*Ursus thibetanus*), musk deer (*Moschus moschiferus*), long-tailed goral (*Naemorhedus caudatus*), red fox (*Vulpes vulpes peculiosa*, rare in Korea),[46] Oriental white stork (*Ciconia boyciana*), black shiner (*Pseudopungtungia nigra Mori*), bull-head torrent catfish (*Liobagrus obesus*), and wind orchid (*Neofinetia falcata*).[47] In Europe, the wisent (European bison, *Bison bonasus*), exterminated in the Carpathian Mountain of central and south-eastern Europe more than 200 years ago, had by 2012 been restored to the Carpathians with 6 free-ranging populations of over 350 individuals.[48]

I remember a conversation with my PhD advisor, the late Dr Rainer Brocke of SUNY-Syracuse. We had gotten onto the subject of animal reintroductions, and Rainer said, "Fred, these animals aren't made of glass. They can thrive in any good place where people don't kill them." Rainer's words were at once blunt, simple, and prophetic. He led the effort to restore the Canada lynx (*Lynx canadensis*) in the Adirondacks, beginning in the 1980s. That effort failed, but prophecies can be fulfilled in other ways.

Nature's Intervention in Conservation – Wildlife's "Do-It-Yourself" Rewilding Program

For all their breadth of effort, biologists in Korea made an intentional omission in their rewilding plan – the tiger (*Panthera tigris*). It was understandable that Koreans were not keen to re-establish a large predatory cat capable of killing not only human livestock but the human livestock keepers. Large predators are not naturally favored in human rewilding initiatives. No matter. They show remarkable initiative of their own.

The mountain lion (*Puma concolor*) was historically one of the most widely distributed species in the Americas, from the edge of the Arctic Circle to the southern extent of Patagonia and from the Atlantic to the Pacific in both hemispheres. In contrast, when I began my studies of them in the late 1970s, mountain lions in the United States, though not endangered, were limited to higher-elevation forests and rangelands of the US Intermountain West, Pacific Coast, the Southwest, and the western portion of state of Texas, with a remnant population in Florida. They were hunted by sportsmen, and shot, trapped, or poisoned by ranchers and predator control specialists when they killed cattle or sheep. But times have changed, and so have lion populations. Today mountain lions have established resident and growing populations in the US Great Plains (North Dakota, South Dakota, and Nebraska) as well as the Upper Great Lakes (Michigan). Only in South Dakota was there a human-initiated reintroduction, and even this effort might not have been necessary.

In Europe, no species has been the subject of more intense persecution than the wolf (*Canis lupus*), exterminated throughout most of the continent by the end of the nineteenth century. But unlike many other extirpated carnivores, wolves did not disappear because of habitat change or loss of prey. They were locally eradicated by active hunting, trapping, and poisoning.[49] Such efforts have largely ceased, concurrent with declines in human populations in many mountainous and hilly areas in Europe, with subsequent increases in natural vegetation[50] and wild herbivores in the same places.[51] Employing their intelligence, behavioral plasticity, and ecological resilience, wolves have increased their numbers and expanded their range across the continent, particularly in the last 30 years. In 2022, a report to the Large Carnivore Initiative for Europe found that all mainland European countries now have wolves, with a total estimated population of 21,500 individuals.[52] The wolf has rewilded the continent through its own efforts, leaving humans to report its success. The wolf and the mountain lion demonstrate the truth of the words of conservationist John Terborgh, which are eerily similar to Rainer Brocke's: "Wolves and cougars don't require pristine conditions to thrive; they require freedom from persecution."[53]

The peregrine falcon (*Falco peregrinus*) in North America, once critically endangered, is now listed by IUCN as a species of Least Concern. Its marked, continent-wide population growth has been especially supported by increases in urban areas where peregrines now make use of skyscrapers for nesting habitat and concentrations of urban rock pigeons (*Columba livia*) for food (Chapter 7). In Australia, the powerful owl (*Ninox strenua*) was until recently considered a rare ecological specialist due to its restricted habitat and dietary requirements and its preference for older growth forests. But in the last decade it has become increasingly abundant in the fragmented urban and urban-fringe environments of major cities along Australia's eastern seaboard.[54] Although the IUCN estimates their population at fewer than 3,000 adults, it judges it as stable and contiguous and places the owl, like the peregrine falcon, in the category of Least Concern.[55] The powerful owl's population growth in urban environments has been supported by its adaptation to Australian cities and suburbs, along with human contributions of nest boxes and enhancements of forest environments along urban rivers, a favorite habitat of the owl.[54]

These are but four of hundreds, perhaps thousands, of "natural rewilding initiatives" by animal species themselves. The rewilding concept has focused human effort on restoring extirpated species to their traditional ecosystems. But it has also produced a transformation of human attitudes when animals "rewild themselves" – leading people to see the presence of wildlife as something to be desired in environments in which wildlife have formerly been extirpated, even in species that were once perceived as dangerous and threatening.

Adaptive Management and Value-added Conservation

Adaptive Management

Whatever strategy of conservation intervention is employed in saving biodiversity, it needs two elements to maximize its effectiveness for conservation policy. One is an approach of *adaptive management*. The other is a method to actually *measure the value added by the conservation action* toward achieving the conservation goal. A formal definition of adaptive management might be expressed as *the structuring of policy or management actions as a set of testable hypotheses to promote learning* ***from policy implementation****, and to allow for greater adaptability when change does inevitably occur within the system.*[56] The action considered is first informed by the data managers already possess, which is then used to create a management hypothesis in some version of *if A then B* (where *A* is the action taken and *B* is the expected result). The management action is then applied in an experimental design, such that the results of the experiment will inform future management actions. In this way, the dual goals of performance and knowledge acquisition are accomplished at the same time so that managers *learn about the system through the implementation of their management actions*. This means that *management actions must be designed as direct manipulation experiments and managers must be willing to change management strategies based on the experimental results.*

These kinds of efforts can be justifiably called "performance-based management." But in order to be "performance based," the management experiment must have at least one measurable variable which reveals a response to the treatment. Sadly, even in cases where so-called adaptive management approaches have allegedly been used, *only 1% have reported any response metrics.*[57] When there is no measured response, there can be no determination of conservation effect. Despite the fact that the stakes in conservation have never been higher, few conservation efforts actually employ an adaptive management approach and even fewer reflect management efforts that *actually evaluate conservation policy decisions*. Some efforts are rightly directed toward the goal of acquiring new knowledge to inform policy, but *conservation effectiveness* must be about the application of conservation policies (organized expressions of management decisions) that actually work.

The woodland caribou (*Rangifer tarandus*) in North America has a range covering 3 million km^2 from Alaska to Newfoundland. The caribou's critical habitat overlaps petroleum deposits and forest stands that are collectively worth billions of dollars.[58] Today most caribou populations are in decline and extirpation of local populations is constantly occurring.[59, 60] Given its range and distribution, caribou conservation efforts are having increasing conflicts of use with extractive industries and natural-resource economies.[58] In these circumstances, we have a wide-ranging species of conservation concern and high relevance to both renewable and non-renewable resource use whose populations may be affected by multiple factors, many of them influenced by management policies. To better evaluate and find a solution to the problem of the caribou's ongoing decline, Robert Serrouyaa and his colleagues designed a comprehensive adaptive management experiment.

The primary hypothesis was that caribou populations were declining because of wolf predation, but Serrouyaa's team created treatments that included not only wolf removal, but wolf removal plus "safe havens" for caribou (predator-proof fencing around maternal birthing and rearing areas) and movement (translocation) of caribou to different areas with no wolf removal. In all cases, populations in treated areas were compared with untreated control populations.[61]

In this experiment the response variable was explicit – change in the size of the caribou population. The strongest response occurred in areas that included both wolf removal and protected maternal areas. Here caribou populations nearly doubled, while those in adjacent, untreated areas continued to decline.[61] What gave clarity to the outcomes were, first, an adaptive management framework with control populations, which were critical for determining if population growth increased after a specific treatment, and second, clear experimental demonstration that a treatment had to be applied intensively to produce a measurable effect. Finally, the skill in experimental design provided demonstration that two treatments applied simultaneously produced an additive effect.[61] Here was a management experiment that could clearly inform both subsequent management actions and conservation policies. This is the essence of effective conservation intervention.

Value-added Conservation – The Case for the Counterfactual

The IUCN has had a Red List of Endangered Species since 1964. More recently it has developed a "Green List," or "Green Status of Species," which assesses how close a species is to being "fully ecologically functional" across its range, and how much it has recovered thanks to conservation action.[62] To answer the last, and more measurable, question, one must compare the *observed or predicted species condition* with the *expected species condition under various scenarios* (alternative realities). Thus, to evaluate the past impact of conservation efforts, we must generate a *counterfactual scenario*: an alternative past in which conservation efforts *did not happen*. Using the best available information to create this parallel conservation universe, we then predict the path that the species would have followed *in the absence of conservation action* (the counterfactual) *compared to the current status of the species*. The difference between the counterfactual status and the current status is called the *Conservation Legacy*.[63]

Molly Grace and her colleagues, chief architects of the Green Status criteria, declare that, to establish the counterfactual, one must: 1) define the population, the action, the conservation starting point, the conservation end point, and the variable to be measured; 2) contextualize the action by determining the value of the variable at its starting point and end point, and identify all other factors which could have affected the variable other than the conservation action; and, finally, 3) estimate the impact of the application of the conservation action by recording the evidence that removal of the conservation action would have caused a change in the variable, estimating the value of the variable at the end point if the conservation action had not been taken, consider the other identified factors and if they would have changed the variable without conservation actions, and then state the value of the variable in the absence of the conservation action.[63] That's a mouthful. Perhaps an example can help.

The Kirtland's warbler (*Setophaga kirtlandii*) (see Figure 10.5) was once one of the most endangered birds in the United States, and one of the first to be placed on the US Endangered Species List.[64] By the early 1970s, the warbler had declined to fewer than 500 adults, all within a small area in northern Michigan.[65] One of the reasons

Figure 10.5 The Kirtland's warbler (*Setophaga kirtlandii*) (A), once one of the most endangered birds in the United States, has been removed from the US Endangered Species List as its populations have recovered. The control of its primary nest parasite, the brown-headed cowbird (*Molothrus ater*) (B), resulted in demonstrable increases in the warbler's nesting success and long-term increases in its population. Kirtland's warbler photo by Erin Hillman. Brown-headed cowbird photo courtesy of US FWS.

for the warbler's decline was nest parasitism by the brown-headed cowbird (*Molothrus ater*). Female cowbirds lay their eggs in Kirtland's warbler nests, and the larger cowbird nestlings monopolize food gathered by the parents. In a parasitized nest, few warbler nestlings survive. Nest surveys during the 1960s determined that 83% of Kirtland's warbler nests were parasitized by cowbirds, and, in such circumstances, fewer than one (0.81) Kirtland's warbler young was fledged per nest.[65]

The Kirtland's warbler belongs to a group of birds known as wood warblers, which have an average lifespan of four to eight years. Taking a middle value of six years for a lifespan, and knowing that many Kirtland's warblers do not breed in their first year of adulthood (second year of life), the average adult warblers would have four reproductively active years, which, from parasitized nests, would yield 3.24 fledged young per pair of adults (4 years x 0.81 fledged young per year). But not all fledged young survive. Using a middle value of 0.6 in a range of published estimates of the survival rate of fledged Kirtland's warblers,[66] we would expect 1.94 adult birds surviving per pair over four years, a value below replacement level.

Such calculation is the epitome of "quick and dirty," but provides a plausible guess in judging just how severely Kirtland's warblers were being affected by cowbirds. Recognizing the urgency of the problem, the US FWS, beginning in the late 1960s, started trapping and removing cowbirds from areas where Kirtland's warblers were nesting. Field surveys of nests from 1966 to 1971 determined that the average number of Kirtland's warblers fledged per nest increased from 0.81 to 2.84 in trapped areas.[65] This means that the "value-added" effect of trapping cowbirds was an annual increase of 2.03 fledged young per nest. Using previous assumptions of fledgling survival, this would generate a lifetime output of 4.87 adult birds per breeding pair of Kirtland's warblers. Here is a quantitative estimate of what would have happened to Kirtland's warbler production if cowbirds were not being trapped in Kirtland's warbler nesting areas – an important dimension of a conclusive counterfactual. Full counterfactual design is more rigorous, best expressed through randomized control trials, before–after control-intervention trials, and matched pair comparisons or other quasi-experimental designs,[63] such as those used in the woodland caribou study. Counterfactual designs should be used in all management actions and policy evaluations, but they are not. To be *credible*, conservation must demonstrate that it is *useful*. When interventions are needed, counterfactual evaluation is essential.

The Anthropocene – A World of Conservation Reliance

The goal of managing endangered populations is their recovery. That is why management plans for endangered species are called "recovery plans." In the United States, the goal of the US Endangered Species Act (ESA) is to bring listed species to "the point at which the measures provided [by the] Act are no longer necessary" (ESA 1973 §3:2). A review of the ESA's success through 2013 found that 98% of listed species still survived, 27 formerly endangered species had been reclassified and were moving toward recovery, and 29 species were being considered for reclassification from endangered to threatened. But only 22 ESA-listed species have been designated as recovered and subsequently removed from the Endangered Species List. That is because, even when the population of an endangered species increases, most listed species will continue to depend on ongoing management action to maintain their path to recovery. Such cases are examples of *conservation-reliant species*, a descriptor that can be applied to more than 80% of species protected under the US ESA.[67]

The Kirtland's warbler was removed from the US Endangered Species List in 2019 and is now the subject of a still-developing post-delisting monitoring plan. If its population should decline in the future, it can be relisted, but if it is stable or increasing, its delisting will become permanent. While such delisting may itself be an encouragement to conservationists, it comes with the loss of both legal protection for the species and US federal government funds designated for such protection, funds that had been used for habitat creation (the Kirtland's warbler is a specialist of young, even-aged jack pine (*Pinus banksiana*)) and cowbird removal. The mechanism that can be used to replace these losses is a Conservation Management Agreement (CMA).

The Kirtland's Warbler Recovery Team (KWRT), a group of scientists and managers dedicated to the Kirtland's warbler's recovery, recognized that population monitoring, research, and public education would remain essential to maintain Kirtland's warbler populations after delisting.[68] Faced with this reality, they began the development of a CMA for Kirtland's warblers. A CMA requires four elements to be effective: 1) a partnership agreement that will continue to enable effective management actions; 2) a management plan informed by understanding critical elements of the species' recovery; 3) sufficient funding; and 4) legal enforcement (see Figure 10.6).

For the Kirtland's warbler, CMA partners will include the US federal government (the US FSW and the US Forest Service, with the latter having jurisdiction over many areas that Kirtland's warblers use as breeding habitat), the state government (the Michigan DNR, which has taken a leading role in creating new warbler habitat), private NGOs, private landowners (who control access to many breeding areas), and private industry, especially the wood products industry, which owns large areas of jack pine forests that could be managed to provide breeding habitat. In this case, the KWRT worked to create a nonprofit conservation organization dedicated to the Kirtland's warbler's continued management and governed by its own Board of Trustees who will receive input from an advisory board of experts from public and private sectors. These boards will form the partnership that will oversee the continuing recovery of the Kirtland's warbler.[68]

To carry out its intentions, a CMA must have financial resources. Here, the partners plan for the development of a private trust fund dedicated to Kirtland's warbler

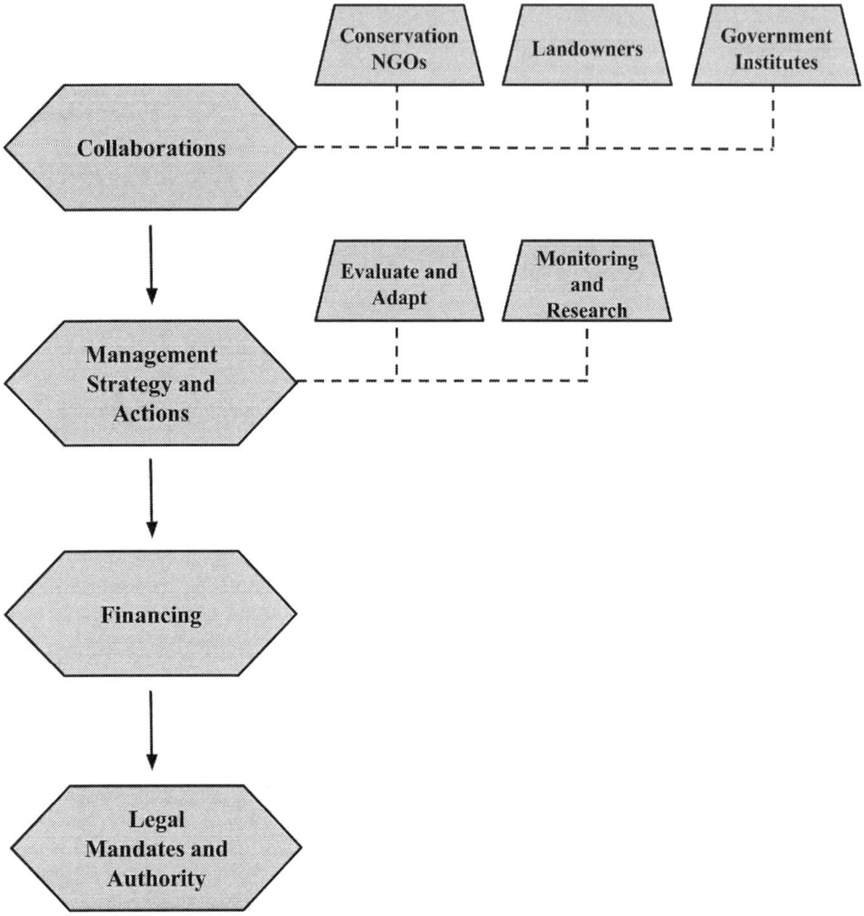

Figure 10.6 Schematic representation of the elements needed for an effective Conservation Management Agreement (CMA) as exemplified for a conservation-reliant species, the Kirtland's warbler (*Setophaga kirtlandii*). Illustration by Katie DeVoss. Reprinted with kind permission from SpringerNature, Clam, Switzerland.

conservation. The trust fund will seek contributions from conservation NGOs, the public, and private industry to continue to finance needed management actions.[68] Finally, the CMA must have authority to makes its management plan and management actions legally enforceable, framed as an enforceable contract or, on private land, a conservation easement giving power to regulate the kinds of use and activity that can take place on lands used as breeding areas by the Kirtland's warbler. In this case, the government management agencies involved will be expected to develop, agree to, and sign legally binding agreements, and ensure that such agreements are implemented.[68]

This example reflects ongoing efforts for one species, imperfect in present efforts and uncertain in future results. But the elements of the Kirtland's warbler CMA offer a blueprint for any species that "recovers" from official endangerment yet remains "reliant" on conservation management, a condition that will be increasingly common for thousands of species in the Anthropocene. This final intervention must be a lasting one, an engagement in which humans reject the binary of "endangered" or "not endangered" and instead embrace an interaction with nature in which every species will be an object of intentional care from the human community.

In this example, conservation intervention was directed to a single species, but sometimes it may require interventions that affect an entire ecosystem. For that example, we end with the story of an ongoing intervention effort in Indonesia.

Story Box 10.1 Restoration of Indonesian Peatlands – Management, Recovery, and Restoration of Conservation-Dependent Species

Hesti Lestari Tata and Bambang Tejo Premono

Research Center for Ecology and Ethnobiology, National Research and Innovation Agency, Jakarta, Indonesia

Tropical peatlands play pivotal roles in maintaining biodiversity and are home to a wide range of plant and animal species (see Figure 10.7). The first time I (Hesti) visited a tropical peat swamp forest in Riau province (Indonesia) more than a decade ago, I realized this ecosystem had unique characteristics. Peat soils are saturated organic soils dominated by undecomposed and partially decomposed plant materials. The primary water source is rain, but its black-tea color in a peat swamp

Figure 10.7 A peat swamp forest in Kalimantan, Indonesia similar to the Bram Itam Forest in Indonesia's Jambi Province, where Hesti Tata and Bambang Premono worked with local residents to restore native vegetation and improve ecosystem function. Peat swamp forests provide many essential ecosystem services, including high levels of carbon sequestration which can reduce the release and concentration of CO_2 into the atmosphere. Ruanda Agung Sugardiman/AusAID. Wikimedia. CC-BY 2.0. <https://creativecommons.org/licenses/by/2.0/deed.ed>.

forest comes from the decomposition of organic materials that are released into the surface waters.

Tropical peatlands have suffered extensive deforestation, mainly due to agricultural development. Converting peatlands to agriculture destroys important ecosystem functions and releases large amounts of stored carbon into the atmosphere, contributing to climate warming. Peatlands have also been extensively drained and fragmented, leaving them more susceptible to wildfires. Peatland degradation leads to the loss of the unique biodiversity and ecosystem services that peatlands provide, especially water regulation and flood control.

We and our colleagues work with local residents to conserve peatlands in the Bram Itam peat swamp forest in Tanjung Jabung Barat regency (Tanjabar) in Indonesia's Jambi Province. Bram Itam has been designated as a "protection forest" by the Minister of Forestry, giving it special consideration for conservation. The local people are resident Malay as well as Buginese, Banjarese, and Javanese immigrants who have lived in the forest since the 1920s. Such immigration has increased the population of Tanjabar, particularly after 1990. As Tanjabar's human population has grown there has been accompanying deforestation, driven by the increasing demand for construction materials, the creation of human settlements, and the collection and production of food.

On the occupied forest land farmers plant the oil palm (*Elaeis guineensis*) because of its high economic value, although the palm is not native to Tanjabar's peat-swamp forest. Much of our reforestation effort has been concentrated on forest land between oil palms plantations, ensuring that such land will remain in native forests characteristic of tropical peat swamps. Some residents have welcomed the forest rehabilitation program, but others have refused to participate because the restoration has threatened to reduce the available land they might be able to occupy. Nevertheless, active replanting and assisted regeneration of native vegetation is being implemented with the help of many people in the local community.

Our restoration is implemented by "rewetting," blocking the human-made canals that have removed water for agricultural irrigation to increase groundwater levels. As groundwater levels rise, we are able to plant more native plant species that advance our peatland restoration effort. We help local farmers rehabilitate the occupied peat forest and peatlands by planting *Dyera polyphylla* (a native peatland tree) and intercropping with galangals (*Alpinia galanga*) and turmeric (*Curcuma domestica*). Ecosystem restoration takes a long time, but ten years after planting, we have begun to see *D. polyphylla* growing well in agroforestry systems with oil palm and *Coffea liberica* (Liberian coffee).

This and other kinds of intercropping are producing a more diverse ecosystem and storing more carbon than monocultures of oil palm. Such rewetting and peat forest rehabilitation can reduce the negative effects of climate change by slowing greenhouse gas emissions to the atmosphere. More directly, farmers receive benefits from peatland restoration that include honey from stingless bees (*Trigona* spp.), black-water fish (such as *Helostoma temminckii*), Areca nuts (*Areca cathecu*), *C. liberica*, and latex taken from *D. polyphylla*. In recent years, villagers have begun to appreciate that planting woody trees species is important not only for the ecosystems, but also to provide timber for their domestic needs. Support from both the government and private sectors has also created a "green market" for these products which ensures that income from them can continue. Initiatives for conservation and restoration, sustainable land management techniques, and laws intended to lessen deforestation and peatland degradation are some of the ways that tropical peatlands are being conserved and restored here in Indonesia using a restoration strategy with specific assignments of roles and responsibilities to carry out work needed to enhance and restore peatlands. We are discovering that we can rehabilitate the ecological integrity, biodiversity, and potential for carbon sequestration for future generations in the peat forest by putting these assisted recovery and intervention management techniques into practice. And we are glad and honored to be a part of it.

References

1. Van Dyke, F., B. W. Van Ee, S. Harju, et al. 2023. Distinguishing and using two species of fish in assessment of stream quality. *Ichthyology and Herpetology* 111:584–596.
2. Ojibwe Treaty Rights. 2018. Great Lakes Indian Fish and Wildlife Commission, Odanah, Wisconsin, USA. <https://glifwc.org/publications/pdf/2018TreatyRights.pdf>. Accessed 2 February 2024.
3. Holtgren, M., 2013. Bringing us back to the river. Pages 133–146 *in* N. Auer and D. Dempsey, editors. *The Great Lake sturgeon*. Michigan State University Press, East Lansing, USA. <https://doi.org/10.14321/j.ctt9qf52k>.
4. Crossman, J. A., K. T. Scribner, P. S. Forsythe, et al. 2018. Lethal and non-lethal effects of predation by native fish and an invasive crayfish on hatchery-reared age-0 lake sturgeon (*Acipenser fulvescens* Rafinesque). *Journal of Applied Ichthyology* 34:322–330. <https://doi.org/10.1111/jai.13558>.
5. Wassink, L., B. Huerta, D. Larson, et al. 2022. Hatchery and wild larval lake sturgeon experience effects of captivity on stress reactivity, behavior and predation risk. *Conservation Physiology* 10(1):coac062. DOI: 10.1093/conphys/coac062.
6. Briggs, A. 2024. Michigan's lake sturgeon rehabilitation strategy. Michigan Department of Natural Resources. <https://www.michigan.gov/dnr/-/media/Project/Websites/dnr/Documents/Boards/NRC/2023/December-2023/Lake_Sturgeon.pdf>. Accessed 25 January 2024.
7. Visvader, J. 1996. Natura naturans: Remarks on the nature of the natural. *Human Ecology Review* 3:16–18, 18.
8. International Union for the Conservation of Nature. 2024. IUCN Red List of Threatened Species. <https://www.iucnredlist.org/>. Accessed 25 January 2024.
9. Hoffman, M., C. Hilton-Taylor, A. Angulo, et al. 2010. The impact of conservation on the status of the world's vertebrates. *Science* 330:1503–1509.
10. Bolam, F. C., L. Mair, M. Angelico, et al. 2021. How many bird and mammal extinctions has recent conservation action prevented? *Conservation Letters* 14:e12762. DOI: 10.1111/conl.12762.
11. Goodall, J. 2010. *Hope for animals and their world*. Grand Central Publishing, New York, USA.
12. Audubon California. 2024. History of Audubon and the condor. National Audubon Society. <https://ca.audubon.org/conservation/history-audubon-and-condor>. Accessed 25 January 2024.
13. United Press International. 1986. US can capture condors to save species, court holds. *Los Angeles Times*. 5 September 1986. <https://www.latimes.com/archives/la-xpm-1986-09-05-

mn-13655-story.html#:~:text=The%20injunction%20against%20capturing%20the,about%20the%20effects%20of%20captivity>. Accessed 25 January 2024.

14. California Department of Fish and Wildlife. 2024. California condor. State of California. <https://wildlife.ca.gov/Conservation/Birds/California-Condor>. Accessed 25 January 2024.
15. Terwilliger, M. 2016. World CA condor update – 2016 population status. National Park Service. <https://www.nps.gov/articles/caco-world-2016.htm>. Accessed 25 January 2024.
16. National Park Service. 2022. World CA condor update – 2022 population status. National Park Service. <https://www.nps.gov/articles/000/caco-world-2022.htm>. Accessed 25 January 2024.
17. Biggins, D. E., B. J. Miller, T. W. Clark, et al. 2006. Restoration of an endangered species: The black-footed ferret. Pages 581–585 *in* M. J. Groom, G. K. Meffe, and C. Ronald Carroll, editors. *Principles of conservation biology*. 3rd edition. Sinauer, Sunderland, Massachusetts, USA.
18. Holt, W. V. 2016. The black-footed ferret recovery program: A strong advocate for establishing semen banking programs as support tools for small population welfare. *Animal Conservation* 19:116–117.
19. Howard, J. G., C. Lynch, R. M. Santymire, et al. 2016. Recovery of gene diversity using long-term cryopreserved spermatozoa and artificial insemination in the endangered black-footed ferret. *Animal Conservation* 19:102–111.
20. Biggins, D. E. 2012. Information on black-footed ferret biology collected within the framework of ferret conservation. *Western North American Naturalist* 72:129–133.
21. Black-footed Ferret Friends. 2023. Reintroduction efforts. <https://www.blackfootedferret.org/reintroduction-efforts.html>. Accessed 25 January 2024.
22. Hylkema, A., A. O. Debrot, E. E. van de Pas, et al. 2022. Assisted natural recovery: A novel approach to enhance *Diadema antillarum* recruitment. *Frontiers in Marine Science* 9:929355. DOI: 10.3389/fmars.2022.929355.
23. Araki, H., B. Cooper, and M. S. Blouin. 2007. Genetic effects of captive breeding cause a rapid, cumulative fitness decline in the wild. *Science* 318:100–103.
24. Guidi J., M. Lucente, N. Sonino, et al. 2021. Allostatic load and its impact on health: A systematic review. *Psychotherapy and Psychosomatics* 90:11–27. DOI: 10.1159/000510696.
25. Lee, S., Y. Iwasaki, and G. Yoshizaki. 2016. Long-term (5 years) cryopreserved spermatogonia have high capacity to generate functional gametes via interspecies transplantation in salmonids. *Cryobiology* 73:286–290.
26. Lee, S., and G. Yoshizaki. 2016. Successful cryopreservation of spermatogonia in critically endangered Manchurian trout (*Brachymystax lenok*). *Cryobiology* 72:165–168.
27. Hagedorn, M., C. A. Page, K. O'Neil, et al. 2018. Successful demonstration of assisted gene flow in the threatened coral *Acropora palmata* across genetically isolated Caribbean populations using cryopreserved sperm. Cryopreservation-Assisted Gene Flow in Elkhorn Coral. bioRxiv 492447.
28. Vitt, P., K. Havens, A. T. Kramer, et al. 2010. Assisted migration of plants: Changes in latitudes, changes in attitudes. *Biological Conservation* 143:18–27.
29. Rout, T. M., C. E. Hauser, and H. P. Possingham. 2009. Optimal adaptive management for the translocation of a threatened species. *Ecological Applications* 19:515–526.
30. Bowgen, K. M., E. F. Kettel, S. H. M. Butchart, et al. 2022. Conservation interventions can benefit species impacted by climate change. *Biological Conservation* 269:109524. <https://doi.org/10.1016/j.biocon.2022.109524>.
31. Linnane, A., C. Gardner, D. Hobday, et al. 2010. Evidence of large-scale spatial declines in recruitment patterns of southern rock lobster *Jasus edwardsii*, across south-eastern Australia. *Fisheries Research* 105:163–171.
32. Green, B. S., C. Gardner, A. Linnane, et al. 2010. The good, the bad and the recovery in an assisted migration. *PLOS ONE* 5(11):e14160. DOI: 10.1371/journal.pone.0014160.
33. Minteer, B. A. 2019. *The fall of the wild*. Columbia University Press, New York, USA, 94–95.
34. Russon, A. E. 2008. Orangutan rehabilitation and reintroduction: Successes, failures, and role in conservation. Pages 327–350 *in* S. A. Wich, S. S. Utami Atmoko, T. M. Setia, et al., editors. *Orangutans: Geographic variation in behavioral ecology and conservation*. Oxford University Press, Oxford, UK. <https://doi.org/10.1093/acprof:oso/9780199213276.001.0001>.

35. International Union for the Conservation of Nature. 2024. IUCN Red List of Threatened Species. Version 2023-1. <https://www.iucnredlist.org/search?query=orangutan&searchType=species>. Accessed 26 January 2024.
36. Rijksen, H., and E. Meijaard. 1999. *Our vanishing relative: The status of wild orang-utans at the close of the twentieth century*. Kluwer Academic Publishers, Dordrecht, the Netherlands. DOI: 10.1007/978–94-010–9020–9.
37. Leiman, A., and N. Ghaffar. 1996. Use, misuse and abuse of the orangutan—exploitation as a threat or the only real salvation? Pages 345–357 *in* V. J. Taylor and N. Dunstone, editors. *The exploitation of mammal populations*. Springer, Dordrecht, the Netherlands. <https://doi.org/10.1007/978-94-009-1525-1_20>.
38. Yeager, C. P. 1997. Orangutan rehabilitation in Tanjung Puting National Park, Indonesia. *Conservation Biology* 11:802–805. <https://doi.org/10.1046/j.1523-1739.1997.95500.x>.
39. Beck, B., K. Walkup, R. Rodrigues, et al., editors. 2007. *Best practice guidelines for reintroduction of the great apes*. International Union for the Conservation of Nature, Gland, Switzerland.
40. Fechter, D., S. Ciuti, D. Kelle, et al. 2019. Spatial behavior in rehabilitated orangutans in Sumatra: Where do they go? *PLOS One* 14(5):e0215284. <https://doi. org/10.1371/journal.pone.021528>.
41. Sherman, J., M. Ancrenaz, and E. Meijaard. 2020. Shifting apes: Conservation and welfare outcomes of Bornean orangutan rescue and release in Kalimantan, Indonesia. *Journal for Nature Conservation* 55:125807, ISSN 1617-1381. <https://doi.org/10.1016/j.jnc.2020.125807>.
42. Cambridge Advanced Learner's Dictionary and Thesaurus. 2024. Cambridge University Press and Assessment, Cambridge, UK. <https://dictionary.cambridge.org/dictionary/english/rewilding>. Accessed 27 January 2024.
43. Noss, R. F. 1990. Indicators for monitoring biodiversity: A hierarchical approach. *Conservation Biology* 4:355–364.
44. Foreman, D. 1998. The wildlands project and the rewilding of North America. *Denver University Law Review* 76:535–553, 536.
45. The Wildlands Project. 1994. *The Wildlands Project*. McMinnville, Oregon, USA.
46. Korea National Park Service. 2021. Achieving co-existence: Restoration of endangered species in the Republic of Korea. Panorama Solutions for a Healthy Planet. <https://panorama.solutions/en/solution/achieving-co-existence-restoration-endangered-species-republic-korea>. Accessed 29 January 2024.
47. The National Atlas of Korea II. 2023. Restoration of endangered wildlife and animal species around the Demilitarized Zone (DMZ). <http://nationalatlas.ngii.go.kr/pages/page_2335.php>. Accessed 29 January 2024.
48. Perzanowski, K., and W. Olech. 2013. Restoration of wisent population within the Carpathian eco-region, Europe. Pages 190–193 *in* P. S. Soorae, editor. *Global re-introduction perspectives: 2013. Further case studies from around the globe*. International Union for the Conservation of Nature, Gland, Switzerland.
49. Zanni, M., R. Brogi, E. Merli, et al. 2023. The wolf and the city: Insights on wolves' conservation in the Anthropocene. *Animal Conservation*. DOI: 10.1111/acv.12858.
50. Navarro, L. M., and H. M. Pereira. 2015. Rewilding abandoned landscapes in Europe. Pages 3–23 *in* L. M. Navarro and H. M. Pereira (editors). *Rewilding European landscapes*. Springer-Nature, Cham, Switzerland.
51. Apollonio, M., R. Andersen, and R. Putman. 2010. *European ungulates and their management in the 21st century*. Cambridge University Press, Cambridge, UK.
52. Large Carnivore Initiative for Europe. 2023. Assessment of the conservation status of the wolf (*Canis lupus*) in Europe. Council of Europe, Strasbourg, France. <https://lciepub.nina.no/pdf/638036032684557257_LCIE%20CoE%20Wolf%20status%20report%202022.pdf>. Accessed 29 January 2024.
53. Terborgh, J. 2015. Protecting the wild: Parks and wilderness the foundation of conservation. Foreword. Pages xi–xvii *in* G. Wuerthner, E. Crist, and T. Butler, editors. *Protecting the wild: Parks and wilderness the foundation of conservation*. Island Press, Washington, DC, USA, xvii.

54. Cooke, R., F. Hogan, B. Isaac, et al. 2018. Powerful owls: Possum assassins move into town. Pages 152–165 *in* C. W. Boal and C. Dykstra, editors. *Urban raptors – ecology and conservation of birds of prey in cities*. Island Press, Washington, DC, USA. <https://doi.org/10.5822/978-1-61091-841-1_11>.
55. International Union for the Conservation of Nature. 2024. IUCN Red List Powerful Owl. <https://www.iucnredlist.org/species/22689389/93229550>. Accessed 30 January 2024.
56. Lamont, A. 2006. Policy characterization of ecosystem management. *Environmental Modeling and Assessment* 113:5–18.
57. Westgate, M. J., G. E. Likens, and D. B. Lindenmayer. 2013. Adaptive management of biological systems: A review. *Biological Conservation* 158:128–139.
58. Hebblewhite, M. 2017. Billion dollar boreal woodland caribou and the biodiversity impacts of the global oil and gas industry. *Biological Conservation* 206:102–111.
59. Hebblewhite, M., C. White, and M. Musiani. 2010. Revisiting extinction in national parks: Mountain caribou in Banff. *Conservation Biology* 24:341–344.
60. Festa-Bianchet M., J. C. Ray, S. Boutin, et al. 2011. Conservation of caribou (*Rangifer tarandus*) in Canada: An uncertain future. *Canadian Journal of Zoology* 89:419–434.
61. Serrouyaa, R., D. R. Seipc, D. Hervieuxd, et al. 2019. Saving endangered species using adaptive management. *Proceedings of the National Academy of Sciences* 116(13):1381–1386. <www.pnas.org/cgi/doi/10.1073/pnas.1816923116>.
62. Grace, M. K., H. R. Akçakaya, E. L. Bennett, et al. 2021. Testing a global standard for quantifying species recovery and assessing conservation impact. *Conservation Biology* 35:1833–1849. <https://doi.org/10.1111/cobi.13756>.
63. Grace, M. K., H. R. Akçakaya, J. W. Bull, et al. 2021. Building robust, practicable counterfactuals and scenarios to evaluate the impact of species conservation interventions using inferential approaches. *Biological Conservation* 261(2021):109259. <https://doi.org/10.1016/j.biocon.2021.109259>.
64. Doyle, M. 2019. After 50-year conservation effort, songbird flies off U.S. endangered species list. E&E News. 8 October.
65. Rapai, W. 2013. *The Kirtland's warbler: The story of a bird's fight against extinction and the people who saved it*. University of Michigan Press, Ann Arbor, USA.
66. Olah, A., C. A. Ribic, K. Grveles, et al. 2023. Low Kirtland's Warbler fledgling survival in Wisconsin plantations relative to Michigan plantations. *Ornithological Applications* 125: 1–14. <https://doi.org/10.1093/ornithapp/duad007>.
67. Scott, J. M., D. D. Goble, A. M. Haines, et al. 2010. Conservation-reliant species and the future of conservation. *Conservation Letters* 3:91–97.
68. Bocetti, C. I., D. D. Goble, and J. M. Scott. 2012. Using conservation management agreements to secure postrecovery perpetuation of conservation-reliant species: The Kirtland's warbler as a case study. *BioScience* 62:874–879.

11 Values to Guide the Anthropocene

Conservation Values in the Anthropocene

The Recognition and Reality of Values

Conservation scientist and ethicist Kyle Van Houtan has written,

> There is no robust hope in the dominant versions of ethics as practiced in the conservation sciences. To succeed as a social cause, conservation needs a hope that academic science itself cannot provide. Conservation needs a cultural legitimacy that inspires enthusiasm, allegiance, and personal sacrifice—in other words, actual changes in human behavior.[1]

The fact that conservation needs cultural legitimacy does not mean it is going to get it. Being hungry does not mean that we shall have bread. But, as the philosopher C. S. Lewis noted, "surely a man's hunger does prove that he comes of a race which repairs its body by eating and inhabits a world where eatable substances exist."[2] Conservation is like that – in search of value, but chronically hungry in its pursuit. Economist Herman Daly called conservation "a policy in the service of a purpose," and, as Daly perceived, "to have purpose means to serve an end, and value is imputed to whatever furthers attainment of that end."[3] Current divisions in conservation are divisions of purpose. In this quest for value and purpose, conservation is a reflection of the human experience. One of history's most thoughtful interactors with nature, Henry David Thoreau, discerned that the pursuit of value is a constant intellectual battle and moral choice. "Our whole life," said Thoreau, "is startlingly moral. There is never an instant's truce between virtue and vice. Goodness is the only investment that never fails."[4] But this conflict is no mere philosophical abstraction. It is a struggle grounded in our own biology and evolutionary experience. As psychologist Jerome Kagan observed,

> What is biologically special about our species is a constant attention to what is good and beautiful and a dislike of all that is bad and ugly. These biologically prepared biases render the human experience incommensurable with that of any other species.[5]

Values provide the fundamental grounding for determining appropriate human intention toward nature, an essential foundation in conservation that is increasingly recognized. But the sources of such values have multiple origins beyond conservation science, and so the legitimacy of these sources must also be recognized. As Wendell Berry noted, "It is not

DOI: 10.4324/9781003404415-11

from ourselves that we will learn to be better than we are."[6] The WWF Beliefs and Values Programme declares that its goal is to "encourage faith communities and conservation partners to work more closely together to protect nature and restore biodiversity."[7] The Society for Conservation Biology's (SCB) Religion and Conservation Biology Working Group's purpose is to "strengthen the collaboration between faith traditions and conservation, and promote awareness of the importance of this collaboration within SCB and the conservation community, based on empirical evidence."[8] The IUCN and the World Commission of Protected Areas (WCPA) are now joint sponsors of a dedicated IUCN Commission Group, The Cultural and Spiritual Values of Protected Areas Specialist Group, whose mission is to "promote the appropriate recognition and inclusion of all relevant cultural and spiritual values, knowledge, and meanings into nature conservation theory and practice; in particular, into protected areas policies, planning, management, and evaluation."[9] When the IUCN and the WCPA declare that they are going to incorporate spiritual values, knowledge, and meanings "into protected areas policies, planning, management, and evaluation," that is the equivalent of saying that such values reflect "real" knowledge about how protected areas (PAs) should be conceived, administered, and assessed. Similarly, Indigenous Peoples stress that traditional knowledge be not merely *recognized* but actively *incorporated* into "real" management of PAs, as is increasingly the case, for example, in managing fire regimes in national parks in Australia.[10, 11] Not surprisingly, one of the most important considerations today in developing enforced agreements for co-management of lands between First Nations People and Parks Canada has been the guarantee of the incorporation of traditional (value-based) knowledge into the policy, planning, and management of PAs.[12]

Nature's Value as Ecosystem Services

An important contribution of the ground-breaking Millennium Ecosystem Assessment (MA) was its premise of the natural value as Ecosystem Services (ES), defined by the MA as *direct and indirect benefits people obtain from ecosystems*.[13] The MA's stated goal was "to meet the needs of decision makers and the public for scientific information concerning the consequences of ecosystem change for human well-being and options for responding to those changes," because "human actions are depleting the Earth's natural capital and putting such strain on the environment that the ability of the planet's ecosystems to sustain future generations can no longer be taken for granted."[13] With many members of the private corporate business community among the over 1,300 experts engaged in writing the MA, the authors sought to integrate the private business sector into the discussion of ecosystem conservation and frame an understanding of the value of nature in terms of services essential to human well-being. The MA included a section, "Opportunities and Challenges for Business and Industry," that spoke directly to an economic understanding of natural resources as "natural capital." "By examining the environment through the framework of ecosystem services," said the authors, "it becomes much easier to identify how changes in ecosystems influence human well-being and to provide information in a form that decisionmakers can weigh alongside other social and economic information."[13] By giving visibility to ES, the MA enabled private industry to strengthen and legitimize its inclination toward various forms of market-based environmental governance.[14, 15]

The MA contributed to an understanding of how deeply human welfare as well as economic growth are wedded to the condition of ecosystems and the services they provide. But the very success of the MA spawned a trend among conservation scientists to express the *goals* of conservation in terms of ES. ES has been repeatedly used by "new conservationists" as the foundation for an advocacy for conservation framed in the currency of nature's instrumental value for humans, not the value of nature in and of itself (nature's "intrinsic value," (IV)) independent of its instrumental value.[16, 17]

Conservation and IV

What IV Is Not

Conservation's contention over whether its motivation should be based on instrumental values or intrinsic values is an old debate. Instrumental value has always offered the appeal of direct benefit (human welfare), commodity markets (things that can be bought and sold), and quantification (unit price times unit quantity equals gross dollar value of the "resource"). In the past, many conservationists unreflectively embraced instrumental quantification as a tool for saving biodiversity, particularly through the mechanisms of "Willingness to Pay" (WTP). WTP is one variant of Contingent Valuation, the creation of "shadow markets" for non-market goods (such as endangered species) by attempting to assess public "preferences" for their survival. But since there is no "market" for the conservation of an endangered species, the "value" of the species is estimated by determining, usually through surveys, how much the respondent would be willing to pay to preserve it. The average WTP times the number of contributing respondents is then used to estimate the endangered species' monetary value expressed as a consumer preference. Such preferences are sometimes referred to as "existence value" (EV) – the value of knowing that something exists, whether you ever see, enjoy, or use it yourself.

WTP suffers from methodological obstacles and logical fallacies, especially as a measure of existence value. Methodologically WTP is unreal. No money changes hands. The respondent can name any level of support for the species without actual financial loss. This problem might be overcome, at least to some extent, by framing the question in terms of a tax or bond, for example, "How much would you be willing to be taxed to preserve the tallgrass prairie at the edge of town?" And if the tax or bond is appearing on the next ballot, the respondent's estimate of worth might have some reality. But the problem with WTP is that it conflates a consumer preference to express EV with a payment supposedly reflecting the actual value of the thing itself. Suppose that a nurse is working for a humanitarian health organization during a civil war in an African nation. Rebel soldiers kidnap one of his patients, a 5-year-old girl, to serve as a hostage in their negotiations with the government. The nurse pays the soldiers US$10,000 of his own money to independently secure the child's release. The nurse will have psychological satisfaction in saving the child's life, but that satisfaction is not a "consumer preference," nor does it in any way reflect the actual value of the child in and of herself. Reducing the value of the child to the psychological satisfaction the nurse experiences for providing the child's rescue would be a moral obscenity and a specious conclusion. And any claim that the US$10,000 paid reflects the actual value of the child is an equally egregious ethical profanity.[18] But non-human nature carries the same objection. A veterinarian who "buys back" a captive infant orangutan from an animal trader in Indonesia for US$400 (the current price on the island of Java to acquire an orangutan in the illegal pet trade)[19] may experience psychological satisfaction in getting the infant to an orangutan center for medical recovery and behavioral rehabilitation (Chapter 10). But is it any less of a moral outrage to declare that the value of the infant orangutan is no more than the veterinarian's "preference" and "psychological satisfaction" in saving it, and no more than the price paid to "buy" it?

We see in both examples that not only is WTP flawed in methodology, but in its confusion about *what* is being measured. WTP might estimate what the respondent is willing to *spend* for a prairie or an orangutan, but that gives us no clue as to why the respondent *would spend it* – and that is the crux of the argument. The motivation, informed by reasons for saving an endangered species or ecosystem, not only fails to be measured, it fails to be recognized. What if the decision to save the prairie, or the orangutan, comes down to a local town hall meeting – one in which people are actually going to have to depend on *arguments* to persuade others? What persuasive power will WTP have then? As economist Marc Sagoff perceived, "One role of applied philosophy is to give defunct theories a proper burial."[20] WTP deserves to be interred.

The fact that WTP cannot provide a good measure of the psychological satisfaction that comes from saving species does not mean that psychological satisfaction for such effort does not exist. It does. New conservationist Michelle Marvier affirms that "being able to enact and live out ... moral convictions" is a psychological benefit that some people receive from doing conservation.[21] If that is the gain that conservation brings, it is correct to declare it an "ecosystem service," a benefit provided by the ecosystem that enhances human well-being. But we cannot construct a persuasive argument for conservation on the premise that engaging in it will benefit me, or even benefit "us," no matter how much money we might be willing to spend for that benefit. Is there anything more to it than that?

The Reality of IV

The ethical examination of nature and our relationship to and interaction with it is even more important than our scientific examination of it. If conservation biologists enjoy studying species, the pleasure they find in knowing them is not what *makes* these species valuable. Rather, such pleasure, because it is appropriate to a growing understanding of the species, is one expression of how they *perceive* its value. But what exactly do we mean by *value*? In general terms, value can be understood as *the basis for an estimation of worth*. And that brings us to the crux of our question. Is this estimation of worth, this "value," a subjective product of thought or an objective property of nature? Some ethicists say it is the first – because value cannot exist outside the mind of the human valuer. This is by definition *anthropogenic* IV – the value perceived in the mind of the valuer who "beholds" the natural entity (for example, a honeybee). In this view of IV, the entity possesses no value in itself, although it may have value in its relation to other things (like, for the honeybee, the flowers it pollinates).[22] It is possible for the entity to have value ("final" value) because of its extrinsic properties (pollinating flowers). It can be an end in itself because of its relations to some other thing(s) through the value it provides to or for them.

The alternative – *autonomous* IV – identifies the human consciousness as the *receptor and perceiver* of the object's value, but not the *locus or origin* of its value. All the attributes of "value" are already present (they are "there") in the object before the human perceives it. As environmental ethicist Holmes Rolston III explains,

> The object [the natural entity, for example, a wildflower] causally affects the subject who is excited by the incoming data [the sight of the flower] and translates this as value, after which the object ... appears as having value ... But nothing is really added

intrinsically to the object at all; everything in the object remains what it was before. Despite the language that humans are the source of value which they locate in the natural object, no value is really located there at all. The only new event is that these properties are registered in – translated into felt values by – the perceptual apparatus of the beholder [see Figure 11.1].[23]

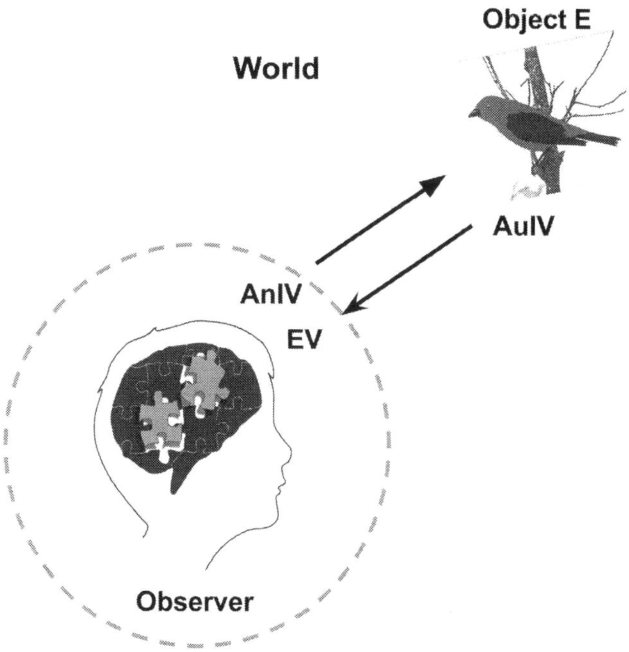

Figure 11.1 The relationship between the perception of autonomous intrinsic value (AuIV) by a human observer and the locus of AuIV in a natural object in which the human consciousness is the *receptor and perceiver* of the object's value, but not the *locus or origin* of its value (Anthropogenic Intrinsic Value, AnIV). Object E possesses an IV perceived by a human observer ("self") and internalized as a personal experience of apprehending the value of E as an "experienced value" (EV). Original drawing by S. H. Ali.

"It is a condition of human flourishing," notes Rolston, "that humans enjoy natural things ... We do not love wildflowers for the pleasure they bring us. They themselves are our pleasure."[24] Rolston's insight reveals the most fundamental vice of anthropogenic IV – its lack of humility. The only value is the human perception, not the creature itself. But what value did that perception bring to the creature, which is itself unchanged by that perception, and possessing in itself everything the human now perceives as valuable?

The Rejection of IV in New Conservation

Substitutes for IV

The values we place on biodiversity conservation affect the actions we choose to advance that conservation. New conservationists stress that the biodiversity crisis should be averted primarily for the benefit of human well-being, that ecosystem protection should be science-based, and that conservation should be accomplished through the collaboration of conservation efforts with capitalist enterprise.[25] The value of nature is what nature contributes to human well-being. Traditional conservation emphasizes that species and ecosystems possess IV, and averting the biodiversity crisis is its own ultimate purpose for the sake of saving all biodiversity – whether or not that satisfies human well-being.[18] New conservation sees the conservation of nature as the preservation of "natural capital," with emphasis on ES-based analyses intended to integrate conservation efforts into economic development, manifested through, among other means, an increasing emphasis on payment for ES (PES) schemes,[26, 27, 28] biodiversity offsetting programs,[29] and corporate partnerships with conservation NGOs to finance conservation efforts.[30] The assumption is that if people do not believe that conservation is in their own best interests, then they will never make conservation a societal priority.

Advocates of this approach see effective future conservation as an expression of enlightened human social and economic development, placing human rights on equal or greater footing with nature's rights. Peter Karieva et al. declare,

> Forward-looking conservation protects natural habitats where people live and extract resources and works with corporations to find mixes of economic and conservation activities that blend development with a concern for nature … we believe that nature can prosper so long as people see conservation as something that sustains and enriches their own lives.[31]

Kareiva, Michelle Marvier, and other new conservationists advocate a vision of conservation that cooperates with capitalism and development.

> Conservation should seek to support and inform the right of development – development by design, done with the importance of nature to thriving economies foremost in mind … Instead of scolding capitalism, conservationists should partner with corporations in a science-based effort to integrate the value of nature's benefits to their operations and cultures.[31]

The concept of IV should be discarded. IV does not motivate conservationists in their own efforts, and IV does not "work" in generating public support for conservation.[17] IV is a subjective state of mind. ES, in contrast, are quantifiable and objective, and therefore socially and politically persuasive as things to be conserved. This view has deep roots, all the way back to the eighteenth-century English philosopher David Hume, that value is only a subjective inference describing one's state of mind toward an object. "When you pronounce any action or character to be vicious," said Hume, "you mean nothing but that from the constitution of your nature you have a feeling or sentiment of blame from the contemplation of it."[32] Statements of value are statements about your feelings.

The Fallacies of Rejection

The Fallacy of Conservation as Moral Proposition

The reasons given by new conservationists for rejecting IV reflect a fallacy of four errors – a failure to understand the nature of the argument for conservation itself, a failure to differentiate reasoning from marketing, a failure to assess the actual results of their argument when it is employed, and a failure to understand the role of IV in their own argument. The first error occurs from the failure to see that the idea of biodiversity conservation is a moral proposition, not a cost–benefit analysis. Michael Soulé, one of the founders of the Society for Conservation Biology, explains that "wild things and places have incalculable intrinsic value, at least as salient as the value of humanity."[33] If one does not agree with this claim, the proposition cannot be dismissed by the accusation that *it lacks popular appeal*. The moral premise that biodiversity conservation has value and should be preserved is precisely the proposition upon which almost all conservation policies are based. A moral premise should be rejected if it is morally inappropriate, as demonstrated by reason. A moral proposition should not be rejected because it is politically unpopular or inexpedient, or alleged to be so.[20] To say that something has IV is to say that it has value beyond anything it does to help us achieve our aims, but that is exactly the kind of value consideration that would be excluded if all conservation endeavors were valued solely because of their conservation of ES.

The Fallacy of Utilitarian Conservation

The second error arises from the first. The claim that IV does not "work" in generating popular support for conservation confuses marketing with morality. The real error, however, is even worse – the error of failing to discern the difference between marketing and truth, and to recognize (or even, apparently, remember) the history of conservation as a social and political movement. In the United States, every major environmental law begins its text with an inspirational and forward-thinking message.[34] These traits are clearly visible in landmark legislation like the US Endangered Species Act (ESA), the National Environmental Policy Act (NEPA), and the Clean Water Act (CWA). The same pattern is apparent worldwide. The constitution of Ecuador makes the rights of nature itself constitutionally explicit, declaring the inalienable rights of *ecosystems* to exist and flourish. It further gives people the authority to petition on behalf of nature and requires the government to remedy violations of nature's rights. The constitution of Bolivia is even broader, explicitly affirming that nature has a "right to exist."[35]

The problem is not that the inspirational message of IV *does not work*. The problem is that conservationists use the argument with such clumsiness and temerity that *they often fail to make it work*. Imagine, for example, that a pharmaceutical company develops a cure for the common cold. They hire an advertising agency to promote their product. After months of intense publicity, sales are abysmal. At this point the company would not discard the product; it would fire the ad agency! Conservationists have long displayed an uncomfortable embarrassment in the advocacy of IV in nature. The remedy is – get over it! The inspirational message of nature's IV has, throughout the history of modern conservation, motivated social, cultural, and political response to nature preservation. We will never improve conservation law or inspire new statutes or better policies through the "inspiration" of shared ES. Conservation law arises on the basis of shared ideals, not shared utilities.

The Fallacy of Conservation as Economic Development

The third error is that while new conservationists argue that moral arguments for conservation, such as those based in IV, do not work, many studies of conservation effectiveness indicate that the opposite is true – economic development does not work in conservation. Brazil, for example, permitted the transition and conversion of many of its traditional national parks to areas that permitted sustainable development through direct extraction of resources. Such "extractive" conservation reserves were also allowed to engage agribusiness, the lumber industry, hydro energy generation, and other forms of use, with the argument and intention that such development would be good for conservation. In fact, there was an overall loss in the acreage of total PAs, not only because of the changes in policy but also because many traditional national parks were reduced in size or simply delisted from protective status.[36] In the Brazilian state of Rondonia, the Rondonia Assembly eliminated 12 state PAs covering over 1 million hectares of land. The state of Mato Grosso also reduced the size of several of its most important parks because, in the words of the Mato Grosso Assembly, "sustainable development is compatible with conservation."[36] Based on his studies of changes associated with the increased institution of extractive reserves, researcher Marc Dourojeanni concluded,

> Success of direct use reserves depends upon unachievable conditions – stabilization of population growth and pressure on natural resources to sustainable levels; and administration of the areas must have the capacity and means to enforce the law and mutually agreed upon standards developed with resident community.[37]

These conditions were invariably lacking, and the amount of land associated with truly protected conservation areas declined.

The Failure to See the Role of IV in "Economic" Arguments

The fourth error is best understood analytically. The argument for conservation expressed as economic development is based on two premises:

> Premise 1. Human well-being is good as an end in itself.
> Premise 2. We should protect things that are good as ends in themselves.
> Conclusion. We should protect human well-being by combining conservation with development.

This argument runs parallel to that of biodiversity conservation based on its IV:

> Premise 1. Biodiversity is good as an end in itself.
> Premise 2. We should protect things that are good as ends in themselves.
> Conclusion. We should protect biodiversity.[18]

The second argument depends on the proposition that *biodiversity* as well as *nonhuman creatures* possess IV. The first argument depends on the proposition that *human beings* possess IV (but not necessarily other species, or to a greater degree than other species). The first argument does not *remove* IV from consideration, but *restricts* its application to humans alone. It also obscures the need for deeper examination and renovation of conservation's tactics and messaging. Environmental scientist Daniel Doak and his colleagues name the problem: "The stance that conservation progress should be driven by transient economic preferences rather than enduring values also hampers recognition of the possibility of even the need for structural and institutional changes in order to achieve and sustain conservation objectives."[38] Conservation is a moral premise advanced by persuasive argument. It cannot be advanced by cost–benefit ratios. It is just as John Muir stated: "Nothing dollarable is ever safe, however guarded."[39]

A Wolf in the Sheep's Clothing

Throughout its history, conservation has been the companion of Western colonialism at the expense of Indigenous Peoples – either through their forced relocation to create national parks and PAs or through their cultural assimilation in education intended to purge them of their cultural values and beliefs (Chapter 8). To assert that conservation must be framed in terms of economic development because people will not otherwise engage in it is not a step forward, but backward. It is simply a new version of colonial conservation, one that displays the same Western bias and blindness of the past, but now cloaked in the sheep's clothing of "development." It assumes that capitalist economic development – which has been the leading edge of nature's destruction – can now be the engine of its restoration. But, logically, the capitalism that generated the ecologic crisis cannot also be its solution. Douglas Tompkins, founder of the outdoor clothing lines *The North Face* and *Esprit*, explains that

> ...my interest in these kinds of "green business" measures soon faded. All profit-oriented corporations, as much as their owners may try to make them responsible, are stuck in their own ditch of contradictions. Ultimately it was too paradoxical to reconcile running a successful company with my motivation to help nature stave off the very impacts of commerce. I could not see anything better to do than to direct my energy toward full-time conservation, and I sold my interest in the business.[40]

The assumption that conservation should, indeed must, continue as an expression of Western economic development reflects a dismissive ignorance of those systems that do not share its premises, particularly the economic systems of Indigenous Peoples. Despite their diverse histories, cultures, and locations, Indigenous Peoples share a common understanding of the purpose of transactions. As Nancy Turner and Darcy Matthews concluded from their study of Indigenous cultures, "Indigenous peoples worldwide, despite their differences, are rooted in cultures of gratitude arising from the tenet of respect."[41] In Western capitalism, the purpose of exchange is to *maximize individual wealth*. In Indigenous cultures, the purpose of exchange is to *optimize mutual benefit*. Hence, Indigenous Peoples place value on giving and receiving gifts, and correspondingly on the virtues of thankfulness and generosity. The purpose of giving such gifts is not to create obligation, but to foster relationships of reciprocity that join the giver to the receiver and, over time, make these roles interchangeable. These principles are followed in exchanges among people, but they are also applied in "exchanges" with nature. Indigenous Peoples consistently show a pattern of "resource use" (although they would not call it that) characterized by nurturing actions toward nature preceding, accompanying, or following their taking of things ("resources") from nature. Invariably the resource is taken in such a manner that the majority of its substance and its regenerative capacities are left in place – not to achieve *sustainable use* (which Western thinking understands as *voluntarily restricted income*) but to achieve *continued use* as well as *continued presence* of the resource as something that sustains an *abundant life* tied to the "nature" around them (like the Ottawa Indians and the lake sturgeon, Chapter 10).

The replacement of such Indigenous economic relationships with Western versions of economic development has often been negative. The Yánesha are an Indigenous People of Peru's upper central Amazon region. International forestry corporations embedded themselves in relationships with the Yánesha through "economic development" that provided employment and services for the Yánesha, but also created an interdependence of debt-based relationships between the corporation and the people. By the end of the

1960s, 90% of Yánesha men owed money to their employers and spent most of their time working off their debts.[42] Traditional Yánesha society did not provide for accumulation of wealth and the institution of private property – a practice imported by private interests who were permitted by the Peruvian government to operate in the Yánesha's traditional homeland areas. The problems that arose were generated by the failure to allow Indigenous Yánesha true independence and sovereignty to purge capitalist practices that eroded their culture, problems facilitated by a corrupt government determined to maintain the benefits of that corruption through ties to resource-based corporations.[42]

The Yánesha are a single example of a worldwide pattern. In a global study of 169 cases of the role of economic development and its interactions with Indigenous People in conservation, Neil Dawson and his colleagues found that market-oriented conservation produces social exclusion and perverse incentives.

> Thirty-one cases of market-oriented conservation interventions, were included in our sample (18.3%), comprising payment for ecosystem service projects, reduced emissions from deforestation and forest degradation (REDD+) programs, alternative livelihood programs, and certification schemes ... closer analysis of the extent of local influence and the roles of local institutions led to all 31 cases being categorized as "externally controlled." Private companies rarely controlled these interventions single-handedly, as NGOs and state agencies also held influence in almost all cases.[43]

But resident Indigenous Peoples did not hold that influence. In contrast, Dawson et al. perceived that "holistic approaches regard a person's well-being as an ongoing process shaped by interactions with others and the surrounding physical environment rather than just a state to be attained by an individual."[43] The "conservation" imposed by Western forms of economic development does not foster those commitments. In the words of Dawson et al., such "imposed conservation" can be "ecologically effective, but at great social and financial cost."[43] It is also increasingly apparent that economic development associated with these kinds of efforts often provides its primary benefits to Western conservation NGOs, not to Indigenous Peoples. A 2008 study by the Rights and Resources organization found that Indigenous-led conservation projects protected the land at a cost of $3.50 per hectare compared with $3,500 per hectare for projects undertaken by US and European-based conservation organizations.[44] Human occupancy works in achieving conservation goals when the area's occupants practice Indigenous livelihoods. It fails when human occupancy provides license for extractive technologies of resource use.

Transforming the Metaphor – Stewardship to Reconciliation

Service not Supervision

In his ground-breaking book *Man and Nature*, George Perkins Marsh was among the first to assert that human beings bore a responsibility to care for the Earth (Chapter 2). "Man has too long forgotten," wrote Marsh, "that the earth was given to him for usufruct alone, not for consumption, still less for profligate waste."[45] This unfamiliar word, "usufruct," is derived from the Latin phrase *usus et fructus*, "for use and fruit" (enjoyment). It is defined in English as "the right to use or enjoy a thing possessed, directly and without altering it" or as "the designated right and responsible use of another's property short of the destruction or waste of its substance." Often used in legal documents, the term unites the separate property interests of two different parties – ownership by one, use and enjoyment by the other. This is the essence of the modern metaphor and understanding of "stewardship."

Stewardship, a concept applied to environmental care by both faith communities and secular voices, has been understood to mean "the conducting, supervising, or managing of something" with an implied concern for its continuance. A concept may be at once popular yet inaccurate. Within the Christian tradition stewardship has often carried the concept of environmental protection, but stewardship is not a biblical metaphor for interaction with nature. The original instruction to the first man, Adam, was to "serve and protect" the Earth (Hebrew – *abad* and *shamar*, often but less correctly translated "rule and subdue"), not "supervise" it. The concept was also susceptible to cooptation by those seeking to make the goal of stewardship not the *care of nature* but the *exploitation of nature* to advance human well-being. *An Evangelical Declaration on the Care of Creation*,[46] published in 1994, declared, from the perspective of evangelical Protestants, that the purpose of stewardship was the care of creation. That purpose was later altered while the metaphor was superficially preserved in the subsequently published *Cornwall Declaration on Environmental Stewardship*, which asserted that the purpose of stewardship was that of promoting "growing affluence, technological innovation, and the application of human and material capital ... to environmental improvement."[47]

Stewardship suffers from the same vice at anthropogenic IV, but in slightly different clothing – creatures are only valuable if humans manage them. More recent efforts in the Christian community and others active in conservation are now framing conservation differently – as "reconciliation ecology," a process requiring recognition of human responsibility for nature's degradation, acts of repentance to remediate human abuse, and renewal of the human–nature relationship expressed in mutually beneficent interactions.[48] The change of metaphor reflects not only a determination to change the current conditions of the Earth, but a change in the human response to the Earth that is required to move toward those conditions. A fundamental transformation in human interaction with nature will demand not only engagement of the intellect, but of the emotions, the will, the body, and the sociocultural relationships of human beings – changes in the whole person.

Right in My Own Backyard

In 2011, I became the Executive Director of a faith-based (Christian) environmental NGO, the Au Sable Institute of Environmental Studies, in northern Michigan. Au Sable's efforts were focused in environmental education through a K-12 program of hands-on experiences of nature study and care throughout northern Michigan, a curriculum of college-level courses in environmental and conservation science serving over 60 colleges and universities in the United States and Canada, outreach and practical application (through a retreat program at the Institute for adult and family groups who wanted to learn more about how to express their Christian faith in and through nature conservation), and a research program to conduct studies to provide original knowledge about the function and management of natural systems and native species that could be applied to conservation. The education and outreach efforts were strong, but the research program was not. It had been oriented to studies of interest to in-residence students and faculty (but of little interest to anyone else), limited to studies on or near the main campus ("sampling by convenience"), rarely produced peer-reviewed publications, and was not supported by external grants (making it an institutional expense, and a drain on a limited institutional budget).

I was determined to change these things and design research that would advance the Institute's work. I realized that, in northern Michigan, we were surrounded by a multitude of organizations, agencies, and communities active in conservation – private NGOs, government agencies, and Indigenous Peoples – yet we were not asking questions of interest to anyone but ourselves. I launched a program of self-education – my own – to find out what these groups were doing, what problems they were trying to solve, and what was keeping them from solving them. I began to schedule interviews with leaders and representatives of all of these groups, and, in the interviews, always asked the same questions – how would you express your mission, and what kind of research could Au Sable do that would genuinely advance it?

I got an earful, but it was a good earful, and extremely educational. I began to understand the scope and particulars of conservation concerns around us, as well as their national and international importance. I was at my desk one morning pondering how to respond to what I was learning when we received a visitor. It was Jerry, the local forester of the Michigan Department of Natural Resources (MDNR). Jerry had been assessing some forests nearby as potential timber sale sites, but it was a cold November morning, and Jerry remembered that Au Sable always kept a pot of hot coffee, and a fascinating assortment of coffee cups, in the main office. He came in to say hello, and warm up.

Jerry was sitting at one of the office tables, drinking his coffee and beginning to appreciate the experience of warmth, when I suddenly realized my opportunity for another interview. I sat down next to Jerry and asked, "Jerry, what kind of research could Au Sable do that would really advance your efforts in conserving these forests?" Without missing a sip of his coffee Jerry answered, "Find out why these abandoned oil well pads never reforest." Oil and natural gas deposits are plentiful in this landscape. They are created by installing well pads in the forest, each taking up about 2 acres of land, and require clearing the site of its trees. At this time there were over 60,000 well pads throughout the state of Michigan, mostly in the northern portion, and disproportionately clustered around the Institute. After the oil and natural gas are extracted from such sites, the pad is left inactive and functionally abandoned.

I took Jerry's response seriously and began to study the pads. I learned that such extractions were contributing to human-initiated destruction of the local northern temperate deciduous forest, a forest I lived and worked in. After the first dozen site inspections with Jerry, I began to get the picture. On these pads, the sites had entered a state that ecologists call "arrested succession" (Chapter 3) (see Figure 11.2), a problem caused by direct or indirect human disturbance, and one that was becoming more and more common worldwide.[49] Here in northern Michigan, removal of the top soil and its moisture, compaction of the surface by heavy machinery, and destruction of the vegetation which held the soil in place created sites that had now remained denuded of trees for 20, 30, and even 40 years after drilling and extraction had been completed. Eventually, I secured a grant for a five-year study of the problem.

Within a 50 m x 50 m area on multiple pads present in our area, we applied four soil treatments (disking, fertilization, disking plus fertilization, and no treatment (control)) in which we planted four species of native forest trees (red oak, white oak, jack pine, and red pine). We also created a counterfactual plot (Chapter 10) by designating an equal-sized area on the same pad in which we applied no treatments or plantings. This counterfactual plot would tell us what would happen if we did nothing at all to these

Figure 11.2 A vacated oil well pad in northern Michigan, USA, 2015. Extractive activities on this site ceased over 30 years earlier, yet natural reforestation has not taken place. Photo by F. Van Dyke.

pads (would they finally reforest on their own?), and then we monitored both the experimental and the counterfactual sites for the next five years.

Research as Reconciliation – Designing Moral Development into Conservation Effort

I designed our study to be an experimental approach rooted in adaptive management, but I also conceived it, as a teacher of students and the leader of a conservation organization, to engage the whole person in Au Sable's whole mission – *Au Sable inspires and educates people to serve, protect, and restore God's Earth*. Our journey in this understanding began with *recognition* (see Figure 11.3). *Recognition* is an engagement of the intellect – in this case, it is a realization that harm had been done to the forest, and that harm, to this extent, at this magnitude, and for this duration, was morally wrong. That recognition of wrong brought an awareness of *moral responsibility*. Human beings did this. We chose to do it. We made no reparation for it. In our extraction of oil and natural gas from each well pad, we had followed the classic pattern of interacting with nature as tap and sink – a tap (resource) from which to remove a market commodity for financial gain, and then a sink of an abandoned site stripped of its trees which we incurred no expense to remediate.

When we fully recognize what we have done to nature, the appropriate response is guilt. Guilt is not something that we should simply "get over." Guilt is an invaluable resource to us. When our awareness of our responsibility to nature comes with the recognition of our failure in that responsibility, our guilt should be followed by *remorse*. Many conservationists think that remorse is a waste of time. They are mistaken. Remorse is the link between the recognition of wrong by both our *intellect* and our *emotions* through which our *values* will begin to affect our *behavior*. Conservationists have consistently ignored, even ridiculed, the role of emotion in human engagement, especially the value of our emotional response to the formation of our own character. But it is through our faculties of feeling that we perceive and appreciate moral and emotional qualities appropriate

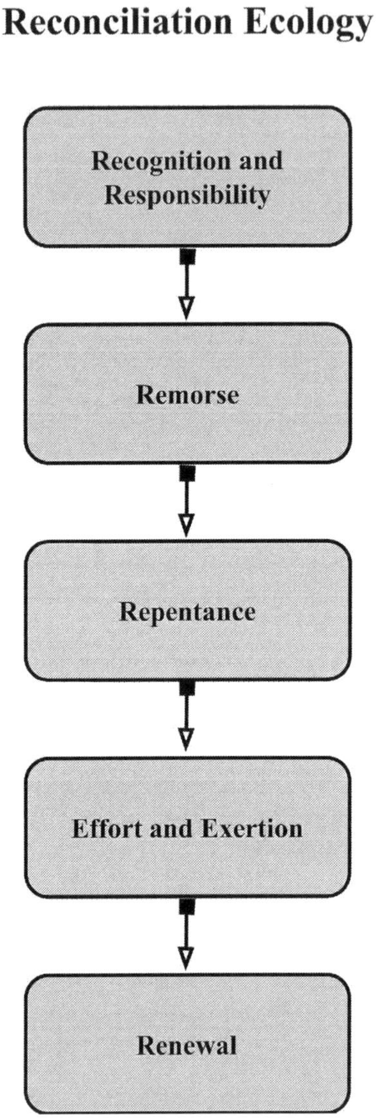

Figure 11.3 A progression of interaction with nature as conceived in reconciliation ecology – a framework for changing human–nature interactions that engages human values at the level of intellect, emotion, will, body, and relationship with nature and with others. *Recognition* of the harm done to nature by human action engages the *intellectual dimension* of our human perception, leading to an ownership of moral responsibility for the harm inflicted and the need to remedy it. *Remorse* engages human *emotional faculties* in response to intellectual perception, leading to original initiatives of *repentance* – new and different behaviors engaging *human will* (creativity) to envision and then produce action that will foster a remedy and remediation of harm inflicted and requiring our *physical effort and exertion* to accomplish the needed work. *Renewal of relationships between people and between people and nature* results from repentant actions that make the relationship different and better than it was previously. Design by F. Van Dyke and S. H. Ali.

to people or events – we *feel* differently toward an act of cruelty than we do toward an act of kindness. By training our capacity for emotional perception and judgment we increase our ability to act out *moral knowledge* and so cultivate a character embedded in *virtue*, especially in our capacity to act with compassion and wisdom. And it is our perceptual faculties for feeling – our virtues, properly trained and guided – that provide insight into the normative properties of the world and, through those perceptions, inform and guide our ethical response.[50]

The emotions of remorse and regret inform the applications of justice. To prosecutors and judges in our court systems, it matters whether wrongdoers are remorseful, for such a person is, however despicable in their actions, someone who could change because in their remorse they may be driven to repent. To begin the process of justice toward the natural world, the proper emotional response to our treatment of nature should be such remorse with accompanying regret. There can never be hope for change in human interaction with nature without the anguish of both. "There seems," wrote George Wuerthner, "to be no expression of loss or grief that we are now pushing many species towards extinction. Humans have survived the Black Plague, the Holocaust and many other losses over the centuries, but one does not celebrate these."[51]

Conservationists have forgotten the need for grief and mourning in our relationship to nature, but it is essential in the Anthropocene, an age marked by rapacious destruction, as well as cruelty, in human interaction with nature and non-human life. Environmental philosopher Thom van Dooren, author of the book *Flight Ways: Life and Loss at the End of Extinction*, poignantly reflects on the essential role of mourning in understanding the cost of the extinctions we have caused.

> Mourning is about dwelling with a loss and so coming to appreciate what it means, how the world has changed, and how we must ourselves change and renew our relationships if we are to move forward from here ... The reality ... is that there is no avoiding the difficult cultural work of reflection and mourning. This work is not opposed to practical action, rather it is the foundation of any sustainable and informed response.[52]

Remorse is not an end point. It must be transformed into intention, intention into decision, and decision expressed in *repentance*. Repentance is the action that engages our *will* – perhaps the most distinctively "human" part of us. It is out of our will that we first desire, then imagine, and then create what did not previously exist. "Our consents and non-consents," said philosopher William James, "...are the measure of our worth as men ... the one strictly underivative and original contribution which we make to the world."[53] Our decision must take the form of an action that we ourselves imagine and then create to address the problem and, in this case, restore a successional process that can lead to the reforestation of an abandoned well pad.

To carry out the repentance of our will we must engage our *bodies*. It may sound strange (or even repulsively unscientific) to believe that our body matters in conservation effort. It does. Not only in conservation, but, as Aristotle taught, because, "For as regards those things we must learn how to do, we learn by doing them."[54] Conservation is no different. It succeeds in changing human behavior toward nature only when it engages us *physically*, requiring not only mental effort but corporal *exertion* directed by *intention* to *do and complete* what we imagine, and by training in *physical competence* to complete the required task. I saw this repeatedly from watching Au Sable's Environmental Education Coordinator, Paul Wiemerslage, engage children in the work of ecological restoration through their physical labor of *doing ecological restoration* – planting the native plants where they belonged and would flourish in places where they had once thrived but then been eradicated.

To devise a bridge that could carry the well pad research from regret to restoration, we created an experimental approach that permitted an evaluation of the four soil treatments with independent combinations of the tree species we planted on the pad – what statisticians call a "split-plot" design – and then determined the best treatment–species combination to achieve reforestation. Our counterfactual plots (where we monitored the vegetation on the pad but applied no treatment at all) gave us insight into what would have happened to the pads "on their own" without management intervention. Our results were complex, but can be summarized this way – disking (breaking up the compacted soil) outperformed all other treatments no matter what tree species was planted in it – red pine outperformed all other tree species no matter what treatment it was planted in. Together a prescription – disk the soil of the pad and plant red pine.[55, 56] The counterfactual "do-nothing" plots, in contrast, showed no significant reforestation by any of the surrounding native trees. Those portions of the well pads went right on looking the same way they had before the study started, the same way many of them had looked since their onsite drilling activities ceased decades ago.[55]

Renewal – The Completion of Reconciliation

This research effort showed that conservation matters. In this case, the reforestation of inactive well pads is both possible and practical. Pad sites could again be contributors to an intact, productive, and healthy forest. But to bring this change to its right conclusion, forest managers must now declare that the reforestation of the well pads is part of the forest management plan and that *the care of the pads must be incorporated into the care of the forests*. If managers make these kinds of changes, they will be the initiators of *renewal* – the final and most essential dimension of conservation, a changed relationship with nature that sees the reality of need, engages human effort to meet it, and forms interactions between humans and nature better than those we had before. Such renewal requires us to address our interactions with nature on legal, moral, and ecological grounds, even if we still have an imperfect understanding of all three.

The well pad research may contribute to stopping arrested succession and initiating the reforestation of abandoned oil wells, but it was also important for my student assistants to understand and experience how to frame conservation as an expression of the whole person – intellect, emotions, will, body, and society – a society of humans in nature, interacting with nature, nurturing the health of nature. Conservationist and essayist Wendell Berry expressed the intimacy of such right interaction this way:

> To live, we must daily break the body and shed the blood of Creation. When we do this knowingly, lovingly, skillfully, reverently, it is a sacrament. When we do it ignorantly, greedily, clumsily, destructively, it is a desecration. In such desecration we condemn ourselves to spiritual and moral loneliness, and others to want.[57]

The landscape artist Alan Gussow, with a trained eye for the beauty of places, said, "A place is a piece of the whole environment that has been claimed by feelings."[58] Many have asserted that the Anthropocene is a time when nature will be what we would make it. But before we presume to make anything of nature, we must first make something of ourselves. We must create connections with nature in which our feelings influence our intentions toward it and our interactions with "places" in it. Our reconciliation with nature is not about making it into whatever we want. It is about making ourselves into the people we ought to be – people who can rightly relate to nature in every dimension of who we are.

An essential foundation of our own reformation is to change how we view ourselves, what we would want for the "nature" around us, and how we put these visions into words. "A part of what we need to do..." says political scientist Jedidiah Purdy, "is just to be bold in voicing the visions of the natural world that we carry and the ways they matter to us."[59] Reconciliation leading to renewal begins with changing our conversation – the verbal channels of our ideas – into a different way of speaking and thinking about "nature conservation." We do not need conservation to express the economic development of nature. We need conservation to embody reconciliation with nature, a reconciliation grounded in habits of interaction enforced by prescribed best practices toward nature and performed for its own good. It must eventually take shape in convention or law, but before that it must first reflect our aspirations for ourselves as well as our aspirations for the natural world.

The value of areas protected by conservation is not to be measured by what people will pay to protect them, but by the values they place on such areas which are reflected in both their expenses and their efforts. The economic resources of conservation must be strategically employed, but the source of those resources must be intelligently restrained. They can be the servant of a greater purpose, but they are not the master of that purpose nor the standard to which it should comply. We must learn the truth of the words of Indigenous American author Robin Wall Kimmerer: "To love a place is not enough. We must find ways to heal it."[60]

References

1. Van Houtan, K. 2006. Conservation as virtue: A scientific and social process for conservation ethics. *Conservation Biology* 20:1367–1372, 1371.
2. Lewis, C. S. 2001. *The weight of glory: And other addresses.* HarperCollins, San Francisco, California, USA, 32.
3. Daly, J. H. 1999. The lurking inconsistency. *Conservation Biology* 13:693–694, 694.
4. Thoreau, H. D. 1946. *Walden: Or life in the woods.* Dodd, Mead, and Company, New York, USA, 191.
5. Kagan, J. 1998. *Three seductive ideas.* Harvard University Press, Cambridge, Massachusetts, USA, 191.
6. Berry, W. 2002. *The art of the commonplace: The agrarian essays of Wendell Berry.* Edited and introduced by N. Wirzba. Counterpoint, Washington, DC, USA, 29.
7. International Union for the Conservation of Nature. 2024. IUCN WCPA Cultural and Spiritual Values of Protected Areas Specialist Group. IUCN, Gland, Switzerland. <https://www.iucn.org/our-union/commissions/group/iucn-wcpa-cultural-and-spiritual-values-protected-areas-specialist>. Accessed 5 February 2024.
8. Society for Conservation Biology. 2024. Religion and conservation biology. Society for Conservation Biology, Washington, DC, USA. <https://conbio.org/groups/working-groups/religion-and-conservation-biology>. Accessed 5 February 2024.
9. Faithplans. 2024. WWF's beliefs and values programme. Faithplans, UK. <https://www.faithplans.org/wwf-beliefs-and-values>. Accessed 5 February 2024.

10. Russell-Smith, J., C. P. Yates, A. C. Edwards, et al. 2015. Deriving multiple benefits from carbon market-based savanna fire management: An Australian example. *PLOS ONE* 10:1–21. DOI: 10.1371/journal.pone.0143426.
11. Santini, N. S., and Y. Miquelajauregui. 2022. The restoration of degraded lands by local communities and Indigenous Peoples. *Frontiers in Conservation Science* 3:873659. DOI: 10.3389/fcosc.2022.873659.
12. Stevens, S., editor. 1997. *Conservation through cultural survival: Indigenous Peoples and protected areas*. Island Press, Washington, DC, USA.
13. Millennium Ecosystem Assessment. 2005. *Ecosystems and human well-being*. World Resources Institute, Washington, DC, USA.
14. Powledge, F. 2006. The Millennium Assessment. *BioScience* 56:880–886. <https://doi.org/10.1641/0006-3568(2006)56[880:TMA]2.0.CO;2>.
15. Hrabanski, M. 2017. Private sector involvement in the Millennium Ecosystem Assessment: Using a UN platform to promote market-based instruments for ecosystem services. *Environmental Policy and Governance* 27:605–618. <https://doi.org/10.1002/eet.1780>.
16. Maguire, L. A., and J. Justus. 2008. Why intrinsic value is a poor basis for conservation decisions. *BioScience* 58:910–911. <http://dx.doi.org/10.1641/B581002>.
17. Marvier, M., and H. Wong. 2012. Resurrecting the conservation movement. *Journal of Environmental Studies and Science* 2:291–295. <http://dx.doi.org/10.1007/s13412-012-0096-6>.
18. Batavia, C., and M. P. Nelson. 2017. For goodness sake! What is intrinsic value and why should we care? *Biological Conservation* 209:366–376. <https://www.sciencedirect.com/science/article/abs/pii/S0006320716307522>.
19. Orang Utan Republik Foundation. 2024. The illegal pet trade. Orang Utan Republik Foundation. <https://www.orangutanrepublik.org/learn/issues/illegal-pet-trade/>. Accessed 5 February 2024.
20. Sagoff, M. 2008. *The economy of the Earth*. Cambridge University Press, Cambridge, UK, 18.
21. Marvier, M. 2014. A call for ecumenical conservation. *Animal Conservation* 17:518–519. <http://dx.doi.org/10.1111/acv.12130>.
22. Callicott, J. B. 1999. Intrinsic value in nature: A metaethical analysis. Pages 239–261 *in* J. B. Callicott. *Beyond the land ethic: More essays in environmental philosophy*. State University of New York Press, Albany, USA.
23. Rolston, H. III. 1988. *Environmental ethics: Duties to and values in the natural world*. Temple University Press, Philadelphia, Pennsylvania, USA, 115.
24. Rolston, H. III. 1988. *Environmental ethics: Duties to and values in the natural world*. Temple University Press, Philadelphia, Pennsylvania, USA, 117.
25. Kareiva, P., and M. Marvier. 2012. What is conservation science? *BioScience* 62:962–969. DOI: 10.1525/bio.2012.62.11.5.
26. Ferraro, P. 2001. Global habitat protection: Limitations of development interventions and a role for conservation performance payments. *Conservation Biology* 15:990–1000.
27. Kosoy, N., M. Martinez-Tuna, R. Muradian, et al. 2007. Payments for environmental services in watersheds: Insights from a comparative study of three cases in Central America. *Ecological Economics* 61:446–455. <https://doi.org/10.1016/j.ecolecon.2006.03.016>.
28. Wunder, S. 2007. The efficiency of payments for environmental services in tropical conservation. *Conservation Biology* 21:48–58.
29. Ives, C. D., and S. A. Bekessy. 2015. The ethics of offsetting nature. *Frontiers in Ecology and the Environment* 13:568–573. <http://dx.doi.org/10.1890/150021>.
30. Tallis, H., and P. Kareiva. 2005. Ecosystem services. *Current Biology* 15:746–748.
31. Marvier, M., P. Karieva, D. Felix, et al. 2019. The benefits of nature exposure: The need for research that better informs implementation. *Proceedings of the National Academy of Sciences* 120(44):e2304126120. <https://doi.org/10.1073/pnas.2304126120>.
32. Hume, D. 2007. David Hume: A treatise on human nature: Volume 1. Page 301 *in* D. F. Norton and M. J. Norton, editors. *David Hume: A treatise on human nature: Volume 1*. Clarendon Press, Oxford, UK.

33. Soulé, M. 2014. Also seeking common ground in conservation. *Conservation Biology* 28:1–2, 1. DOI: 10.1111/cobi.12293.
34. Rodgers, W., Jr. 1994. The seven statutory wonders of U. S. environmental law: Origins and morphology. *Loyola of Los Angeles Law Review* 27:1009–1021. <https://digitalcommons.law.uw.edu/faculty-articles/258/>.
35. Kotzé, L. J. 2014. Human rights and the environment in the Anthropocene. *The Anthropocene Review* 1:252–275. DOI: 10.1177/2053019614547741.
36. Dourojeanni, M. J. 2015. Human impact on protected areas of the Peruvian Amazon. Pages 215–225 *in* G. Wuerthner, E. Crist, and T. Butler, editors. *Protecting the wild: Parks and wilderness, the foundation of conservation*. Island Press, Washington, DC, USA.
37. Dourojeanni, M. J. 2015. Human impact on protected areas of the Peruvian Amazon. Pages 215–225 *in* G. Wuerthner, E. Crist, and T. Butler, editors. *Protecting the wild: Parks and wilderness, the foundation of conservation*. Island Press, Washington, DC, USA, 222.
38. Doak, D. F., V. J. Bakker, B. E. Goldstein, et al. 2015. What is the future of conservation? Pages 27–35 *in* G. Wuerthner, E. Crist, and T. Butler, editors. *Protecting the wild: Parks and wilderness, the foundation of conservation*. Island Press, Washington, DC, USA, 33.
39. Muir, J. 1908. Hetch Hetchy damming scheme. Congressional Record "San Francisco and the Hetch Hetchy reservoir." Hearing before the Committee on the Public Lands of the House of Representatives, December 16, 1908.
40. Tompkins, D. R. 2015. Afterword. Pages 277–282 *in* G. Wuerthner, E. Crist, and T. Butler, editors. *Protecting the wild: Parks and wilderness, the foundation of conservation*. Island Press, Washington, DC, USA, 278.
41. Turner, N. J., and D. Matthews. 2020. Serving nature: Completing the ecosystem services cycle. Pages 3–29 *in* H. Bai, D. Chang, and C. Scott, editors. *A book of ecological virtues: Living well in the Anthropocene*. University of Regina Press, Regina, Saskatchewan, Canada, 8.
42. Smith, R. C. 1982. Liberal ideology and Indigenous communities in post-independence Peru. *Journal of International Affairs* 36:73–82.
43. Dawson, N., B. Coolsaet, E. Sterling, et al. 2021. The role of Indigenous peoples and local communities in effective and equitable conservation. *Ecology and Society* 26(3):19. <https://doi.org/10.5751/ES-12625-260319>.
44. Adamson, R., and S. Klinger. 2009. Alliance Action Prize. Indigenous Peoples Climate Action Fund. <https://www.alliancemagazine.org>. Accessed 9 November 2009.
45. Marsh, G. P. 1864. Man and nature. Page 36 *in* D. Lowenthal, editor. 2003. *Man and nature*. University of Washington Press, Seattle, USA.
46. Hatch, N., J. Aldrich, G. Brushaber, et al. 1994. An evangelical statement on the care of creation. *Christianity Today*, April 4.
47. Cornwall Alliance. 2000. The Cornwall Declaration on environmental stewardship. <https://cornwallalliance.org/landmark-documents/the-cornwall-declaration-on-environmental-stewardship/>. Accessed 7 February 2024.
48. Warners, D. P., and M. K. Heun, editors. 2019. *Beyond stewardship: New approaches to creation care*. Calvin College Press, Grand Rapids, Michigan, USA.
49. Goldsmith, G. R., L. S. Comita, and S. C. Chua. 2011. Evidence for arrested succession within a tropical forest fragment in Singapore. *Journal of Tropical Ecology* 27:323–326. <https://doi.org/10.1017/S0266467411000010>.
50. Murdoch, I. 1970. *Sovereignty of the good*. Schocken Books, New York, USA.
51. Wuerthner, G. 2018. Anthropocene boosters and the attack on wilderness conservation. *Ecological Citizen* 1:161–166, 163.
52. van Dooren, T. 2013. Keeping faith with death: Mourning and de-extinction. <https://www.thomvandooren.org/2013/11/02/keeping-faith-with-death-mourning-and-de-extinction/>. Accessed 25 July 2024.
53. James, W. 1918. *The principles of psychology, volume 2*. MacMillan, London, UK, 578–579.
54. Aristotle. 2011. *Aristotle's Nicomachean ethics*. Translated with an interpretive, notes, and glossary by R. C. Bartlett and S. D. Collins. University of Chicago Press, Chicago, Illinois, USA, 26.

55. Van Dyke, F., S. Wilderman, S. Harju, et al. 2022. Effects of soil treatments and tree species on reforestation of well pads. *Restoration Ecology* 30(8). DOI: 10.1111/rec.13658.
56. Van Dyke, F., S. Wilderman, S. Harju, et al. How can we restore forests removed at vacated oil drilling sites? SciAni. <https://bit.ly/3QJiLa9>.
57. Berry, W. 2003. The art of the commonplace: The agrarian essays. Page 304 *in* N. Wirzba, editor. *The art of the commonplace: The agrarian essays*. Counterpoint, Berkeley, California, USA.
58. Gussow, A. 1972. *A sense of place: The artist and the American land*. Friends of the Earth, Washington, DC, USA, 27.
59. Purdy, J. 2015. *After nature: A politics for the Anthropocene*. Harvard University Press, Cambridge, Massachusetts, USA, 265.
60. Kimmerer, R. W. 2015. *Braiding sweetgrass*. Milkweed Editions, Minneapolis, Minnesota, USA, 345.

Index

Kolibri 360 GmbH
Ferdinand-Jühlke-Straße 7
99095 Erfurt, Deutschland
produktsicherheit@kolibri360.de